W0037300

Facing Up to Global Warming

N.F. Gray

Facing Up to Global Warming

What is Going on and How You Can Make a Difference?

 Springer

N.F. Gray
Centre for the Environment
Trinity College
University of Dublin
Dublin, Ireland

ISBN 978-3-319-20145-0   ISBN 978-3-319-20146-7   (eBook)
DOI 10.1007/978-3-319-20146-7

Library of Congress Control Number: 2015943590

Springer Cham Heidelberg New York Dordrecht London
© Springer International Publishing Switzerland 2015

Printed on acid-free paper

Springer International Publishing AG Switzerland is part of Springer Science+Business Media
(www.springer.com)

Preface

In 1972, a startling book was published called *Limits to Growth*. This book became hugely influential in the environmental movement, and while it alerted us to the fragility of our future on planet Earth, it also, inadvertently, helped to eventually undermine the credibility of environmentalism. The book predicted when certain nonrenewables, including fossil fuels and metals, would become exhausted. The predictions were based on the best available knowledge at that time, but what it never envisaged in the early 1970's, was that within a decade humans would be extracting oil, gas, and minerals in some of the remotest, extreme, and fragile places on Earth ... a process that has continued and expanded to the present day. So the predictions proved incorrect in practice, but it reinforced the idea that all resources are limited and are slowly being exhausted.

This book became a driving force for many environmental scientists who realized that we have to act both collectively and individually to preserve our home, planet Earth, with its unique biosphere and which is home to millions of different living organisms of which we are just one species. For me personally, being an environmental scientist has been a long and often disappointing journey and at various times I have been shocked, scared, and often depressed by the unfolding of the current crisis which is so intertwined with global warming. But, to my surprise, in recent years I have begun to feel more hopeful that perhaps we can deal with our climate and resource problems to create a sustainable planet. So in this book, I have attempted to explain what the problems are and suggest some solutions. However, the book comes with a warning. During the 15 chapters that follow, I am going to make a lot of you really annoyed and possibly upset, I apologize in advance. I am not trying to shock; I am simply putting the facts before you so that you can make up your own mind. Neither am I telling anyone that their lifestyle is wrong, or alternatively, that they are better than the next person because they have invested in green energy or a hybrid car. The book is an overview; it is not a text on the theory of sustainability or population dynamics; it simply looks at what the individual should know and addresses some of the issues closest to our everyday lives. There are hundreds of academic and specialist texts on sustainability, but they fail to link sustainability to tackling global warming, especially at the individual level. Adopting any form of sustainable actions in your life will cause significant effects both direct and indirect. Such actions will lead to changes that will influence economic and social norms ... so sustainability if properly applied will mean socioeconomic change. I begin the text by giving a brief overview of the problems of climate change and the real difficulties that having such a rapidly growing population is placing on the idea of a sustainable and equitable planet. Discussions on population is always a very emotive issue and so I have simply given some basic facts, and shown that as population grows our ability to live sustainably on planet Earth becomes more challenging. So this is not a comfortable book.

Is the text political? I have tried not to be, but if you advocate changing people's lifestyles, then it will appear to be political. Then, we have the concept that everyone on the planet matters and that the concept of global justice and human rights is important when assessing the sustainability of our own lifestyles. So trying to avert these climate-mediated crises by ensuring that everyone has enough for their needs without being wasteful is a good starting point. However, that starting point has to be an acknowledgment that all people are equal, and that we should all have the right to pursue happiness and well-being. Is this naïve? Of course it is, but what else are we to strive for in a truly global and fair society. Global warming raises serious social as well as economic questions and many of these are going to be very difficult to deal with in practice, and my aim is to try and make you think about these issues from a personal perspective. Can we have finite economic growth? Can we have finite consumerism? Unfortunately, the answer has to be no to both of these questions, which means that both economists and social geographers or planners have a lot of work to do and that we are going to have to eventually reinvent our economy and social environment to achieve these goals.

There are also many other important environmental issues that we also need to consider and many of these are also linked to global warming such as deforestation, exploitation of new fossil fuel reserves, intensification of agriculture, and overexploitation of water resources. However, many of these issues such as pollution, are less important in the context of global warming, as we now have them largely under control. We have made huge strides in dealing with air, water, and land pollution over the past 40 years, and there are scientists and regulatory bodies all dealing with these issues on an ongoing basis. Such issues are predominately local or at worst regional, but rarely global, and what is important is that we have the technology and infrastructure to deal with them. But controlling carbon dioxide and other greenhouse gas emissions must now be everyone's priority. If we have to reduce carbon emissions by 80 % by 2050 to avert a global crisis, this will mean using significantly less energy in the developed world than we currently use, and while this does not necessarily mean an immediate and huge change in our everyday lives, it does mean changes to our current lifestyles. This is not going to be easy and the burden has to be shared by everyone. However, the fact is that people feel very threatened by the idea of altering their lifestyle, even when change can be for the better.

We also need to understand that some sectors of society are using more than their fair share of global resources, but that in the context of global warming everyone must act responsibly if we are to succeed in mitigating climate change. Those in developing countries also desire the technology, food, travel, etc. we enjoy, and to break this cycle we in the developed world need to begin to pull back from our current high-energy lifestyle while allowing the poorer nations to develop and become sustainable.

This book will never be welcomed by those who are pretty happy with the status quo and who have not become genuinely concerned, possibly scared, by the possibility of what global warming may do to our home, planet Earth. This is a very general text that looks at different aspects of our lives which we, as individuals, have control over. It is simple things like travel, food, recycling, using resources … all those things which we are all involved in on a daily basis; and how our actions affect the future of planet Earth and our ability to sustain that ever growing human population. I hope that this book will help you think and act from a position of knowledge and reassurance.

I hope that the majority of you will be reassured that we are beginning to successfully tackle global warming, but in order to succeed in stabilizing our new climate, and I believe we can, we need your help through direct action. You really can make a difference. This book is a personal journey and during it I will be asking you to do various things. Some are critical others will be just things that I hope you will try, but to work I need a commitment from you. The journey is not free, it comes at a cost, and you have to decide just how much you are willing to pay for your planet. This is about your future.

Dublin, Ireland   Nick Gray
Spring, 2015

Contents

Part I   The Concept of Living Sustainably

1   Defining the Problem 3
   1.1   Introduction 3
   1.2   Population 5
      1.2.1   The Mechanism of Population Growth 9
      1.2.2   The Consequence of Population 11
      1.2.3   The Dilemma of the Malthusian Catastrophe 13
   1.3   Global Warming and Our Climate 17
      1.3.1   Temperature 18
      1.3.2   Precipitation (Rainfall and Snow)................................ 24
      1.3.3   Wind.. 26
   1.4   Conclusions... 29
   Homework! 30
   References and Further Reading 30

2   What Is Sustainability? 33
   2.1   Only One Earth 33
   2.2   What Do We Mean by Sustainable?................................... 37
      2.2.1   Environmental Sustainability................................... 43
      2.2.2   Stern 44
   2.3   So Where Are We Now Regarding Sustainability?.................. 48
   2.4   Conclusions... 49
   Homework! 49
   References and Further Reading 50

3   The Concept of Resources 53
   3.1   Renewable and Non-renewable Resources........................... 53
      3.1.1   Renewable Resources 54
      3.1.2   Non-renewable Resource 56

3.2   Key Extractable Resources 58
    3.2.1   Crude Oil/Petroleum 58
    3.2.2   Coal/Lignite 62
    3.2.3   Natural Gas 63
    3.2.4   Helium 66
    3.2.5   Uranium/Nuclear Energy 66
    3.2.6   Metals 67
3.3   Land as a Finite Resource 69
    3.3.1   Soil and Processes 69
    3.3.2   Deforestation and Land Degradation 70
    3.3.3   Action to Protect the Land Resource 73
3.4   What Are We Supposed to Do? 74
3.5   Conclusions 75
Homework! 76
References and Further Reading 77

Part II   Greenhouse Gases and Global Warming

4   Global Warming and CO_2 81
4.1   The Greenhouse Effect 81
4.2   Radiative Forcing 85
4.3   Global Warming Potential 87
    4.3.1   Albedo 89
4.4   Effects of GHG Emissions and Models 90
    4.4.1   Who Is the IPCC? 91
    4.4.2   Predicting Emissions 93
4.5   Proposed Limits 95
4.6   Conclusions 100
Homework 101
References and Further Reading 101

5   Measuring CO_2 Emissions 103
5.1   Introduction 103
5.2   Total Carbon Footprint 109
5.3   Embedded and Secondary Emissions 117
    5.3.1   Embedded Energy 117
5.4   Examples of How We Use Energy 119
    5.4.1   Driving 119
    5.4.2   Lights 121
    5.4.3   The Internet 124
    5.4.4   Mobile Communication 124
5.5   Making the Right Choice 126
    5.5.1   Plastic vs. Paper Bags 127
5.6   Rebound Effect 129
5.7   Conclusions 130
Homework! 131
References and Further Reading 133

6   The Real Cost of Carbon 135
  6.1   How Do Government's Tackle Climate Change? 135
  6.2   Background to Emissions Trading 136
      6.2.1   Emissions Trading Scheme (ETS) 138
      6.2.2   Joint Implementation (JI) and the Clean Development
              Mechanism (CDM) 138
      6.2.3   The Cap and Trade Mechanism 140
  6.3   Emissions Trading 140
  6.4   The Cost of Sequestration 143
  6.5   Carbon Taxation 145
  6.6   The Real Cost of Carbon Offsetting 147
      6.6.1   How Does Offsetting Actually Function
              and Does It Work? 148
      6.6.2   Carbon Sequestration in Agriculture and Forestry 149
      6.6.3   Offsetting as a Mechanism of Controlling Emissions 153
  6.7   So Where Do We Stand on Carbon Pricing? 154
  6.8   Conclusions 155
  Homework! 156
  References and Further Reading 157

7   Ecological Footprint 159
  7.1   Action and Reaction 159
  7.2   Ecological Footprint 160
      7.2.1   Calculation of Ecological Footprint 163
  7.3   Global Living Planet Index 170
  7.4   One Planet Economy Network 171
  7.5   Setting Sustainability Targets 173
  7.6   Conclusions 174
  Homework! 174
  References and Further Reading 175

Part III   Our Use of Resources

8   Energy: Green or Otherwise 179
  8.1   How Much Energy Do We Use? 180
      8.1.1   Electricity 184
      8.1.2   All Fuels 188
  8.2   Renewable Energy 192
  8.3   The Nuclear Debate 198
  8.4   Household Energy Use and CO_2e Emissions 201
      8.4.1   Home Energy Measurements 203
      8.4.2   Is Standby Really a Problem? 204
      8.4.3   Turning Off Desktop PCs 206
  8.5   Energy Targets 207
      8.5.1   Personal Targets 209
  8.6   Conclusions 210
  Homework! 211
  References and Further Reading 212

9   Travel: Here, There, Everywhere 215
   9.1   Introduction.. 215
   9.2   Travel as Part of Our Carbon Footprint 216
   9.3   Aviation... 217
           9.3.1   Emissions from Flying 219
           9.3.2   Contrials 222
   9.4   Travel by Other Means 224
           9.4.1   The Car... 225
           9.4.2   Commuting... 229
           9.4.3   Are Modern Cars Really That Efficient?........................ 231
           9.4.4   Embedded Footprint of Car Manufacture 236
           9.4.5   Alternatives 236
   9.5   Conclusions.. 238
   Homework! 239
   References and Further Reading.. 239

10   Having Enough to Eat 241
   10.1   Introduction.. 241
   10.2   Climate Change and Agriculture... 243
   10.3   Who Will Be Affected Most by Food Scarcity? 245
   10.4   The Food We Eat and GHG Emissions.. 251
           10.4.1   Food Miles.. 251
           10.4.2   Local Is Best?.. 252
           10.4.3   Organic Food: Is It the Sustainable Option? 253
   10.5   The Food Footprint 255
           10.5.1   Calculating the Food Footprint 259
           10.5.2   Examples of Food Calculators Are................................ 261
   10.6   Can We Reduce Our CO_2e Emissions in Our Food? 262
           10.6.1   Food Waste 264
   10.7   Conclusions.. 269
   Homework! 270
   References and Further Reading.. 270

11   Where Does Water Fit in? 273
   11.1   Introduction.. 273
           11.1.1   Water Use 277
           11.1.2   Water Scarcity 277
           11.1.3   Water Conflict... 281
   11.2   Water Demand Management.. 283
           11.2.1   Water Conservation 283
           11.2.2   Water Efficiency Labelling.. 284
           11.2.3   Metering Supplies 286
           11.2.4   Household Water Use and CO_2 Emissions...................... 286
   11.3   Peak Water 287
           11.3.1   Desalination.. 293

	11.4	Water Footprints	294
		11.4.1 Water Diary	298
	11.5	Conclusions	299
		Homework?	299
		References and Further Reading	300
12	**Waste Not Want Not**	303
	12.1	Introduction	303
	12.2	Recycling—The Science of Signs	304
	12.3	Waste Production	311
	12.4	Waste Hierarchy Is Pivotal to Sustainability	314
	12.5	Facts About Recycling	316
	12.6	At a Personal Level	319
		12.6.1 Electronic Items	321
		12.6.2 Someone Somewhere Wants It	322
		12.6.3 What Is in Your Bin?	323
		12.6.4 The Way Forward	325
	12.7	Conclusions	327
		Homework!	328
		References and Further Reading	328

Part IV Responding to the Impact of Global Warming

13	**The Planet's Health**	333
	13.1	Whose Planet Is It Anyway?	334
		13.1.1 Biodiversity	335
	13.2	Maintaining Earth's Current Organic Balance	336
		13.2.1 The Carbon Sink	336
		13.2.2 Volcanoes	338
		13.2.3 Other Warming and Cooling Effects	339
	13.3	Wildfires	339
	13.4	Ice Cover	342
	13.5	Sea Level	344
	13.6	Permafrost	347
	13.7	Methane Hydrate (Methane Clathrate)	350
	13.8	Sea Acidification	351
	13.9	Tipping Points in Planet Health	353
		13.9.1 Should We Care?	353
	13.10	Conclusions	355
		Homework!	355
		References and Further Reading	356
14	**Your Health and Wellbeing**	357
	14.1	Health	357
		14.1.1 Temperature-Related Illness and Death	359
		14.1.2 Vector-Borne and Rodent-Borne Diseases	359

	14.1.3	Waterborne Diseases	363
	14.1.4	Extreme Weather-Related Health Effects	363
	14.1.5	Air Pollution-Related Health Effects	364
	14.1.6	Who and How Many Are at Risk?	365
14.2	Positive Health Benefits of Climate Change		367
	14.2.1	Cooking	367
	14.2.2	Electricity Generation	368
	14.2.3	Transport	368
	14.2.4	Eating Less Meat and Dairy	368
14.3	Wellbeing and Sustainability		368
14.4	Conclusions		372
Homework!			372
References and Further Reading			373

15	**In Your Hands!**		**375**
15.1	Introduction		375
	15.1.1	Global Warming	376
15.2	Revisiting the Previous Chapters		380
	15.2.1	Defining the Problem	380
	15.2.2	What Is Sustainability?	381
	15.2.3	The Concept of Resources	381
	15.2.4	Global Warming and CO_2	382
	15.2.5	Measuring and Offsetting CO_2 Emissions	382
	15.2.6	The Real Cost of Carbon and Offsetting	383
	15.2.7	Ecological Footprint	384
	15.2.8	Energy: Green or Otherwise	385
	15.2.9	Travelling Here, There, Everywhere	385
	15.2.10	Having Enough to Eat	386
	15.2.11	Where Does Water Fit in?	386
	15.2.12	Waste Not Want Not	387
	15.2.13	The Planet's Health	388
	15.2.14	Your Health and Wellbeing	388
15.3	The Next Step?		389
15.4	Implementing Personal Action		390
	15.4.1	The Personal Plan	391
	15.4.2	More on Setting Targets	394
	15.4.3	Checklist	396
15.5	In Conclusion		396
References and Further Reading			397

| **Index** | | | **399** |

Part I
The Concept of Living Sustainably

Chapter 1
Defining the Problem

In a global context there are two major pressures affecting planet health and these are population and global warming induced climate change. In this chapter we examine how they are related. Image James Cridland. Reproduced under Creative Commons Licence

1.1 Introduction

Climate change? Heard about it? Of course you have. There is hardly anyone from the arctic or Antarctic to the equator who will not nod his or her head in acknowledgement. Let's face it there is hardly ever a day when your newspaper will not have some feature or news article on climate change. It's news; it is very big news, bigger

© Springer International Publishing Switzerland 2015
N.F. Gray, *Facing Up to Global Warming*, DOI 10.1007/978-3-319-20146-7_1

than anything else that you are going to read in today's or tomorrow's newspapers. Occasionally the headlines are quite alarming. For example The main headline of the front cover New Scientist magazine of the 17th November, 2012 read:

CLIMATE CHANGE
FIVE YEARS AGO WE FEARED THE WORST.
BUT TODAY IT'S LOOKING EVEN WORSE THAN THAT

Examples of other headlines:

Worst ever CO₂ emissions leave climate on the brink
Font page headline The Guardian, 30th May, 2011

Oceans on brink of catastrophe
Font page headline The Independent 21st June, 2011

OECD warns of catastrophic climate change: Governments urged to 'break out of national mindsets'
Irish Times 24th November, 2011

UN climate science panel issues starkest warning yet
Irish Times 23rd September, 2013

But sometimes the headlines can be quite contradictory and confusing. The massive headline on the front cover of New Scientist magazine, this time on the 7th December, 2013 declared:

CLIMATE SLOWDOWN
IS IT TIME TO STOP WORRYING ABOUT GLOBAL WARMING?

So if it is such big news why do we largely ignore global warming? Let's be honest, has it actually changed your life or stopped you doing anything … of course not. Not yet anyway. So why has climate change become something that is just in the background like some irritating elevator music, it's non-news really, just the same old doom and gloom. We have come to accept climate change just as we except the movement of the moon in our sky. It seems to wax and wane in intensity and urgency at regular intervals, but for the vast majority of us, especially those living in Northern Europe and North America, nothing actually seems to change.

Yet the vast majority of scientists are agreed that global warming induced climate change is a reality, indeed Governments from around the world have largely agreed with these experts and have decided to act to curb greenhouse gas (GHG) emissions in an attempt to slow and limit global warming.

The aim of this book is to inform you of the problems we are all facing due to global warming and to find solutions from the perspective of personal action. The book is written for the non-scientist, and while it does contain some difficult concepts, they are hopefully explained in an accessible and easily understood manner. What is more difficult, is to see ourselves as actual players in this drama, for we are all in part to blame for global warming but at the same time we are also, individually, the solution.

In this book we are going to look at different aspects of the problem. Each chapter will hopefully explain to you in simple terms what the causative effects are, how they affect us and the planet we all share, and what we can do about it. Each chapter has lots of links if you want to explore things in more detail or follow up on major points. Importantly the book attempts to alert you to the changes that we will all experience due to global warming and how they will affect you personally. The core of the book is about how you can make a difference to climate change problems and how to adapt a little better to the changes that may occur. The final chapter brings together these main points and explores in more detail what you need to do in order to survive climate change. Within the text there are grey boxes that either highlight key points or give further information, including examples of useful calculations. At the end of each step there is a summary statement that I want you to consider and hopefully accept.

The book is about creating a personal plan to deal with global warming and at the end of each chapter there is a section which I have called 'homework', which is designed to help you do this. These sections allow you to explore in a more detail some of the key points raised in the book from your own perspective, which hopefully you will find both interesting and exciting to do. Each output from the homework section should be compiled together to form a portfolio of information about yourself and your family which will form the core of your personal plan. You don't need to do the homework sections, they are just there in case you feel you want to explore the areas more. Nor do you have to develop a personal plan; these are just options available to you.

So let's make a start. We are all familiar with pollution, albeit it contamination of our water, soil or air; but pollution is generally a local, or less frequently, a regional problem. However, in a global context there are serious pressures facing not just you and me but the planet as a whole. So what are these key pressures? Well we can break them down into two key mega-issues to start with: **population** and **climate change**. These two pressures lead to new problems or exacerbate existing problems relating to pollution, loss of biodiversity and habitat, as well as shortages of food, fuel and water. In turn these affect communities and individuals by causing conflict, social and political instability, loss of communities and culture which leads to migration, loss of wellbeing, and ultimately illness and death. In this first chapter I want to introduce you to the underlying problems of both population and climate change before going on to look at more specific issues in detail. This brief overview will help you understand the following chapters more clearly, although all the main issues raised will be explored in more depth later on in subsequent chapters.

1.2   Population

Most of us rarely experience large numbers of people. We may feel that a room or train carriage is crowded, but that is more to do with density per unit area than actual numbers. Sometimes when we are stuck in a traffic jam or are travelling on the subway (whether it's the New York Metro or the London Underground) during rush hour we begin to get that feeling of being just a one of a large group of people, all separate individuals, each with a family, relatives and all the trappings and aspirations that you have. The image at the start of this chapter shows a large crowd leaving a concert in Paris. But in practice it is hard to appreciate just how many people we share the planet with.

There are 70 people on a crowded bus, perhaps a 120 on a crowded train carriage, a 1000 at a Rave or perhaps 10,000 at a large concert. At a Premier Division football match or a rugby international there could be 20–80,000 people and up to a 100,000 at a major demonstration. There were estimated to be 1,000,000 on the streets of London at the recent Royal Wedding in London.

The global population is currently over 7,000,000,000. That is approximately the same as the maximum number that can squeeze into Croke Park (Dublin) (80,000) multiplied by the number that can be seated in either Wembley Stadium (London) or the Rose Bowl Stadium (Pasadena, CA) (90,000).

The population of Dublin is 1.1 million with an average density of 2950 people per square kilometre which compares pretty favourably to Mumbai with its population of 14.5 million people giving an average density of 29,905 people per square kilometre. But population density can be misleading as in New York the population density is lower than in Dublin at 2050 people per square kilometre even though its total population is in excess of 18 million. In Dublin City where 0.6 million people live we have an average density of 4588 people per square kilometre rising to 19,500 people per square kilometre where we have high rise accommodation. So the concept of population is difficult to perceive and like the current financial crises we find it difficult when we start talking in tens or hundreds of millions or in case of global population thousands of millions, which are of course billions (Fig. 1.1).

In November 2011 the world population reached **seven billion**. University students attending College in 2015 were largely born when there were around five billion people on the planet. Let's assume you were born on the 16th January, 1994, then you would have been the 5,608,680,165th person alive on earth at your moment of birth just past midnight. When I was born there was a global population of 2,648,162,381 and when my father was born there were 1,924,614,475 people alive. However, when his father was born the population was not that much smaller with the death rate and birth rate very similar so that the average life span had not increased significantly over the intervening period.

A child born at this very moment that I am writing this sentence (1st November 2014 at 16.30 h) would be the 7,226,610,234th person alive on the planet. Today 379,056 children were born and after all deaths are subtracted that is a net increase in the global population of 217,222 in a single day. Since 50,000 BC a staggering total of 84 billion people have lived on earth up to and including today. In 10 years time the global population will be eight billion and in 25 years time nine billion using the most accurate estimates. For those born since 1994 the world's population has already increased by almost a third. What we are experiencing is an unprecedented increase in growth rate. Controlling factors such as disease, poverty and starvation have been significantly reduced; which coupled with our ability to

Fig. 1.1 For most of us, urban commuter train networks will be the greatest experience of population density that we regularly experience. A London Underground train is designed to carry a maximum of 4 passengers per square metre of carriage. The system has 3.4 million passengers each working day compared to 7.2 million on the Mumbai local train network, that is 2.5 times more passengers than it is designed to carry. *Source*: Oxyman. Reproduced under common licence http://commons.wikimedia.org/wiki/File:1972_Stock_at_Kilburn_High_Road_4.jpg

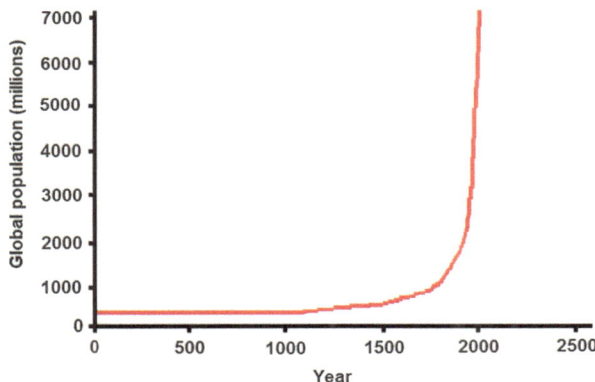

Fig. 1.2 The hockey stick (exponential) growth curve of global population. *Source*: IPCC WG1-AR4 Report (IPCC 2007). Reproduced with permission of the Intergovernmental Panel on Climate Change, Geneva

increase fertility, plus the increasing longevity and improved health of the elderly, has resulted in a global population that has rapidly expanded over the past half century and continues to grow. It is depicted in Fig. 1.2 and is commonly known as the population growth hockey stick due to its characteristic exponential shape.

It is hard to predict what the population will be in 30–40 years time. Over the period 1950–2050 the global population will have at least tripled from 3 to 9 billion adding six billion in only one century assuming a significant reduction in birth rate.

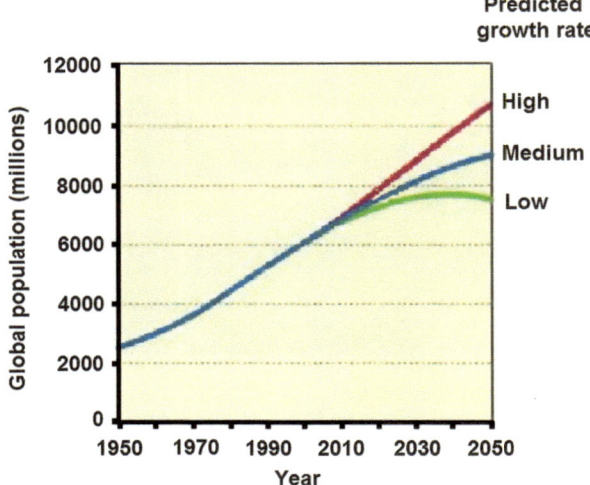

Fig. 1.3 World population predictions 1950–2050 produced by the United Nations. We are closely following the high growth rate line (*red*) which could mean a global population as high as 10–11 billion by 2050! *Source*: IPCC WG1-AR4 Report (IPCC 2007). Reproduced with permission of the Intergovernmental Panel on Climate Change, Geneva

Table 1.1 Countries with the largest populations expressed as millions at the end of October, 2014

Country	Population (millions)	Country	Population (millions)
China	1397	Mexico	124
India	1273	Philippines	101
U.S.A.	323	Ethiopia	97
Indonesia	254	Vietnam	93
Brazil	203	Egypt	84
Pakistan	186	Germany	83
Nigeria	180	Iran	79
Bangladesh	159	Turkey	76
Russia	142	Congo	70
Japan	127	Thailand	67

Some projections are as high as 11–12 billion, others predict a fall with global population stabilizing out at 6–7 billion, which is now generally thought to be unlikely. The latest UN predictions suggest we are on course for a global population of 11 billion or more by 2050 (Fig. 1.3).

The population of the top most populated countries is listed in Table 1.1 with China and India collectively the home to 37 % of the world's population. Global and regional trends in population growth can be explored using the Worldometers® information link: http://www.worldometers.info/world-population/

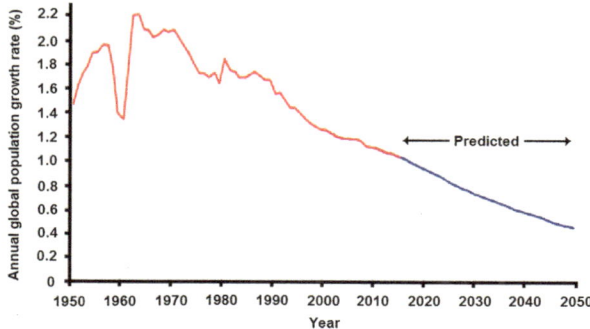

Fig. 1.4 Annual global population growth rate (1950–2050). The global population rate has been steadily falling (*red*) since the mid 1960s and using past trends is predicted to continue to fall (*blue*) but not fast enough to prevent the global population exceeding ten billion by 2050

Table 1.2 Examples of population growth rate (%)

Qatar	4.93 %	Russia	−0.1 %
Zimbabwe	4.36 %	Romania	−0.26 %
Niger	3.36 %	Latvia	−0.63 %
Uganda	3.30 %	Ukraine	−0.63 %
Nigeria	2.55 %	Estonia	−0.65 %
Ireland	1.10 %	Bulgaria	−0.8 %
USA	0.90 %	Syria	−0.8 %
UK	0.55 %	Moldova	−1.1 %
Sweden	0.18 %	Cook Islands	−3.14 %

Those below 0.00 % show a reduction in population growth. Countries on the left show a net increase in population growth while those on the right a net decrease, although both are highly affected by immigration and migration respectively. *Source*: The World Fact Book (CIA 2012 https://www.cia.gov/library/publications/the-world-factbook/rankorder/2002rank.html. Reproduced with permission of the Central Intelligence Agency, US Government

1.2.1 The Mechanism of Population Growth

Global population growth rate is determined by calculating change in population over unit time expressed as a percentage of the number of individuals at the beginning of that period. The rate may be positive or negative. Global growth rates are declining overall (Fig. 1.4) but still remain high in many areas (e.g. for the period 1990–2010: Africa 55 %, Middle East 51 %, Asia 35 %, North America 24 % Europe 9 % non-OECD Europe −11 %); and per country (e.g. Nigeria 62.4 %, Pakistan 55.3 %, Bangladesh 41.3 % India 40.2 %; USA 22.5 %, China 17.2 %, Japan 3.5 %, and Russia, −3.6 %). Population expansion is expected to rise most in Asia and Africa, although it is Asia where stabilization of growth rate is most likely to occur. Currently the global average population growth rate is 1.1 % with the USA at 0.90 % and the UK at 0.55 % (Table 1.2). However, 1.1 % is simply too high and we need to get the global population growth rate as close to zero as possible in order to eventually stabilize the population.

> **Definition of Population growth rate**: The average annual percentage change in the population, resulting from a surplus (or deficit) of births over deaths and the balance of migrants entering and leaving a country.

> **The challenge is to stabilize our global population**. To achieve this we must reduce our global population growth rate to zero which means the number of people who are born equals the number who die each day. For a reduction in global population we would need a negative growth rate (i.e. that is more people must die each day than are born).

The total fertility rate (TFR) is quite a complicated concept and is based on the assumption that every woman needs to give birth to one daughter in her lifetime to maintain the capability of a population to replace itself in order to sustain the population at the same level. If every woman lived to the end of her reproductive life then the TFR would be around 2.1 in most developed countries. It is slightly higher than 2.0 as there are slightly more boys born than girls. In developing countries, due to higher risks of mortality this ranges from 2.5 to 3.3 (Fig. 1.5). In a global context the

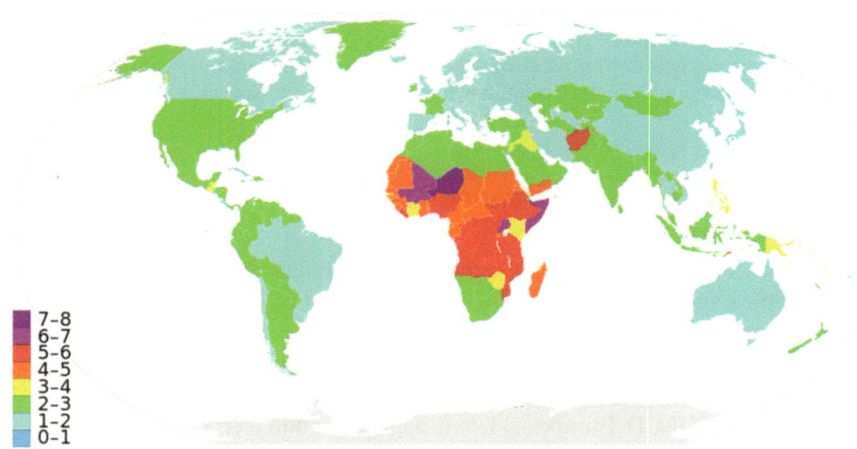

7–8
6–7
5–6
4–5
3–4
2–3
1–2
0–1

Fig. 1.5 Total fertility rates (TFR) per country in 2012. *Source*: The World Fact Book (CIA 2012). Reproduced with permission of the Central Intelligence Agency, US Government

TFR is 2.33, and if this rate could be achieved then the global population growth rate would become zero.

> **Definition of the total fertility rate (TFR)**: The average number of children born to each woman in a population.

Changes in TFR take a long time to affect population due to a lag effect known as population momentum, which is getting longer as life expectancy increases. In the UK the TFR is currently 1.98, so excluding the effects of migration and immigration, the population should eventually stabilize. The TFR in the United States has generally been lower than in Europe peaking at 3.8 in the 1950s. In the mid 1970s it fell below 2.0 reaching a low in 2001 at 1.63. Since then it has risen but remains below the replacement TFR level at 1.89.

The global average TFR has fallen steadily from 4.9 in the 1950s to 4.2 in the 1970s to 2.9 in the 1990s. It is currently around 2.4. But **the TFR needs to fall to at least 2.3 in order to see the global population stabilize out which should occur about 40–50 years after we have reached that point, but at what level the global population will stabilize at we just don't know**. To reduce the global population, if this was necessary to achieve a sustainable population size, then a period of below replacement TFR would be needed after the global population had stabilized.

More information: https://www.cia.gov/library/publications/the-world-factbook/ rankorder/2127rank.html
https://www.cia.gov/library/publications/the-world-factbook/rankorder/2127rank. html

In 1972 the average life expectancy of a person born and living in Bangladesh was under 50 years with women having on average 7 children (i.e. the TFR was 7). Today, the life expectancy for those living in that country has risen by over 20 years to 70 and the birth rate per woman has drop to 2.2. This is reflected by a general global shift from short lives and large families to longer lives and smaller families which is echoed around the world, although this transition has been slower in some parts of Africa. So the very good news is that we are getting closer to a TFR which could then begin the stabilization of our population in about 60 years time probably at around 11 billion.

1.2.2 The Consequence of Population

As the population increases then pressure on resources, especially food, water and our ability to deal with waste increases. Other factors such as life quality, wellbeing, health are also compromised (de Sherbinin et al. 2007). This can be measured in

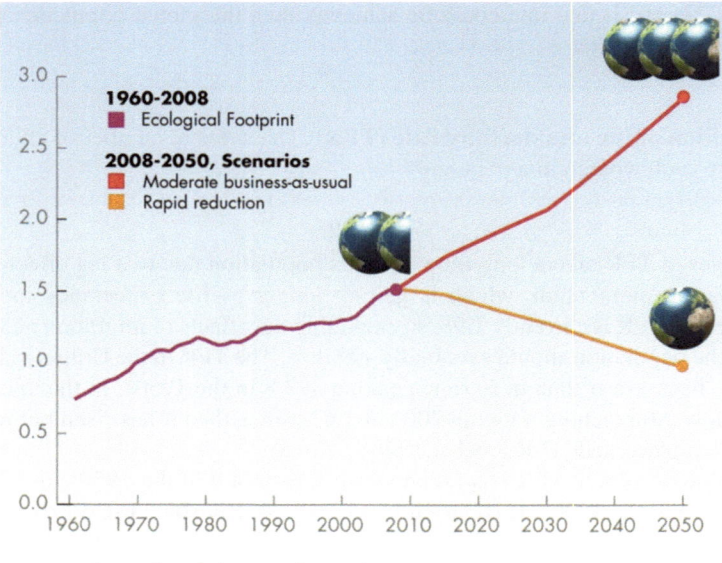

y-axis: number of planet earths, x-axis: years

Fig. 1.6 Trend in the global average ecological footprint measured in Earths per person 1970–2050. Two predicted scenarios, the *red line* is moderate growth (i.e. medium growth as shown in Fig. 1.3), and the *yellow line* a rapid decline in population growth which is now thought to be very unlikely. *Source*: Global Footprint Network. http://www.footprintnetwork.org/. Reproduced with permission of the Global Footprint Network, Geneva

terms of global hectares of land available to maintain each person on the planet. This is called the ecological footprint and is expressed in global hectares per person which can be converted to the number or fraction of Earths needed to support the lifestyle of individuals (Sect. 7.2). As the population rises and we also lose productive land through climate change, especially desertification, forest loss and other factors, then the resources needed to support the human population far outstrips what the planet can sustainably provide. In 1900 we had 7.9 hectares (ha) of land per person but by 1950 this had fallen to 5.7, by 1987 it was 2.6 and in 2005 just 2.20 ha. Currently (based on 2012 estimates) it is just 1.79 ha and is still falling. A report by the World Wildlife Fund (WWF) in 2010 shows that in 2007 the global ecological footprint was 18 billion hectares which means that the global population needed this area of productive land to provide everyone with the resources they needed to support their lifestyles and absorb their waste. The problem is that there was only 11.9 billion hectares available at that time. Since then population has continued to rise and the area available to sustain us has continued to fall.

Currently the human race uses the equivalent of 1.5 Earths to provide the resources it needs and to absorb its waste (Fig. 1.6). By 2030, assuming the UN estimates of moderate population growth, we will need the equivalent of two Earths to support us. Obviously the goal is for everyone to live within 1.0 Earths per person, but at the moment the average for Ireland is 3 Earths and a massive 5 Earths for every person in the US.

WWF Report: http://awsassets.wwf.org.au/downloads/mc035_g_living_planet_report_2010_14oct10.pdf

More information: http://www.footprintnetwork.org/en/index.php/GFN/page/footprint_for_nations/

Our population is now vast compared to other large vertebrate species. When you visit a zoo or watch a programme about wildlife we rarely appreciate just how far we have pushed once common species to the brink of extinction by hunting and more importantly through the destruction of their habitats for farming and urbanization. Today, for every Northern White Rhino there are 200,000,000 people, for every Black Rhino 2,600,000, and for each Mountain Gorilla, 10,000,000 people. Even for each Asian Elephant and Giraffe, both of which are still thought to be relatively common, there are 175,000 and 35,000 people respectively. We are everywhere and there is nowhere on planet Earth where we haven't visited and left our mark. As we grow in numbers all the other species we co-exist with, rely on for food or for pollinating our crops and other ecological services, are put at risk as is seen by the rapid loss in biodiversity which is being accelerated by global warming (Sect. 7.3).

1.2.3 The Dilemma of the Malthusian Catastrophe

The consequence of population has been subject to intense debate for centuries. In 1798 the Reverend Thomas Malthus (1766–1834) wrote a treatise entitled *An Essay on The Principle of Population*. Malthus was to become highly influential as a political economist, although today many of his ideas are rather outdated. However, this socio-political work introduced the idea that unchecked population growth would eventually exceed the growth in food supply which would inevitably lead to catastrophic shortfall in supplies at some time in the future (Fig. 1.7) (Malthus 1798). This Malthusian catastrophe is seen in terms of a radical increase in food prices

Fig. 1.7 The Malthusian catastrophe. Malthus believed that if unchecked, population growth which is seen as an exponential curve would eventually exceed the growth in food supply, which he considered to be more of a linear progression. This he believed led to natural checks (e.g. famine) to further population growth

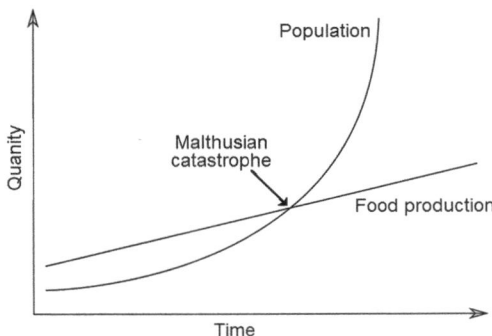

driven by shortages due to increased demand leading to political and social chaos. Leaving aside some of the misguided views of the period, the underlying idea that there is a sustainable population size that can be supported by agriculture and resources became established. The work of Malthus was based on the elimination of poverty, however, his original ideals have been replaced with a more modern interpretation (Neo-Malthusianism) based on the prevention of severe famine, environmental damage caused by human pressure and to ensure sufficient resources for future generations.

The issue of population and resources became a major issue again in early 1960s after the publication of a report *The Growth of World Population* commissioned by the US National Academy of Science (1963). With the population growth rate at 2 %, scientists were becoming increasingly concerned over the relationship between population growth, food production and the impact on the environment. However, it was Paul Ehrlich's book *The Population Bomb* published in 1968 that inspired scientists from a broad range of disciplines to begin to examine the effects that population growth would have on resources and ecological processes in general (Ehrlich 1968). However, the Malthusian catastrophe where population growth exceeds agricultural output has so far not happened. More land has been brought into production, while better scientific and technological advances have made agriculture increasingly productive keeping pace with increasing demand from a continually expanding population. Linked with an efficient and cheap global transportation network, we are able to feed our global population of seven billion. Likewise we have been able to exploit new sources of other critical resources such as fossil fuels and metals, once thought to be inaccessible. However, to a great extent our ability to feed ourselves is dependent on a range of potentially environmentally damaging practices, such as the use of chemical fertilizers as well as chemical pesticides, a high dependency on fossil fuels and the use of genetically modified crops. These practices are creating other environmental pressures that in themselves may become limiting. As we will see in Chap. 10, global food shortages do occur, often linked to climate related crop failures, leading to increased global food prices which in poorer parts of the world has led to famine and social unrest. So in theory the Malthusian catastrophe is still a possibility, although currently there is plenty of food to feed the world population. But increasing global population is not just about food, it is about the demand for other resources, the impact of people directly and indirectly on the environment, on environmental processes and ecological services on which food production and our very survival are dependant. It is about our ever increasing levels of waste that is exacerbating air, water and soil pollution, as well as adding to GHG emissions.

So the question we have to ask ourselves is: Will population growth at some time in the future exceed resources or our capacity to deal with our waste; and if it does when will this happen? (Fig. 1.8) In terms of available biocapacity per person we know that this is decreasing annually and that we are, according to Ecological Footprint analysis, currently living beyond what is sustainable. That extra capacity is leaving behind a legacy of environmental damage much of which is irreversible.

Fig. 1.8 In practice food
production and availability to
resource has kept pace with
demand. Inevitably there will
be a time (t) when that
demand is not met as output
slows and eventually levels
out or even begins to fall

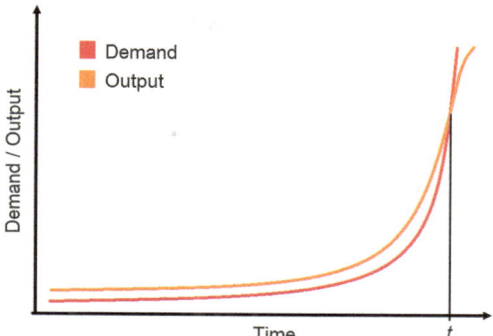

Just as Malthus stated we tend to rely on natural disasters and factors such as famine and disease or manmade factors such as war, to control population. In our modern world this is unacceptable and we rightly respond internationally to prevent and alleviate such disasters. However, it is important that we accept that there is a maximum population that can be sustained while preserving wellbeing and health for each and every one of us that make up the global family, while at the same time protecting our environment as a viable ecosystem. All people rightly desire the same standard of living of those living in the richest countries such as the US or in Western Europe. What is clear is that as population growth continues that this universal standard of living is not possible. There is a difference in having enough to eat in terms of rationed basic foodstuffs and our desire to eat high protein diets and being able to choose what we eat.

In this book, the concept of Neo-Malthusianism is solely concerned with the idea of maintaining environmental support systems, optimizing resources and preventing further losses of biodiversity. This is discussed further in the next chapter on sustainability.

More information: de Sherbinin, A., Carr, D., Cassels, S. and Jiang, L. (2007) Population and Environment. *Annual Review of Environmental Resources*, 32, 345–373. http://www.ncbi.nlm.nih.gov/pmc/articles/PMC2792934/

1.2.3.1   Limits to Growth

I mentioned in the preface how in 1972 a report titled *Limits to Growth* was commissioned by the Club of Rome, an independent think tank, which explored the effects of population growth and increased demand on social and economic stability (Meadows et al. 1972). This report was perhaps the most import consequence of Paul Ehrlich's book *The Population Bomb* published 4 years earlier. Apart from predicting tends in population growth, industrial output, resource depletion and wealth using a range of different scenarios, they also predicted that some key resources such as oil and certain metals would run out within 20–30 years and warned of a [Malthusian] catastrophe. As we shall see in later chapters, predictions about when resource depletion would occur were inaccurate due to the data available at that time (pre 1970). They were also unable to foresee the speed of scientific and technological advances, especially in the areas of oil and gas extraction and to what lengths companies would go to extract other fossil fuels and minerals. While the concept was sound we are still here with almost double the global population and still enjoying an energy-rich lifestyle. For that reason the report has been largely dismissed as a doomsday fantasy, based largely on the fact that we did not run out of these key resources when predicted.

Written by Donella and Dennis Meadows, who were scientists based at the Massachusetts Institute of Technology, the report is based on the predictions from a computer model (World 3) they constructed. The task was herculean with historical data on population, industrial development, pollution, and resources used to predict the future in terms of resource depletion using a number of different scenarios depending on how these resources were managed. Using the business-as-usual scenario (BAU), the model predicted 'overshoot and collapse' by 2070. Another problem was that the central hypothesis of the book was similar to that of Malthus, in this case that planet Earth is finite and so unlimited growth in population and exploitation of resources will eventually lead to a collapse of the socio-economic structure of society. However, in his report '*Is Global Collapse Imminent*', Graham Turner form the University of Melbourne has re-examined the predictions from 1972 and compared these with current data up to 2014. He plotted data for the intervening years against that predicted for the same period in 1972 and found that the BAU scenario in the *Limits to Growth* Report was pretty much on target.

> **If the present growth trends in world population, industrialization, pollution, food production, and resource depletion continue unchanged, the limits to growth on this planet will be reached sometime within the next one hundred years. The most probable result will be a rather sudden and uncontrollable decline in both population and industrial capacity.**
>
> *Limits to Growth*, **1972**

In this discussion on population, we have not considered the negative effects that global warming induced climate change will have on our ability to support our global population. What is clear is that many critical resources, especially food production and water availability have already been adversely affected by climate change reducing the population that can be supported in many areas leading to famine and migration. The impact of global warming will inevitably alter the balance between output and demand.

Link: *Is Global Collapse Imminent*? http://www.sustainable.unimelb.edu.au/files/mssi/MSSI-ResearchPaper-4_Turner_2014.pdf
More information: http://www.clubofrome.org/

- Growing population is putting natural resources and ecosystem processes at risk
- Our natural resources are both finite and renewable. Those that are finite are rapidly being exhausted or are being extracted with increasing environmental impact.
- Those that are renewable including water and food are also being exhausted by shear demand

There is a finite global population that is sustainable. Evidence suggests that we may have already exceeded it.

1.3   Global Warming and Our Climate

When we discuss climate change it is often confusing to know exactly what is meant by the term climate. It seems obvious, but climate is actually defined using the concepts of average weather which in turn is described in terms of average temperature, rainfall, wind direction and speed. It is usually averaged over a standard period normally the past 30 years and is based on detailed metrological records that have been meticulously recorded since 1850 at a large number of fixed locations throughout Ireland and the UK and more globally for the past 100 years. In an unchanged climate these parameters are expected to remain within average ranges, although extreme events can occur based on the probability of their occurrence once in every 10, 25 or 100 years. Other parameters are also used to assess the effects of global warming including atmospheric carbon dioxide concentration, snow and ice cover, sea level and sea pH.

What will global warming mean to our climate? Unfortunately we are already seeing it's consequences in the form of more frequent and intense precipitation events leading to flooding; more severe and devastating storms leading to destruction of homes and infrastructure as well as storm surges; and increased global temperatures leading to wildfires, heat waves and drought. These effects differ not only regionally but also locally with many of us experiencing unusual or extreme weather patterns and events. Each of these key climate parameters are summarized below but are considered in greater detail later in the book.

More information: http://www.weather.gov/

1.3.1 Temperature

The rate of increase of global temperature goes up and down, largely due to solar
activity. The average global temperature has risen by 0.8 °C since 1850, and while
this doesn't seem very much, any alteration in temperature can have significant
consequences for established weather patterns. There have been two periods of temperature rise. Between 1910 and 1940 the global temperature increased by 0.35 °C,
which was followed by slight cooling during 1940–1970, followed by another
increase during 1970–2011 of 0.55 °C. Currently the rate of increase has stabilized
due to a reduced level of solar activity and other factors, but locally temperatures are
continuing to rise with Australia, central Africa, the Arctic and Antarctic all experiencing significant temperature rises. Average temperatures are set to increase worldwide with increasingly more extreme heat events. The number of days with
temperatures greater than 32 °C (90 °F) is expected to increase throughout the
Northern hemisphere, especially areas already experiencing heat waves. In the US
the Southeast and Southwest are currently experiencing an average of 60 days per
year above this threshold, but this will increase to at least 150 days each year by
2050 causing significant effects to crop production. Another problem is that these
heat events will get increasing hotter as time goes.

Today the measurement of global temperature is based on thousands of land and
sea monitoring stations from around the world with temperatures collected every day.

Figure 1.9 shows average global temperatures since records were available. As
technology has improved the variability associated with these mean values has

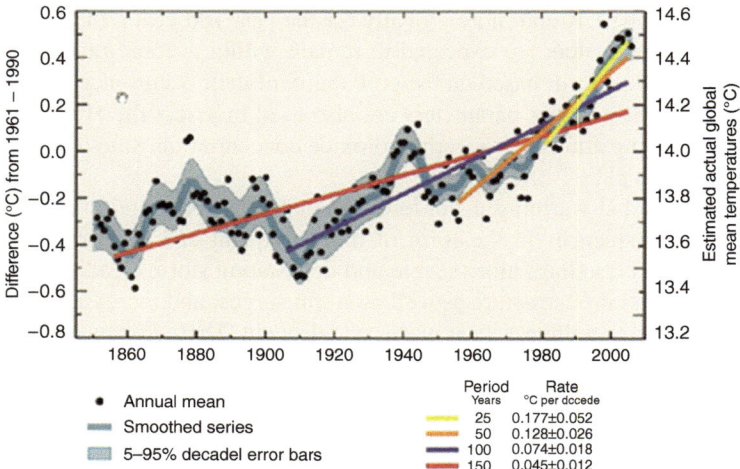

Fig. 1.9 Average global temperatures. The graph shows the rate of increase over different time
periods, each demonstrating that global temperatures are rising and the rate of rise is increasing.
Source: IPCC WG1-AR4 Report (IPCC 2007). Reproduced with permission of the Intergovernmental Panel on Climate Change, Geneva

become much less making these average values far more reliable and precise (i.e. the error bars shown by the shaded area are becoming smaller either side of the average (mean) value). What is interesting is that overall the rate of rise is rapidly increasing as can be seen by comparing the steepness of the slope when using the past 150, 100, 50 and the last 25 years of data from which the current rate of increase of 0.18 °C per decade has been calculated.

So why has the rate of warming slowed down and if global warming is still happening where is all this energy going to? There is no doubt that the past decade has been the hottest on record for many countries, but the average rise in global surface temperature (per decade) has been the lowest since 1951. The highest rate of increase was seen in the 1990s at 0.28 °C per decade but fell during the 2000s to 0.09 °C per decade. The current rate is just 0.04 °C per decade. So does this slowing down mean that our temperature predictions for the future are no longer valid because at this rate surface temperatures would be around 1 °C by 2100 and nowhere near the critical temperature of 2 °C we are so concerned about (Sect. 4.5). So what's going on?

The energy entering the Earth's atmosphere is absorbed as heat (Fig. 4.1). What we often forget is that over 70 % of the planet is covered by oceans and that it takes 3000 times more energy to heat water by 1 °C than the equivalent volume of air. So over the past 40 years it has been estimated that 94 % of the energy from the sun has been absorbed by the oceans, with 4 % absorbed by land and ice, and just 2 % causing the recorded rise in surface temperature. What has happened to cause this apparent slowdown in the rise of surface temperatures is a mixture of events, although the total amount of energy being absorbed by the planet has continued to rise as expected. First the oceans are absorbing more energy than normal and a rapid increase in the burning of coal, especially in China, over that period combined with increased volcanic activity leading to increased atmospheric sulphur dioxide levels and also particulates, has all contributed to a cooling effect (Sect. 4.1). **Together this has slowed the rate of surface temperature increase, although almost certainly this is a temporary effect**.

Apart from the temperature of the oceans, especially the deeper oceans which are warming faster than normal, the rate at which heat is absorbed and released from the surface of the oceans into the atmosphere has also been influential in reducing the rate of global surface temperature rise over the past 10–15 years. Heat transfer between the oceans and the atmosphere is what causes the natural variation in surface temperatures with the El Niño effect increasing and La Niña effect reducing surface temperatures.

During an **El Niño** east winds spreads warm water over the surface of the equatorial Pacific Ocean situated off the western coast of South America, resulting in more energy than normal being released into the atmosphere. This warms the entire planet as well as creating high air surface pressure as was the case in 1998. In turn this influences the climate globally causing weather extremes such as droughts and floods during November to February. The worse impacts are felt along the coasts of Chile, Peru and as far as New Zealand and Australia. It is also linked with increased hurricane intensity in North America. Conversely **La Niña** is caused by westerly winds spreading cool water over the surface of the same area of the Pacific Ocean reducing temperatures by 3–5 °C below normal. This results in more heat being absorbed by the water which rapidly cools the atmosphere above causing global surface temperature to drop. This is accompanied by low surface air pressure in the western Pacific area again affecting global climate. In the past 10–15 years there have been a number of prolonged La Niña events but no major El Niño which has caused a drop in surface temperatures that has contributed to masking the rise in global warming of the planet. Climate change is expected to double the frequency of extreme El Niño events which currently occur on average every 23 years. However, 2014 was the warmest on record even though there was no El Niño, establishing a long-term temperature rise trend of 0.16 °C per decade.

More information on El Niño: http://www.elnino.noaa.gov/
More information on La Niña: http://www.elnino.noaa.gov/lanina.html

The sun also plays an important role in surface temperatures. The sun goes through an 11 year cycle of activity when the sun varies in terms of activity. These relatively small variations in activity (less than one tenth of one percent 0.1 %) during the solar cycle can have significant effects on our terrestrial climate through complex reactions in the upper atmosphere. The sun releases extreme ultraviolet (EUV) radiation, which peaks during the period of maximum solar activity. NASA has recently discovered that the sun's output of EUV radiation is much greater than the level of solar activity measured by, for example, by sun spot activity and varies by a factor of 10 or possibly even more thereby affecting both the chemistry and thermal structure of the Earth's upper atmosphere. Dips in solar activity are often associated with very cold spells in Europe and North America. The solar cycle has just gone through a period of minimum activity (2006–2011) which means that the current low solar activity may be masking the real rate of heating of the planet. So during this decade solar activity will increase reaching a peak during 2018–2022 which should correspond to a significant increase in the rate of temperature rise.

Other factors such as the effect of rising concentrations of atmospheric sulphur dioxide and particulates over the past decade from burning coal, especially in China, and from increased volcanic activity have reflected the sun's energy back into space thereby having a cooling effect. The Intergovernmental Panel on Climate Change (IPCC) in its latest report has identified the oceans as being responsible for half of the reduction seen in surface temperatures with sun activity and extra volcanic activity making up the rest. These factors are considered further in Sect. 4.1.

We should really define global warming in terms of the total heat absorbed by the land, oceans as well as the atmosphere. When we do this then global warming is still accelerating even though the rise in surface temperature has slowed.

So the apparent slowdown in surface temperature rise is actually just a temporary phenomenon. The energy reflected back into the atmosphere is no longer a problem while the fate of the extra energy stored in the seas is more difficult to predict. Some will be retained causing shifts in the biological balance of the oceans, with concern being expressed about the effects to the deep ocean not only in terms of its ecology but also in accelerating the release of methane from frozen reserves (Sects. 13.6 and 13.7). Of course when the sun reaches its peak in maximum solar activity and part of the stored energy in the oceans is released due to an El Nino event, then we will see a rapid increase in surface temperatures again. The energy is still there and it will be transferred back into climate activity causing more unprecedented weather events.

Climate sceptics often say that we are just as likely to have another ice age as global warming. Is this true? The orbit of the Earth around the sun is critical to the amount of solar energy that reaches the surface of the planet. Ice ages in the Northern hemisphere are only able to occur when three separate events coincide. These events are actually cycles. The first is the perihelion which is the time of year the Earth is closest to the sun. Currently it occurs in January but it changes over thousands of years. So now, as it occurs during the winter then our summers are cooler; but when the perihelion occurs in the summer months then our summers become much warmer as a consequence. The second cycle is the angle of the tilt of the Earth. This ranges from 22.0° to 24.5° to the vertical. The smaller or shallower the angle the cooler the summers become and *vice versa*. At the moment the Earth is about midway at 23.4°. Finally the third cycle is the shape of the orbit of Earth around the Sun. The more elliptical the orbit then the lower the summer temperature, while the less elliptical the warmer the summers. Once again we are currently in mid phase. So are we heading for another ice age? It is only when all three cycles line up that summer temperatures become cool enough to trigger an ice age. The good news is that it will be another 60,000 years before all three cycles do coincide and trigger another ice age. There is bad news of course. Each one of these cycles can significantly exacerbate the effects of global warming during the summer months especially when they coincide with one another, creating a seriously warm period.

Every climate record in Europe has been broken in the last decade, and repeatedly so in the British Isles, with 11 of past 12 years exceeding the previous maximum average temperatures. Temperatures are now higher than at any time in the last 1300 years. In the UK, 2012 was the wettest summer in 100 years, while 2011 was the coldest for 20 years. This seems at odds when you consider that it is now on average 2 °C warmer than in the 1950s in many parts of the country, especially the south and southeast England. The number of nights with a temperature below freezing have fallen by 10 % in these areas. The weather is certainly changing but what is interesting is that even in somewhere small like the UK the local weather will be different depending on where you live. This creates a problem in trying to accurately predict how local weather will be affected by global warming. In Europe the area where extreme temperatures are most likely to occur is southern England, Germany, Denmark and the Low Countries, who have all experienced the greatest increase in temperature on the hottest days for over 60 years. The greatest response is found in an area from Northern France to Denmark, with temperature increases on the hottest days of at least 2 °C, over four times the global mean change over the same period. In winter the coldest nights are also getting warmer, which is particularly seen in Scandinavia.

More information: Stainforth, D.A., Chapman, S. C. and Watkins, N.W. (2013) Mapping climate change in European temperature distributions. *Environmental Research Letters,* 8, doi:10.1088/1748-9326/8/3/034031

The US Geological Survey (USGS) has produced a unique online database which allows users to see projections of both temperature and precipitation trends over the coming century. Developed by NASA the database has been created by downscaling all 33 climate models which were used the Intergovernmental Panel on Climate Change 5th Assessment Report (IPCC AR5). The USGS has allows users to explore changes in climate not only at State level, but down to County level, providing visual as well as data summaries. The individual models can also be compared directly. As well as allowing future projections to be explored it also allows historical data from 1950 onwards to also be examined. Using the average model projections then the maximum expected rise in average temperature is expected in areas such as Minnesota (7.2 °F), Utah (7.0 °F) and Montana (6.5 °F) with more southern coastal area having the minimum increase such as Georgia (5.2 °F) and Florida (4.5 °F). Seasonally this represents an average increase from 87.1 °F (average 1980–2004) to 94.3 °F (average 2050–2074) which is an 8 % increase during the summer, but a much larger relative increase in temperature in the Winter increasing from 41.5 to 47.3 °F over the same time frames which represents a 14 % increase (Fig. 1.10).

More information: http://www.usgs.gov/climate_landuse/clu_rd/nex-dcp30.asp

The latest report on the effects of global warming on Ireland, '*Irish climate: The road ahead*' published in September 2013 gives the best estimation of how my own climate will change in the short to medium term. The effect of climate change will see an increase of summer temperatures of 2 °C on average with a reduction in overall precipitation of about 20 % making summers both hotter and drier. In contrast the autumn and winter will get milder and wetter with about 14 % more rainfall than now with much of that falling as intensive episodes resulting in an increase in the frequency and severity of flooding and storm surges. Winter day time temperatures will rise by 2 or even 3 °C overall with night time temperature also warmer. The problem is warm air carries far more moisture which results in more rainfall and possibly on occasions snow. However, as the Arctic Ocean warms then occasionally cold polar air will move southwards causing severe cold periods as seen in 2010 and 2011.

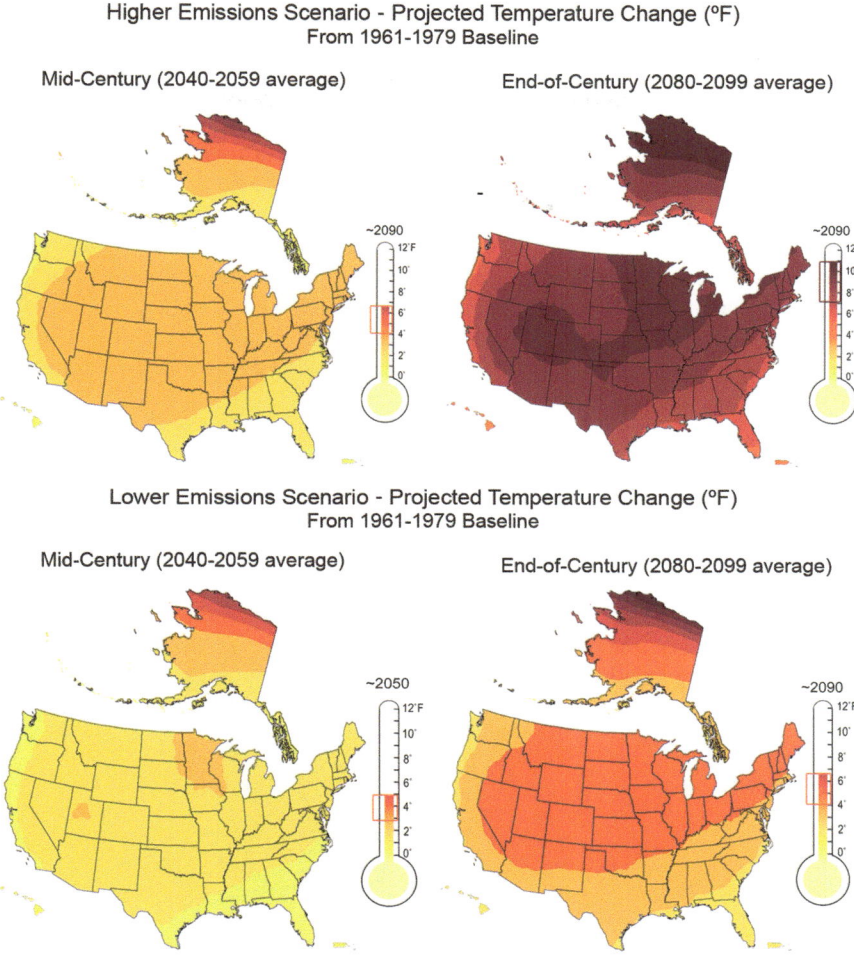

Fig. 1.10 Predicted changes in surface temperature in the United States for the years 2040–2059 (*left*) and 2080–2099 (*right*) for different emission scenarios. The thermometers at the side of each map indicate the range of model projections (in *brackets*). *Source*: USGCP http://globalchange.gov/. Reproduced with permission The US Global Change Research Program, Washington, DC, USA

More information: *Report by Met Éireann (2013)* Irish climate: The road ahead http://www.met.ie/publications/IrelandsWeather-13092013.pdf

Recorded temperature trends are outside what can be considered extreme events in an unchanged climate, so it is clear that our climate is changing and this change is occurring quite quickly.

1.3.2   Precipitation (Rainfall and Snow)

There have been significant changes in rainfall pattern, volumes and intensity in the
past 20 years, with heavier precipitation events as a consequence of more water
vapour in the atmosphere due to warmer temperatures. It is a misnomer to think that
global warming will lead to less rain, instead climate will become increasingly varied and less predictable in the medium to long term. Some trends are becoming
evident. For example, the Mediterranean region, the Sahel, southern Africa and
southern Asia are all becoming drier. While east, north, and southern America,
northern Europe, and both northern and central Asia are all becoming wetter. In fact
we are seeing rainfall migrating north and south from the equator which is shown
very clearly in the Sahel region of Africa, where the desert is moving south with
forests in retreat due to a lack of rainfall (Figs. 1.11 and 1.12).

More   information: http://oceanworld.tamu.edu/resources/environment-book/
desertificationinsahel.html; http://www.millenniumassessment.org/documents/doc-
ument.355.aspx.pdf

So while many areas are having to deal with water shortages, in most of Northern
Europe, including Ireland and the UK winters will become wetter as a consequence
of global warming with more intensive storms releasing heavy volumes of rainfall
over relatively short time periods. It is this rapid release of rainfall which is particularly damaging, especially in areas where land has been largely covered by building,
roads and paving. In Cities and towns throughout the UK gardens have been paved

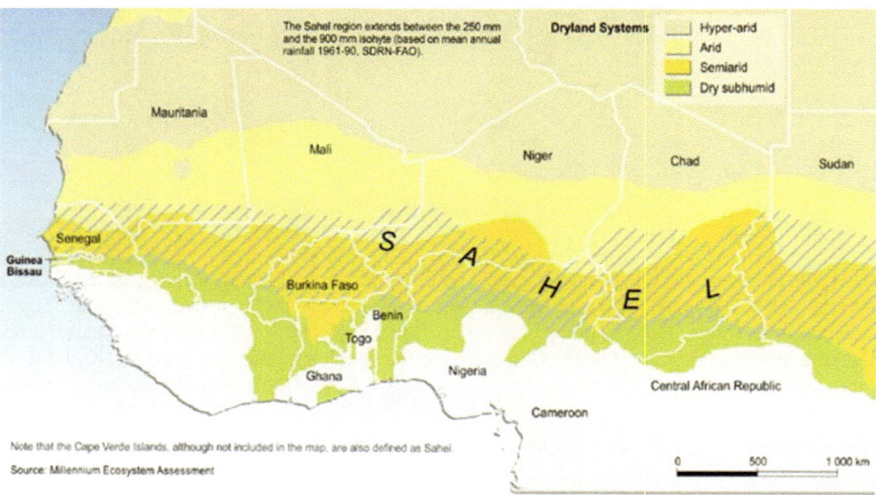

Fig. 1.11 The Sahel area of Africa. Global warming is driving desertification southwards and
damaging forests and grasslands. *Source*: Millennium Ecosystem Assessment http://www.millen-
niumassessment.org. Reproduced with permission of the United Nations Environment programme,
Nairobi, Kenya

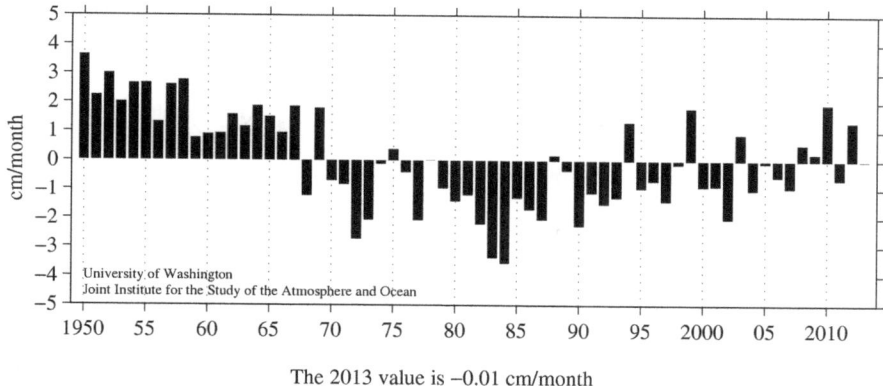

The 2013 value is −0.01 cm/month

NOAA NCDC Global Historical Climatology Network data

Fig. 1.12 Rainfall in the Sahel as a deviation in cm per month (June to October) from the average over the period 1950–2013. Global warming is a considered major factor in precipitation patterns in this region. *Source*: JISAO http://www.jisao.washington.edu/. Reproduced with permission of the Joint Institute for the Study of the Atmosphere and Ocean, University of Washington, Seattle, WA, USA

over, especially front gardens to provide off street parking resulting in much less natural percolation area for water to seep away into the ground. Instead this water rapidly flows into surface drains and very quickly finds its way into local water-courses that often flood. It is not only rain swelled rivers and streams that are causing flooding, but the normal road and combined sewer systems that simply can't cope with the massive downpours. Storms like these also wash vast quantities of debris that block drains and culverts, as well as smaller streams and rivers that can cause unexpected flooding.

Over 1.5 million homes in the UK are built on river flood plains requiring significant protection from expensive flood defences. Of these about half a million are at a high risk from flooding. As storm events become more common and severe then it is becoming increasingly difficult to protect homes as was seen during the most recent flooding during the winter of 2013/2014. Recent floods have highlighted the need to protect important infrastructure such as electricity sub-stations and water treatment plants from flooding, and even relocate them out of high-risk areas. This problem is not just confined to Europe, but flooding is a worldwide problem that will require all stakeholders to prepare and plan for increasing threats of flooding as global temperatures rise.

In Ireland and the UK people whose homes are at risk of flooding are beginning to become much more resilient to the problem. On advice from insurance companies they are redesigning their homes to minimize damage in the future. Apart from using portable flood barriers for doors, the floors and walls are waterproofed to protect foundations; walls are painted rather than wallpapered; carpeted and wooden floors are being replaced with tiles; all the electrical sockets are raised above the expected water level including the internal and external junction and meter boxes; and ensuring that all valuables that are not waterproof are kept upstairs. At a community level with the help and advice of the Environment Agency and the local authorities people are also working together to improve surface water drainage in the area by removing garden paving, ensuring culverts are kept clean, and that litter and other material that could be mobilized during a storm and cause blockages to drains doesn't occur.

It is not possible to eliminate all the damage when extreme events occur. For example, when Hurricanes Christian and Xavier passed over Northern Germany in October and December 2013 respectively causing immense damage and the most severe and extensive flooding on record. But with better planning both at the national and local level, and some acceptance of the inevitability of future vulnerabilities of some areas, then damage can be minimized.

1.3.3   Wind

Winds are generated by differences in sea and land temperatures and these are changing due to surface warming of the oceans causing changes in wind patterns and intensity. As the planet heats up these changes in wind patterns and intensity will become more intense requiring different local and regional responses. The jet stream is also affected which causes localized cold spells in northern Europe. Wind patterns in some areas have altered due to changes in storm tracks and while the number of hurricanes in the North Atlantic have not increased in number it is certain that their intensity has (Fig. 1.13). It is predicted that the intensity of Atlantic

Fig. 1.13 Hurricane Sandy: Monday 29 October, 2012. It's not only severe winds that cause damage but also intense rainfall and coastal storm surges. *Source*: NASA Earth Observatory http://earthobservatory.nasa.gov/. Reproduced with permission, NASA. Greenbelt, MD, USA

hurricanes will increase as ocean temperatures rise and with each 1 °C it is predicted that the precipitation rate will increase by 6–18 % and the wind intensity by 1–8 %.

More information: Karl, T.R., Melillo, J.M. and Peterson, T.C. (eds.) (2009) United States Global Change Research Program. Cambridge University Press, New York, NY, USA

The effects of these storms can be immense. Eleven million people were affected by Super Typhoon Haiyan which hit the Philippines in November, 2013 (Fig. 1.14) which is just one in a series of super strong storms that have wreaked havoc over the past 5 years. Tropical cyclones like this are caused by warm air rising off the sea. Tropical storms are common in the Philippines, and Super Typhoon Haiyan (category 5 at landfall) was the 25th such storm that season. It had an average wind speed of 233 km/h. In December 2012 Typhoon Bohpa also hit the Philippines with higher average winds of 261 km/h (also category 5 at landfall) but with less strong gusts making it less impactful although well over 1000 peoples died in that storm. In comparison Hurricane Katrina was only a category 3 with average winds of 205 km/h although the death toll was in excess of 1800. What made Super Typhoon Haiyan so special and so utterly devastating was just the amount of energy in the

Fig. 1.14 Super Typhoon Haiyan as it approached the Philippines on 7 November. The storm was 600 km in diameter wide enough to engulf Europe from London to Berlin. *Source*: NOAA http:// www.noaa.gov/. Reproduced with permission of the National Oceanic and Atmospheric Administration, Washington, DC, USA

winds, with sustained winds of some 320 km/h with gusts of up to 378 km/h, that is 199 and 235 mph respectively making it probably the strongest winds ever recorded.

Global warming is now rapidly heating the oceans and so there is simply more energy being generated creating these incredibly fast cyclones.

Although it is too early yet to be able to definitely link storms such as Hurricane Katrina (2005), Tropical Storm Sandy (2012) and now Typhoon Haiyan to global warming, it is looking increasingly likely that their intensity is being driven by the extra heating of the oceans. So we have to start thinking how we are going to prepare and mitigate such disasters in the future and that means taking climate change seriously when it comes to building homes, living in low lying areas susceptible to storm surges and also having practiced and effective emergency plans in place (Fig. 1.15). Of course there have been many other natural disasters such as earthquakes and tsunamis, which have caused massive death tolls (Indian Ocean 2004 and Japan 2011) and they too would benefit from similar actions.

Climate modellers in Ireland predict a small overall increase in the energy content of the wind by up to 8 % during the winter and a decrease of between 4 and 14 % in the summer months. With an increasing reliance on wind power this may

Fig. 1.15 It is not just the wind that causes the devastation but the storm surges and flooding associated with these super storms. Super Typhoon Haiyan was particularly devastating due to the low lying land and islands that makes up much of the Philippines. *Source*: Reuters. Reproduced under licence Reuters Thompson

have a significant impact on generation potential. This does not exclude the real possibility of severe storm events associated with hurricanes occurring in the Western Atlantic seaboard of America affecting the country.

> **We are part of a planet that has always changed over time. It is currently happening at such a fast rate that we (and other species) don't have enough time to naturally adjust and adapt.**

1.4   Conclusions

- The global population is continuing to rapidly expand and will result in a global population of between 11 and 12 billion by the end of the century. However, the total fertility rate is very close to the point where population growth rate will stabilize.
- There is significant evidence to support the concept that there is a finite population level which is sustainable without irreversibly damaging the planet. Beyond this limit irreversible damage will be caused to Earth's ecosystems and its associated ecological services. Many scientists believe we are very close to or have already exceeded that limit.
- Climate is changing with more frequent and intense precipitation events, more severe and devastating storms, increased global temperatures, increased desertification and wildfires, increased risk of inland and coastal flooding.
- We all need to look very critically at where we live and the predicted risks from climate change whether that is increased wind speeds, flooding or water scarcity. We can help ourselves by careful planning and being prepared for these seasonal and often unexpected events. At the Local, Regional and National level much can be done in terms of better physical planning on the ground, building more appropriate homes and having emergency response action plans ready.
- Climate change will result in major migration from dry areas and due to more pressure on limited and dwindling resources. As our climate changes there will be a gradual repositioning of population within and between continents.

> **Don't Panic**
> It sounds bad .... it is bad
> But it is not the end ... it's simply a new beginning
>
> **The first step is to accept that our climate is changing and that there is a finite global population that is sustainable.**

Homework!

Try and put yourself into a population context. This is actually really hard to do. I suppose standing in the middle of Mumbai is one way of feeling population pressure, but there is an easier way. I would like you to examine the websites below that allow you to explore population trends:

http://www.ined.fr/en/pop_figures/countries_of_the_world/
http://www.census.gov/popclock/

Then do a couple of simple tasks to see where you and your family fit into the global family. Using the website below, or another population calculator, calculate the number of people on the planet when you were born, and then do this for the rest of family including your parents and grandparents. Try and create a simple plot of population (vertical or y-axis) against the year (horizontal or x-axis) using your family dates to generate the global population at that time, but use today's global population for the current year.

http://populationaction.org/Articles/Whats_Your_Number/Summary.php

When you are ready then move onto the next step which poses the question '*what is sustainability?*'

References and Further Reading

CIA. (2012). *The world fact book*. Washington, DC: Central Intelligence Agency. Retrieved from https://www.cia.gov/library/publications/resources/the-world-factbook/
de Sherbinin, A., Carr, D., Cassels, S., & Jiang, L. (2007). Population and environment. *Annual Review of Environmental Resources, 32*, 345–373. Retrieved from http://www.ncbi.nlm.nih.gov/pmc/articles/PMC2792934/
Ehrlich, P. R. (1968). *The population bomb*. New York: Ballantine Books.
IPCC. (2007). *IPCC fourth assessment report. Working Group I Report: The physical science basis* (WG1-AR4). Geneva, Switzerland: Intergovernmental Panel on Climate Change. Published on behalf of the IPCC by Cambridge University Press, Cambridge, England.
Malthus, T. R. (1798). *An essay on the principle of population as it affects the future improvement of society*. London, England: Johnson.
Meadows, D. H., Meadows, D. L., Randers, J., & Behrens, W. W. (1972). *The limits to growth: A report for the Club of Rome's project on the predicament of mankind*. New York, NY: Universe Books.
US National Academy of Science. (1963). *The growth of world population*. Washington, DC: US National Academy of Science.

CIA World Fact Book

https://www.cia.gov/library/publications/the-world-factbook/index.html

Population

http://ngm.nationalgeographic.com/7-billion
http://www.un.org/en/development/desa/population/
http://www.unfpa.org/pds/
http://www.worldometers.info/world-population/

National Oceanic and Atmospheric Administration

http://www.nhc.noaa.gov/
http://www.youtube.com/watch?v=dmLYjs0kwnc&feature=youtu.be
http://www.youtube.com/watch?v=N4SCe_YCw_s

Solar Activity

http://science.nasa.gov/science-news/science-at-nasa/2013/08jan_sunclimate/

Arctic Ice Data

http://nsidc.org/arcticseaicenews/

General

http://climate.nasa.gov/evidence/
http://globalchange.gov/home.html
http://www.climate.gov/

Chapter 2
What Is Sustainability?

In this chapter we explore the meaning of sustainability and how it can help and hinder our response to dealing with global warming. Image by Leah-Anne Thompson. Reproduced under licence

2.1 Only One Earth

Growing up as a teenager in rural Gloucestershire I seemed to have missed the swinging sixties. The decade was in fact a heady period of cold war, rapid technological and industrial expansion, and the beginning of consumerism after the long period of post war austerity. Perhaps I was still a little too young to appreciate all of this, I like to think so. However, one thing I do remember as being exciting was the

© Springer International Publishing Switzerland 2015   33
N.F. Gray, *Facing Up to Global Warming*, DOI 10.1007/978-3-319-20146-7_2

space race between the USA and Russia, and the birth of telecommunication satellites such as Telstar in 1962. Telstar was also the name of a hit record by the Tornados later the same year. It was also the period when we first began to see grainy images of our planet from space.

In 1969 Life magazine reproduced the first picture taken by man of planet Earth, taken during the Apollo 8 mission (Fig. 2.1). That picture showed us that while the planet seems vast for those of us on the ground, it is in fact finite which means that all our resources are finite as well. This was a major point in the environmental movement, and the picture of planet Earth with its green land and blue seas become an iconic symbol of environmentalism.

The picture tells us quite bluntly that this is all we have in terms of space and resources, and it has to last humankind forever. Regardless of what the science fiction writers may suggest, once these resources are exhausted or our natural ecosystems are destroyed then there is nowhere else to go. These resources have to last us all on planet Earth forever. So it is important to understand that the word environment is not

Fig. 2.1 The first image of the whole planet Earth taken by man that featured on the cover of Life magazine. Taken at a distance of 30,000km with south at the top with North America in the bottom right. *Source*: http://history.nasa.gov/ap08fj/photos/a/as08-16-2593.jpg. Reproduced with permission of NASA, Washington DC, USA

an abstract term but describes our one and only home. Unfortunately its meaning has become weakened through general use becoming an intangible entity such as the terms arts, heritage etc. But the environment is the place and system which keeps us, and all species that we share the planet with, alive. **Quite simply, without a healthy and well managed environment we can't survive**.

Our environment is in crisis and has been for a long time, so long in fact that we have become immune to the numerous and often quite stark warnings (Sect. 1.1). Pressing environmental concerns include: the hole in the ozone layer, acid rain, accumulation of toxins in the food chain, loss of biodiversity, loss of topsoil and desertification, pollution and acidification of the seas, lakes and rivers, unsustainable exploitation of non-renewable and renewable resources (which can also be depleted) including forests, fish stocks and freshwater. None of these problems have gone away, but we now have a greater problem … this is global warming induced climate change.

> *Global warming will alter the very nature of the planet's surface on which we live in terms of water availability, food production and also how and where we can live.*

According to ecological footprint analysis, if everyone lived as we live here in Ireland or the UK then we would need at least three Earths to support our current lifestyle (Sect. 7.2). Increase that to five Earths for the USA. The problem is that we only have one Earth which we all have to share as equal stakeholders. So how does that work? It's quite simple. It is only poverty of others that has allowed us to live the way in which we do and has possibly stopped the Earth already plummeting into ecological meltdown (Fig. 2.2).

Fig. 2.2 This iconic book by Susan George first published in 1976 explores the inequality between developed and developing nations and led to the concept of global justice. Reproduced with permission of Penguin Books, London

Our lifestyles have evolved largely through the past colonization of developing countries, the exploitation of which has continued in many countries through corporate exploitation and sometimes corruption. **Everyone is entitled to a fair share of the Earth's resources … aren't they**? China and India are both booming economies emerging from extensive poverty, and they have their eyes set on a similar lifestyle to the west. Would this lead us to the brink of ecological disaster? Yet it is inconceivable that others should be denied the lifestyle that we have enjoyed here for so long. So something must be done to make human life (collectively) on Earth both equitable and sustainable.

Sustainable

Adjective

1. *Able to be maintained at a certain rate or level.*
2. *(esp. of development, exploitation, or agriculture) Conserving an ecological balance by avoiding depletion of natural resources.*

Life began 3.6 billion years ago with bacteria and photosynthetic algae extracting carbon dioxide from the atmosphere and releasing oxygen (a waste product) back into the atmosphere. Plants evolved and continued to remove CO_2 and storing it over millennia as coal, natural gas, and peat. Likewise small creatures removed the CO_2 stored in seawater as carbonate to build shells and exoskeletons and as they died and sank to the bottom of the ocean they built up boundless layers of sedimentary carbonate rocks. So bioforms have changed the planet from its original lifeless state to what we see around us today. The atmosphere, oceans and that thin terrestrial layer on which we all live has all been changed, some may say engineered, by evolving diversity of living species.

The Earth today has evolved into a hugely complex interrelated life form, with the millions of species that comprise the planet ecosystems (including humankind which is just one of those species) linked to each other through numerous delicate relationships. These relationships are also highly dependent on the climate and other physical processes. Gaia was the Greek goddess of the Earth, the mother of all. In 1979 James Lovelock published a book *'Gaia: a New Look at Life on Earth'* where he used the term to explain the concept that our planet was in fact a highly complex interrelated system in which all life forms are an important part creating an interdependent giant life form—Earth. Many scientists have dismissed the concept of Gaia as simply a metaphysical description of Earth's inorganic and biological processes. His second book presented the scientific evidence for his theory, but what is clear is that the Earth is still evolving and all life forms are part of this continuing evolution (Lovelock 2000, 2007, 2010).

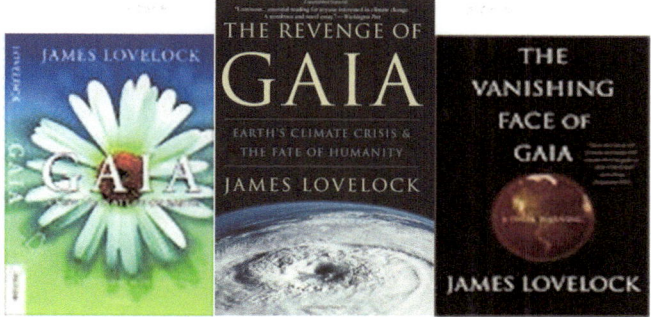

Covers reproduced with permission of Oxford University Press, Oxford, UK and Basic Books, New York, USA

> *The Gaia hypothesis states that temperature, oxidation state, acidity, water are all kept constant automatically and unconsciously by the biota through self-regulating homeostasis which is regulated by active feedback mechanisms.*

It seems bizarre in this consumerist society that we currently get our oxygen free, our light free, the air cleaned for free, our heat and energy that drives the planet and its' ecosystems all for free. The energy that grows our food is free, our food is pollinated for free to produce fruit, nuts and seeds, fish are free, and all this relies on a healthy balanced planet which we take completely for granted. James Lovelock concludes that **it is too late to reverse global warming and argues that mankind must prepare to adapt to a very hot future**.

2.2   What Do We Mean by Sustainable?

For our continued existence on planet Earth to be sustainable we need to ensure that our lifestyle does not prevent future generations from also experiencing a full and meaningful life. This doesn't necessarily mean the same wealth or consumerism levels as we have today. Wealth and consumerism are not really prerequisites to a full and meaningful life and many people are happy even at comparatively low consumption levels. Research has supported this idea, as we will see later, but of course a certain level of income and support is needed to prevent poverty and to sustain wellbeing. However, the question is **at what level does this need end and**

consumerism itself becomes the goal rather than wellbeing? This is explored further in Sect. 14.3.

Sustainability and sustainable development are often used interchangeably but they are actually fundamentally different.

- *Sustainability* *is the endpoint where civilization can thrive within the limits posed by only having one planet.* Where we are going with this is trying to identify what our individual share is and learning how to survive in a meaningful and complete way within its confines.
- *Sustainable development* *is the process of getting from here and now to a point of sustainability.* This book explores your journey to living within your equal share of a single planet Earth.

I suppose that sustainability is the nirvana for an environmentalist. However, it is interesting to look at synonyms for the word nirvana. These include paradise, heaven, illusion and fantasy. So the next important question we have to address is **whether global sustainability could be a reality or is just a fantasy**?

There are hundreds if not thousands of definitions of sustainable development and one of the things I always get my students to do is to create a unique personal definition of their own. The most famous definition is that produced by the Brundtland Commission in 1987 and is without doubt the most quoted environmentally related definition: *Sustainable development is development that meets the needs of the present without compromising the ability of future generations to meet their own needs.*

In fact the definition in the report is subtly different: '**Humanity has the ability to make development sustainable**—*to ensure that it meets the needs of the present without compromising the ability of future generations to meet their own needs*' (World Commission on Environment and Development 1987).

This iconic definition has received a lot of criticism with many seeing it as weak and ill defined, while others regard it as condescending and paternalistic. It is certainly more survivalist than environmental. Yet sustainability has become to be seen by all stakeholders, whether they be environmentalists or industrialists, as the nucleus on which the environment and we ourselves can live in harmony while both remaining mostly intact. Yet how can this be achieved as sustainability lacks precise structures or systems to achieve the desired outcomes, even if we knew exactly what those outcomes should be? So it remains a largely abstract concept, even though nearly all the discussions we read or hear relating to the environment, biodiversity and even economics have become a discourse on sustainability. So all our discussions about conservation, climate change, population and the environment in general, have become a sort of do-loop, with everything coming back to sustainability. So much so, that the term sustainability is now as widely used as the term environmental, both being equally vague and perhaps today increasingly meaningless. The weakness of the definition has led to cosmetic environmentalism (i.e. promoting unsustainable activities as sustainable) as well as the inappropriate and misleading use of the term.

> *"Few development interventions or research initiatives these days can successfully attract funding unless the words 'sustainability' or 'sustainable' appear somewhere in the proposal to the funding agency"* (**Bell and Morse** 2008).

So what precisely are the problems with sustainability as a concept? Currently the terms sustainability and sustainable development are closely linked in our minds to global economic, environmental and social crises. So in some sense they have quite negative connotations. Economic growth results in an increase in the rate of production and consumption of both goods and services. This in turn leads to an increase in use of resources, and an increase in the production of waste, by-products and a wide range of pollutants. This will be increasingly evident as we begin to exploit the vast reserves of fossil fuels associated with oil shales and fracking for gas (Sect. 3.2.1.1). Therefore, if the mechanisms of economic growth are not controlled or altered they impact on all of us in an increasingly negative manner through the over exploitation of natural resources, the ability of natural systems to assimilate waste, and an increasingly degraded environment (physical, chemical and biological).

Let's summarize:

- Sustainability addresses the relationship between economic development, its impact on the physical, institutional and intellectual structure of society and the natural world as a whole (i.e. the environment).
- It defines the relationship between dynamic human economic systems and slower changing ecological systems.
- Its objective according to many is to create a system whereby human individuals can flourish, human cultures can develop and diversity, complexity and function of ecological life support systems are protected (Khalili 2011).
- Sustainability is the economic state in which the demands placed upon the environment and natural resources by people and commerce can be met without reducing the capacity of the environment to provide for future generations (Gladwin et al. 1993).

Does this get us any further? Not really, so perhaps it is useful to go back to the very beginning of the concept.

The Nobel Economist Sir John Hicks first conceptualized the concept of sustainability in terms of income in 1946 as *'the amount, whether natural or financial capital, one could consume during a period and still be as well off at the end of that period.'* I suspect that many of us would recognize this basic economic concept from Mary Poppins: expenditure exceeds capital—result misery, expenditure within capital—result happiness. It was not until 1972 that it was first used in context of the future of humankind in the book *Blueprint for Survival*. But it would be another

15 years before the concept took on global significance with the publication of the Brundtland Report (World Commission on Environmental Development 1987). This resulted in a global discourse on what sustainability was and how to define it. For me, it was a definition in 1991 by Solow that has come closest to what I feel sustainability is or could be: '*an obligation or injunction to conduct ourselves so that we leave to the future the options and the capacity to be as well off as we are, not to satisfy ourselves by impoverishing our successors.*' I like this definition as it uses the word obligation and with it brings the moral responsibility that we all have to use our planet wisely, fairly and unselfishly. The concept of sustainability still continues to evolve as our understanding of the complex relationship between economic development and the environment unfolds. The need to define and pursue sustainability is increasingly urgent as the environmental crisis deepens.

'*Human influence on the climate system is clear and growing, with impacts observed on all continents. If left unchecked, climate change will increase the likelihood of severe, pervasive and irreversible impacts for people and ecosystems. However, options are available to adapt to climate change and implementing stringent mitigations activities can ensure that the impacts of climate change remain within a manageable range, creating a brighter and more sustainable future.*'

Intergovernmental Panel on Climate Change (IPCC)
Copenhagen 2nd November, 2014.

More information: http://www.ipcc.ch/index.htm

So where are we right now? Sustainability is currently perceived to be comprised of three interdependent systems the so called economy–ecology–social nexus. All three systems have to be addressed simultaneously if sustainable solutions to the environmental crisis are to be found. **Economic Sustainability** focuses on the portion of natural resources (both renewable and non-renewable) that provides the physical input into the production process for goods and services (i.e. economically the maintenance of the man-made capital). **Environmental Sustainability** focuses of the maintenance of environmental services. Often referred to as the life support system but it is much more than this. **Social Sustainability** addresses poverty and human development. The maintenance of the life support systems is the predominant prerequisite for social sustainability.

The relationship between these three sustainability systems was illustrated at the 2005 World Summit by three interlocking circles (United Nations General Assembly 2005). Note that the social-economic interactions should be equitable, the economic–environmental relations must be viable and that the environmental–social relationship must be bearable. The theory is that sustainability is an equal balance with each sector of equal importance. This is clearly untrue and quite misleading, perhaps even dangerous, as the environment is vital to our survival. This nexus suggests

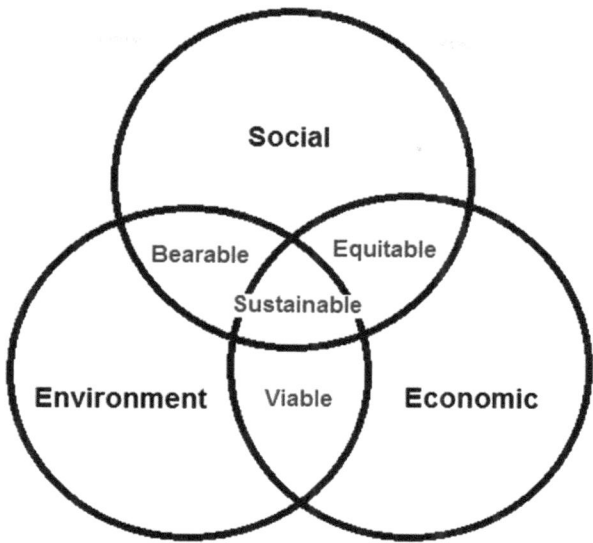

Fig. 2.3 The economy–ecology–social nexus formed the basis of early environmental sustainability theory

Fig. 2.4 The economy–ecology–social nexus has become distorted controlled primarily by economic and social expansion without regard to the biocapacity of the Earth to support it

that there are no limits to growth and that there is always more free resources and capacity to assimilate waste on which to create further growth, which is not the case (Fig. 2.3).

In reality the economic dimension is dominating through continuous growth with the environmental dimension being rapidly depleted. As the environment is limiting and its resources cannot be expanded, Society must flourish within these limits and the economy must then reflect and service the needs of society within those limits. **To create a sustainable society the environmental dimension must gain more importance and for it to be reliably protected** (Fig. 2.4).

Fig. 2.5 A more sustainable
economy–ecology–social
nexus design

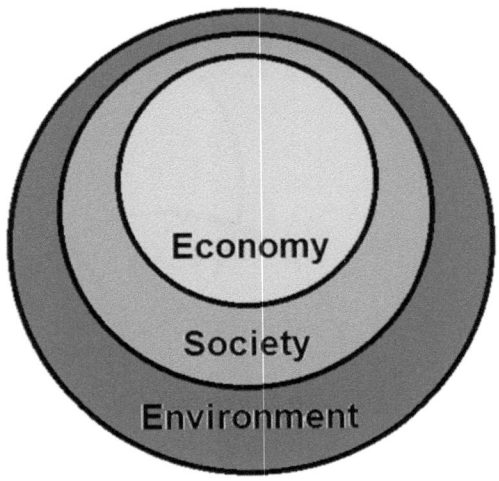

The reality of the economy–ecology–social nexus is that the economy and
social demands far exceeds the Earth's biocapacity … what we need is to
radically adjust our understanding of how this relationship really works

A better model of the economy–ecology–social nexus requires the economy to
operate within limits set by society (e.g. to reflect values such as fairness, justice
and liberty). Society flourishes within limits set by the environment, so that the
three elements are not equal but each serves the other (Fig. 2.5). This is not unique
and the concept has been widely adopted in water supply using a new management
approach known as demand-side management where expansion of water demand
has to be satisfied within a limited available volume of water so that any expansion
has to be achieved through the conservation of supplies and their better management
(Sect. 11.2).

We need to decide on the limits that humankind can exploit the Earth without
destroying its ability to be self-sustaining and self-regulating. Limits are needed
globally, regionally, nationally, locally and individually. The problem is that we
are personally not setting any targets at all, with the ability to pay the only constraint for most of us. Everyone is demanding their rightful share, from the developing nations to industrial manufacturers. **Here lies the conundrum … what is
our share**?

2.2.1   Environmental Sustainability

How we view and relate to the environment is often seen as two opposing theories both of which have their routes in the seventeenth century.

- **Technocentrism** (also known as cornucopianism, expansionism, shallow environmentalism or weak sustainability)
- **Ecocentrism** (i.e. neo-Malthusianism, preservation, steady-stateness, deep ecology or strong sustainability)

Technocentrism centres almost entirely on human wellbeing. Here sustainability is reached if enough investment in manmade and human capital is made to compensate for the degradation of natural capital. It relies heavily on technology solving our environmental problems without causing us to deviate from economic growth. For example, whole planet engineering solutions such as global dimming could in theory allow us to overcome the problem of global warming associated with carbon dioxide emissions, by reducing the energy from the sun getting to the surface of the planet, without having to consider reducing our use of fossil fuels. While technological and scientific advances are critical to dealing with global warming, can they also solve all the problems we now face? Can man actually create an entirely mechanistic planet, rather like a space station, where natural processes are all replaced by computer driven technological systems? Personally I don't think so, and while the environment has absorbed technological mistakes in the past, it is unlikely that it could recover from major damage to whole environmental processes caused by whole planet engineering projects that go wrong. However, many people strongly believe that the fate of humankind should not be left to natural processes.

In contrast, ecocentrism, normally referred neo-Malthusianism (Sect. 1.2.3), is based on the assumption that natural capital should be maintained and nurtured. Natural capital is sustained when renewable resources are used according to their regeneration rate and impact on the ecosphere. Importantly humankind should not exceed the assimilative capacity of planet Earth. Strict adherents to strong sustainability believe that non-renewables are so valuable that their use should be restricted.

Today we tend to accept a middle-of-the-road approach … **Sustaincentrism**. This recent concept accepts that resources are finite and defines the extent to which natural systems can absorb and equilibrate human caused disruptions to Earth's ecological processes. This theory accepts that the global ecosystem is finite, nongrowing, materially closed, vulnerable to human interference and limited in its regenerative and assimilative capacity. Therefore in order for an economic system to provide goods and services to humanity it must sustain all ecological systems, since a change in one significantly affects the other.

> *Sustainability has become very discipline biased with different classifications, definitions and functions, making the transfer of policy into action very difficult and often confused. To some extent we have stalled in our attempts to be proactive by uncertainty as what is the best action to take.*

There is serious concern over the sustainability of consumption as the result of increasing evidence of long-term damage being done to global environmental and ecological processes. Previously impacts from pollution tended to be local, now they are having regional and possibly global effects.

Significant disagreement developed between environmentalists and industrialists in the 1970s. Environmentalists believe that we have to preserve the natural systems of our planet whatever it takes, and that humankind has no more right to the planet's resources than any other species (i.e. Ecocentrism). This was a very unpopular ideology at that time and coincided with the publication of the book *Small is Beautiful* which gave rise to the idea that we were all doomed to live a low-level alternative existence in order to achieve a sustainable world. The book was even more poignant having been written by a leading industrial economist. However, it was during this period that environmentalism was seen to be, quite wrongly, as against economic development and growth.

> *'If we squander the capital represented by living nature around us, we threaten life itself.'*
> Peace is threatened by the desire for wealth which '... *depends on making inordinately large demands on limited world resources ...'*
> *'Localization rather than globalization'*
> **Schumacher, E.F. (1973) Small is Beautiful: Economics as if People Matter,** *published by Penguin Books*

Sustainability is an opportunity to give us a middle way. We cannot simply give up our existing economic model to solve our environmental crises without this leading to the total collapse of society as we know it. **We need a slow ordered transition to a low-energy economy not only to stabilize global warming, but to sustain our ever growing global population and protect them from the increasing threats of, hunger, water shortages, pollution, disease including antibiotic resistant bacteria and many other global threats**.

2.2.2   Stern

The Stern Committee looked at just this problem, how to alter our current global economy without derailing it. The Stern Review on the *Economics of Climate Change* (2006) was carried out for the UK Government (Stern 2007). The review was not primarily about solving climate change, much to the disappointment of some environmentalists, it was largely about how the economic market and economic development would be affected by these changes and how these could be minimized. To a great extent it is about how do we make an ordered transformation from our current resource rich society where energy is plentiful, still relatively cheap and its use unregulated, to a resource-limited society, generally referred to as a **low-carbon economy**.

> The Stern Review states that *'climate change is the greatest and widest-ranging market failure ever seen, presenting a unique challenge for economics.'*

The report is large and complex, but the key findings are summarized below in bold and the comments that have been added are mine and not those of the committee:

- **The benefits of strong, early action on climate change outweigh the costs**.

  - One of the failings in our attempts to deal with climate change at the national level is that we have tried to make it cost effective. Climate change is perceived to be an economic opportunity where businesses can grow, create jobs and make profits. This is just not feasible where fossil fuel derived energy is cheaper than sustainable options. Tackling climate change should be seen in the same way as other infrastructural development or emergency planning.

- **The scientific evidence points to increasing risks of serious, irreversible impacts from climate change associated with business-as-usual (BAU) paths for emissions**.

  - The review clearly tells us that we have to change both in the way we do business and how we live our lives.

- **Climate change threatens the basic elements of life for people around the world including access to water, food production, health, and use of land and the environment**.

  - There is no scepticism here, but a clear and bold statement of fact.

- **The impacts of climate change are not evenly distributed—the poorest countries and people will suffer earliest and most. And if and when the damages appear it will be too late to reverse the process. Thus we are forced to look a long way ahead.**

  - The problem with this and many other types of global problems is that as long as our own weather is okay and farmers are able to sow and harvest their crops, and our water and electricity supplies remain in good order, then we are lured into a false sense of security. We don't tend to go to those areas most affected by climate change for holidays, so to a great extent it's out of sight and out of mind. But people are suffering on a daily basis from the effects of climate change through severe changes in weather patterns and local climate change. Most of these global warming induced changes are not reversible, so once we lose productive land to desertification, for example, it is essentially lost for centuries or millennia to come. What you and I emit today in terms of greenhouse gases (GHGs) will continue to have a direct effect on global warming for at least 100 years from now (Sect. 4.2), so we have to start dealing with this problem now.

- **Climate change may initially have small positive effects for a few developed countries, but it is likely to be very damaging for the much higher temperature increases expected by mid-to-late century under BAU scenarios.**

  - There will be a shift in food production from the American mid west to more northern areas. Cooler countries in the northern latitudes will attract more business as it develops a more temperate climate.

- **Integrated assessment modelling provides a tool for estimating the total impact on the economy; our estimates suggest that this is likely to be higher than previously suggested.**

  - The truth is that the current economic model that has evolved was developed in a different era and is no longer suitable for a world in crisis; where resources are rapidly depleting and our environment is on the verge of system collapse from over exploitation. We need a new economic model and this will require a significant rethink about growth and profit, as well as a change in the way we as consumers live our lives.

- **Emissions have been, and continue to be, driven by economic growth; yet stabilisation of greenhouse gas concentration in the atmosphere is feasible and consistent with continued growth.**

  - Economic growth is undoubtedly the primary cause for GHG emissions. Our problem is that the simplest way of sustaining a rapidly growing population is through economic growth. Demand creates employment and sustains communities. So the challenge is to decouple economic growth from emissions or find alternatives to this simple relationship.

- **Central estimates of the annual costs of achieving stabilisation between 500 and 550 ppm CO_2e are around 1 % of global GDP, if we start to take strong action now in 2006/2007. It would already be very difficult and costly to aim to stabilise at 450 ppm CO_2e. If we delay, the opportunity to stabilise at 500–550 ppm CO_2e may slip away.**

  - The reality of us stabilizing the planet's atmospheric CO_2e emissions at 450 ppm is now improbable and we are resetting targets to more realistic goals (Sect. 4.5). So we know that global warming is inevitable and will continue to increase in the short to medium term resulting in significant climate change. What we must do now is centre all our efforts into reducing emissions regardless of whatever these goals might be and simply to mitigate against higher global temperatures.

- **The transition to a low-carbon economy will bring challenges for competitiveness but also opportunities for growth. Policies to support the development of a range of low-carbon and high-efficiency technologies are required urgently**.

  - A lot of work has been going on behind the scenes to develop new technologies, although often linked with promises of new growth markets, especially in the renewable energy sectors. Again we need clear direction about what needs to be done not only at the industrial and commercial levels, but also in the state sectors. Of course the individual will drive this transition.

- **Establishing a carbon price, through tax, trading or regulation, is an essential foundation for climate change policy. Creating a broadly similar carbon price signal around the world, and using carbon finance to accelerate action in developing countries, are urgent priorities for international co-operation**.

  - A stable and realistic price for carbon is a prerequisite for reducing emissions. We cannot expect new innovations without investment and for companies to be able to manufacture and supply them at a profit; also alternative low-carbon energies must be competitive and this requires carbon taxation at a realistic level (Sect. 6.5).

- **Adaptation policy is crucial for dealing with the unavoidable impacts of climate change, but it has been under-emphasised in many countries**.

  - We are so lucky living in northern Europe where climate change so far has had little impact. However, it is not going to be possible to control problems such as flooding by simply building higher and higher defences. We need to build into our planning at every level the potential effects of climate change that may occur quite unexpectedly. We need to prepare ourselves for the changes that will occur both economically and socially not only regionally, but locally and personally,

- **An effective response to climate change will depend on creating the conditions for international collective action**.

  - We are all part of the problem as well as the solution. We are quick to highlight those countries that have the largest carbon footprints, however, we are all consumers and hence emitters of greenhouse gases. Therefore this is a global problem requiring a global solution, which means that everyone is a stakeholder in solving the issue.

- **There is still time to avoid the worst impacts of climate change if strong collective action starts now**.

  - Even a cynical old environmentalist like myself has to believe that we can deal with this issue. It is possible but it is going to require significant changes over the decades to come in our lifestyles and the framework of our society. Some of these changes will be very challenging as we will see in later chapters.

These conclusions from Stern clearly and equitably summarizes where we stood in 2006 in relation global warming and climate change. Yet in all the intervening years our progress in tackling these issues in both developed and developing countries has been painfully slow. But these conclusions are fundamental to how we should respond to the global dilemma of climate change.

> **The challenge is to decouple economic growth from GHG emissions or find alternatives to this simple relationship.**
>
> *Stern Committee*

2.3   So Where Are We Now Regarding Sustainability?

We seem to have come a long way from our early simple definitions of sustainability. Personally I remain uncertain as to what sustainability is, what its objectives should be, or how these objectives are to be achieved. One problem is that environmentalists are generally suspect of the idea of sustainable development seeing it as an oxymoron, as development inevitably leads to environmental degradation (Redclift 2005).

What does the term sustainability mean now? Has it simply become another buzz word like environmental? What will it mean in the future? Is it simply a way to maintain business as usual in the future, or is it about equality, liberation, and most importantly self-determination? What we need to start considering is taking more control over the rate of economic growth and making it less environmentally damaging. Remember, that ultimately individual consumers control growth. The Earth Charter describes sustainability as "*a sustainable global society founded on respect for nature, universal human rights, economic justice, and a culture of peace*" (The Earth Charter Initiative 2000).

Any definition must be factual, scientific, have a defined endpoint and be quantifiable. Perhaps, the need for an Irish or US constitution shows us that a simple phrase such as '*love they neighbour*' is just not up for the job. So perhaps we will need a global sustainability constitution giving precise agreed actions and endpoints. We all feel we know what sustainability means … it's a personal concept which differs from person to person … but can we actually set a rigid definition? The answer is perhaps we don't have to. Perhaps it is actually impossible to do, and that our inability to agree on a single 'catch all' definition is one of the stumbling blocks that is actually stopping us dealing with the challenges of global warming. What is important is that we all know what is required of us in order to deal with the problem of global warming and how to survive whatever climate change has in store for us individually and regionally.

> *It is probably impossible to have a universally acceptable definition of sustainability and sustainable development. It can be as simple or as complex as you want ... as long as it personally motivates you to act proactively to deal with the problems of global warming.*

2.4 Conclusions

- We must see ourselves as part of the natural system and we cannot exclude humanity in our vision of planet Earth nor must we see humanity in isolation.
- Any resolution of the environmental crisis must ensure continued economic stability, otherwise society will break down and we will enter a global dark age caused by famine and conflict.
- The concept of sustainability is the best mechanism that we have to ensure global stability and fairness, but it needs to have clear aims and objectives.
- We all have a moral responsibility to use our planet wisely, fairly and unselfishly.
- This is a global problem requiring a global solution, which means that everyone is a stakeholder in solving the issue.

The first step was to accept that our climate is changing and the planet does not have the capacity to sustain an unlimited population.

> **The second step is accepting Solow's definition of sustainability as** *'an obligation to conduct ourselves so that we leave to the future the options and the capacity to be as well off as we are, not to satisfy ourselves by impoverishing our successors'* **and personally agreeing to individually act to help achieve this.**

Homework!

Although we have had repeated conferences on climate change we have singularly failed at the national level to really come to grips with the problems, and in part this is because of the difficulty of seeing what precisely has to be done at the regional or local level. So it is down to you and me to solve this problem from the bottom up;

it will be anyhow when Governments eventually decide what exactly needs to be done. Therefore, let's make a start right now. We have seen that the development of a universal definition of sustainability is proving extremely difficult to achieve. It is, however, much simpler to write a personal definition. Such a definition should be personally inspirational and remind us why we are trying to make a difference by tackling global warming.

So what I would like you to do is to write your own definition of sustainability in no more than 50 words. I would like you to put this along with your population data in a personal portfolio. This can be anything from a computer file to a cardboard folder … you could even use the fridge if you have enough magnets. What is important is that all this material is kept together as it will form part of a personal plan.

To get you started have a look at some personal definitions of sustainability by my undergraduate students from Trinity College Dublin:

http://ournewclimate.blogspot.ie/search/label/Definition%20of%20sustainability

When you are ready then move onto step 3 which looks at the science and evidence for global warming.

References and Further Reading

Adams, W. M. (2006). *The future of sustainability: Re-thinking environment and development in the twenty-first century*. Report of the IUCN Renowned Thinkers Meeting, January 29–31, 2006. Gland, Switzerland: The World Conservation Union (IUCN).

Aligica, P. D. (2009) Julian Simon and the 'limits to growth' neo-Malthusianism. *The Electronic Journal of Sustainable Development, 1*(3), 73–84. Retrieved from http://mercatus.org/uploadedFiles/Mercatus/Publications/JULIAN_AND_THE_LIMITS_TO_GROWTH_NEO-MALTHUSIANISM.pdf

Bell, S., & Morse, S. (2008). *Sustainability indicators: Measuring the immeasurable?* (2nd ed.). London, England: Earthscan.

Campbell, T., & Mollica, D. (Eds.). (2009). *Sustainability*. Surrey, England: Ashgate.

Ehrenfeld, J. R. (2008). *Sustainability by design: A subversive strategy for transforming our consumer culture*. New Haven, CT: Yale University Press.

Gladwin, T. N., Kennelly, J. J., & Krause, T. (1993). Shifting paradigms for sustainable development: Implications for management theory and research. *The Academy of Management Review, 20*, 874–907.

Khalili, N. R. (2011). *Practical sustainability: From grounded theory to emerging strategies*. London, England: Palgrave Macmillan.

Lovelock, J. (2000). *Gaia: A new look at life on earth*. Oxford, England: Oxford University Press.

Lovelock, J. (2007). *The revenge of Gaia: Earth's climate crisis and the fate of humanity*. New York, NY: Basic Books.

Lovelock, J. (2010). *The vanishing face of Gaia: A final warning*. New York, NY: Basic Books.

Neumayer, E. (2010). *Weak versus strong sustainability* (3rd ed.). Cheltenham, England: Edward Elgar.

Redclift, M. (2005). Sustainable development (1987–2005): An oxymoron comes of age. *Sustainable Development, 13*(4), 212–227.

Robinson, J. (2004). Squaring the circle? Some thoughts on the idea of sustainable development. *Ecological Economics, 48*, 369–384.

Stern, N. (2007). *The economics of climate change: The Stern review.* Cambridge, England: Cambridge University Press.

The Earth Charter Initiative. (2000). The Earth Charter, San Jose, Coats Rica. Retrieved from http://www.earthcharterinaction.org/content/pages/Read-the-Charter.html

United Nations General Assembly. (2005). 2005 World Summit Outcome, Resolution A/60/1, adopted by the General Assembly on September 15, 2005.

World Commission on Environment and Development. (1987). *Our common future: The report of the World Commission on Environment and Development.* Oxford, England: Oxford University Press. Retrieved from http://www.un-documents.net/wced-ocf.htm

Chapter 3
The Concept of Resources

Are resources really going to run out and does renewable really mean that they are inexhaustible? In this chapter we explore the problems of ensuring that we have enough resources for all future generations. **Image: Jan Schäfer http://www.coci.org/**. Reproduced with permission

3.1 Renewable and Non-renewable Resources

When that first image of planet Earth was taken from space in 1968 it became clear that the amount of natural resources on the planet was limited (Sect. 2.1). It was a kind of watershed or some might say a wakeup call. Evidence of man's activities was clearly visible from space and what became evident was that once these resources had been used then there were simply no more. Just as many species have become extinct through our over exploitation, we are also coming dangerously close to eradicating many of our non-living natural resources as well. Also, as these resources

© Springer International Publishing Switzerland 2015
N.F. Gray, *Facing Up to Global Warming*, DOI 10.1007/978-3-319-20146-7_3

become increasingly scarce and more valuable, then ever larger risks are taken to locate and extract them, often resulting in unprecedented threats to the natural environment by putting biosystems under even greater pressure.

The majority of non-renewable resources have been formed over millions of years and are considered as non-renewable if either their quantities are limited or they cannot be replaced as fast as they are used. Non-renewable resources are used worldwide to create electricity, heat homes, to power vehicles and manufacture goods. I think it is fair to say that the majority of people, when referring to non-renewable resources visualize those resources used as energy or in the production of energy. However, this is no longer the case with a wide variety of minerals and other materials now falling into this category. The demand for non-renewables is closely linked to economic growth and consumerism, which in turn is linked to greenhouse (GHG) emissions.

3.1.1   Renewable Resources

Before we can fully understand what a non-renewable resource is, it may be helpful to consider those resources that are thought to be inexhaustible … the renewable resources. For example, a natural resource is considered renewable if it has the ability of being replaced through biological or other natural processes and replenished within a relatively short period of time. Paul Weiss, the Chairman of the Renewable Resources Study in 1962 defined a renewable resource as '*the total range of living organisms providing man with food, fibres, drugs, etc.*' Therefore soil, water, plants and animals are all renewable resources and are part of the natural biological processes involving all ecosystems (marine, terrestrial and freshwater). So we can consider agricultural food production, fish stocks, forestry for building and fuel all as renewable. Similarly energy generated from the sun, wind, waves, tidal movement, biomass and geothermal resources are all considered renewable.

Increasingly, however, renewable resources are also under threat with their sustainability threatened by industrial development, pollution, over-exploitation and global warming. So they must be carefully managed to avoid exceeding the natural world's capacity to replenish them. The problem of course is that in many cases these resources are not managed at all, they are just there, often living in a very precarious relationship with other species. A simple example is foraging. In the UK there are a large number of professional foragers who collect wild plants and mushrooms for restaurants and for markets. This is different to collecting blackberries, which has little impact on the habitat and in fact may be useful in controlling this problem plant. Foraging involves usually gathering quite rare and unusual plant material, and their collection and removal is having a significant impact on fragile coastal, woodland and other habitats. Just trampling over these areas causes significant damage to species diversity. Those who gain from their exploitation do not

Fig. 3.1 The theory of maximum sustainable yield (MSY) showing the sustainable yield curve

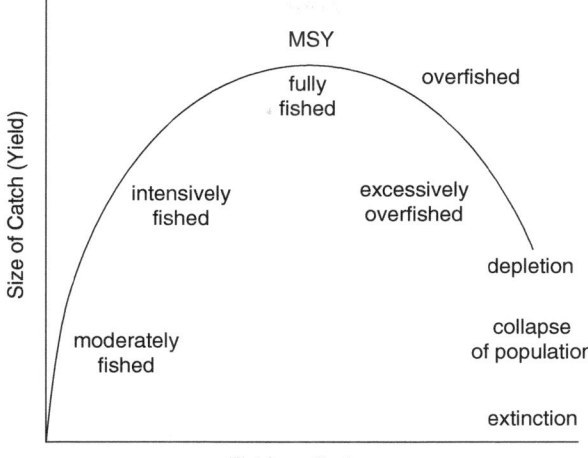

invest in sustaining them for the future and are often deluded into believing that this is somehow a very environmental and sustainable thing to do. So it poses the awkward question, **who do these resources belong to**?

Many of the natural renewable resources that we have taken for granted for generations are now under threat, with increasing numbers of common fish species in particular now close to extinction due to over exploitation. Biological and renewable resources need to be managed to prevent their over exploitation and eventual destruction. Management is loosely based on the concept of **sustainable yield** which is a common concept in agriculture. *Sustainable yield is the amount that can be taken from natural capital without reducing the sustainability of that capital* (Fig. 3.1). It is the amount that should not be harvested to ensure that the ecosystem service being utilized (i.e. the animal or plant being removed) is maintained at the same or increasing level over time. Fishing is the obvious example, so fish stocks must be able to replenish themselves to maintain the volume of stock from year to year. The problem in natural systems is that the yield is not constant and will vary from year to year due to factors such as climate conditions, disease, and natural disasters such as flooding and wildfires. Thus during lean years more of the capital must be left to ensure sustainable yields in the future. The same applies to groundwater used for drinking water supplies or irrigation. The rate of extraction must not exceed the rate of replenishment, and if it does then you begin to reduce the total volume of water in the aquifer which will slowly run out or the quality with be affected when older more mineralized water starts to be abstracted from lower depths.

The open seas are largely a free-for-all with huge fishing vessels able to use an array of satellite and radar technology to identify shoals and then to literally suck them up. Very often shoals are unique populations or communities, and each time this happens whole populations are made extinct, reducing genetic biodiversity as well as reducing viable breeding populations for the future. What makes it worse is that only a fraction of the fish will end up being eaten directly by humans, the majority being turned into cheap protein for livestock and pets. So in Europe cod, mackerel and many other species are verging on extinction simply because of greed. More globally tuna and sardines are also under threat. There could be plentiful fish for everyone if we simply followed some basic management rules in relation to yields. This has been demonstrated in Cornwall where the lobster was almost extinct due to overfishing. Through a co-operative approach a management programme has been put into place. This is based on three principles: (i) returning all female lobsters with eggs, (ii) returning small or immature individuals that do not conform to a minimum size standard (i.e. are not large enough to have bred), and (iii) controlling the total volume of lobsters removed each year. This has also been applied to other species such as line caught tuna for example. The success of the Cornish lobster has also been due to a captive breeding programme. When lobster eggs hatch the tiny lobster is easy prey for other species, so large mortalities are inevitable. By catching female lobsters with eggs and maintaining them in hatcheries, almost 100 % of the young lobsters survive and can then be returned to the sea when they have a grown to a size that they makes them less easily predated on. This has rapidly replenished stocks off the Cornish coast creating a sustainable fishery once more allowing the maximum sustainable yield to be increased, so fishermen are catching more and bigger lobsters than ever before.

> **The maximum sustainable yield or MSY is, theoretically, the largest yield/catch that can be taken from a species' natural stock over a prolonged period.**

3.1.2   Non-renewable Resource

In contrast, *a non-renewable resource is a natural resource that cannot be produced, grown, generated, or used on a scale which can sustain its consumption rate.* So once depleted there is no more available for future needs. The key examples are fossil fuels as the time scale and conditions for the formation of new fossil fuels is just not available. Other non-renewables are rare or non-recyclable metals and helium gas. However, what we have discovered in recent times is that a renewable resource which is consumed much faster than nature can create them can also become a non-renewable resource, as mentioned above examples include certain water resources especially aquifers and fish stocks.

Key Resources at Risk of Extinction

- Energy

  - Oil
  - Gas
  - Coal
  - Uranium

- Metals/ores

  - Most rare earth metals

- Gases

  - Helium

- Fisheries

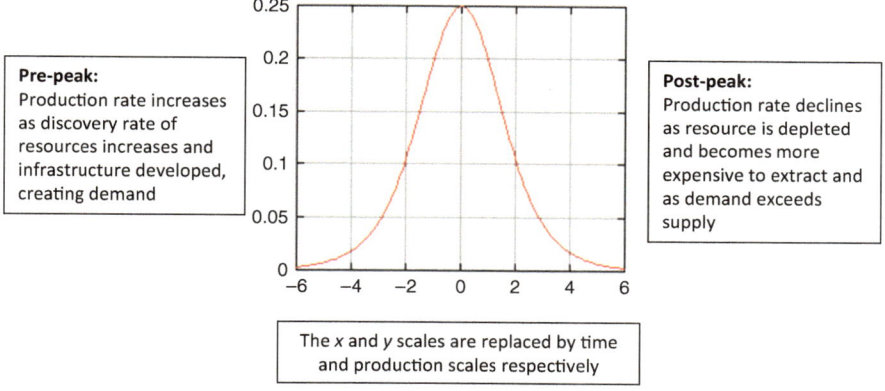

Pre-peak:
Production rate increases as discovery rate of resources increases and infrastructure developed, creating demand

Post-peak:
Production rate declines as resource is depleted and becomes more expensive to extract and as demand exceeds supply

The *x* and *y* scales are replaced by time and production scales respectively

Fig. 3.2 The Hubbert curve of peak production

3.1.2.1   Concept of Peak

The Hubbert peak theory was first published in 1956 and is a simple economic model that describes for any geographical area, such as a single country or the entire planet, that the rate of production of a resource (e.g. oil, metal etc.), if plotted over time, follows a bell shaped distribution (Fig. 3.2). This is based on the assumption that for any area the amount of that resource is finite. Therefore the rate of discovery which initially increases quickly must eventually reach a maximum and then decline. Peak theory can be applied to a variety of finite resources such as coal, oil, natural gas and uranium, but not normally to renewable resources such as water.

3.2   Key Extractable Resources

3.2.1   *Crude Oil/Petroleum*

The most sought after non-renewable resource and the key driver of the global economy is crude oil. It is also the scarcest resource in terms supply and demand. Like all fossil fuels, crude oil is the liquefied, fossilized remains of plants and animals that lived hundreds of millions of years ago. So once oil resources are depleted, they cannot be replaced. It is difficult to know exactly how much crude oil is still available but the best estimates currently suggest only 40 years of reserves are left at the current rate of consumption. By November 2011, global oil supplies had risen to a record high of 90.0 mb (million barrels) per day, and being a major driver of GHG emissions, the increasing use of oil and other fossil fuels is bad news for those trying to control global warming. Most oil producing countries and regions have already reached peak production, including Norway and the USA which includes the giant fields of Alaska and Texas (Figs. 3.3 and 3.4). In Europe the bulk of oil is

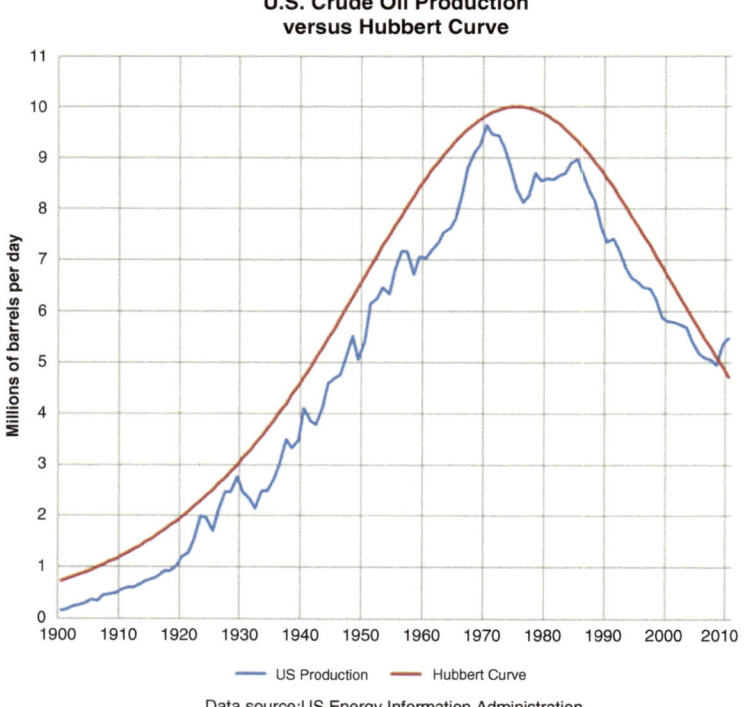

Fig. 3.3 US Crude Oil Production 1900–2010 matched against a typical Hubbert Curve. *Source*: http://www.eia.gov/. Reproduced with permissions of the US Energy Information Administration

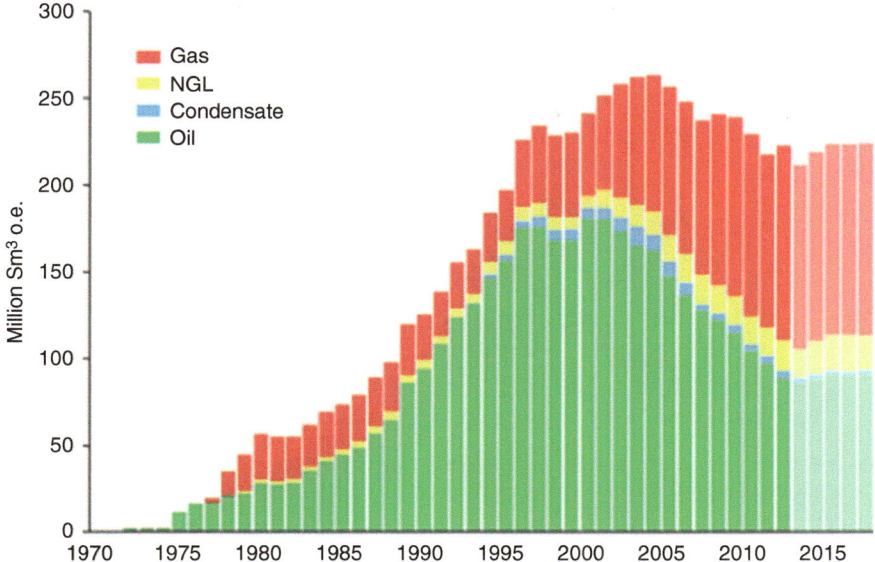

Fig. 3.4 Oil production off the coast of Norway has also reached peak. Two hundred and twenty-four million Saleable Standard cubic metres of oil equivalents (Sm³ o.e.) were produced in 2012. NGL are natural gas liquids. *Source*: http://www.npd.no/en/news/News/2013/The-Shelf-in-2012--press-releases/Petroleum-production/. Reproduced with permission of the Norwegian Petroleum Directorate

imported, creating a dependency on sources that are unpredictable and costly, so that at some stage crude oil will probably become the most expensive bulk resource on the planet.

As I said in the preface, peak oil has been repeatedly delayed due to new discoveries in the remotest, most hostile and most fragile environments. The environmental cost of oil exploration, extraction and transportation is vast. There have been some devastating accidents due to oil extraction in ecologically fragile areas. The Exxon Valdez sank in Prince William Sound in Alaska on March 24th 1989 with over half a million barrels of crude oil lost releasing 210,000 m³ (equivalent to 55,000,000 US gallons) of thick black toxic crude oil into the sea. In excess of 2100 km (1300 miles) of coastline was devastated and a further 28,000 km² (11,000 mile²) of sea affected. Oil spillages are not always caused by ships sinking. Platforms and extraction wells can also cause serious pollution as was seen at The Deep Water Horizon platform in the Gulf of Mexico. A major fire and explosion caused oil to escape from the well head between 20th April until 19th September, 2010 with 4.9 million barrels of oil escaping deep under water, equivalent to 780,000 m³ (210,000,000 US gallons) of crude oil was released. The coastlines of Louisiana, Mississippi, Alabama and Florida were devastated for over 790 km (491 miles) causing irrevocable damage to some of the most pristine and ecologically important marine waters in the US. Huge areas of the sea bed were affected by oil with tar balls still being washed up onto the coastline 4 years later.

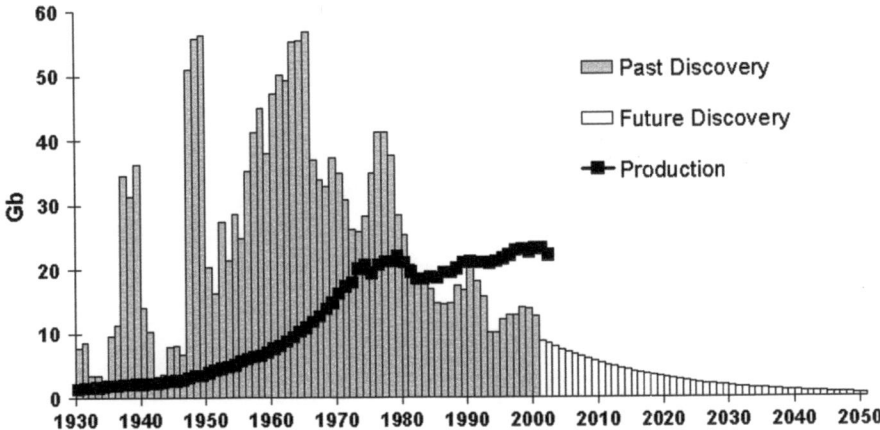

Fig. 3.5 Discovery of new oil wells and oil reserves are falling although demand continues to rise. *Source*: www.hubbertpeak.com. Reproduced with permission of Colin Campbell, HubbertPeak.com

Oil production mimics demand. Global demand for oil peaked in the late 1970s early 1980s and rapidly fell with production levelling off. Since then we have seen a steady rise in demand once again. Figure 3.5 suggests that it would be increasingly difficult for production to meet demand, even with new oil fields being discovered. However, existing reserves and new oil fields are being increasingly replaced by the extraction of oil sands and shales.

> *"All the easy oil and gas in the world has pretty much been found. Now comes the harder work in finding and producing oil from more challenging environments and work areas."*
>
> **William J. Cummings, Exxon-Mobil 2005**

3.2.1.1 Oil Sands and Shales

Unconventional sources, such as heavy crude oil, oil sands, and oil shale have not been traditionally counted as part of global oil reserves as they have generally been considered just too polluting and uneconomic to extract. The sands and shale are recovered by strip mining and the oil extracted by a thermal process which is three times more expensive than conventional drilling technology. This high energy intensive process also requires extensive refining with the result that three times more GHG emissions are produced per barrel of oil which is equivalent to 45 % more GHG missions overall after use.

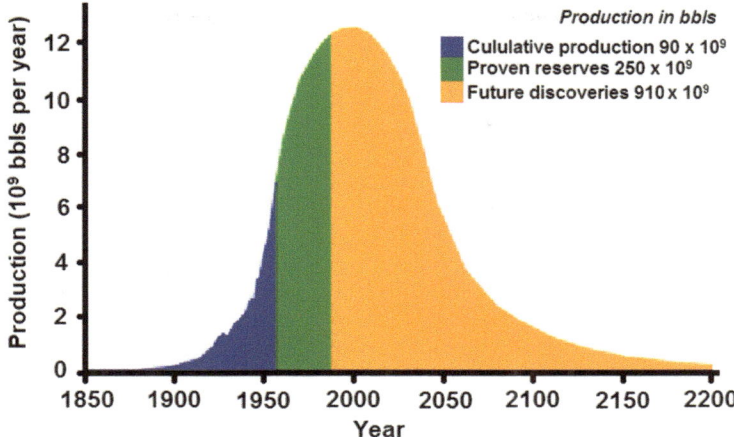

Fig. 3.6 Global reserves of all oil resources including oil sands and shales in barrels (bbls)

The majority of oil sands and shales are also heavily contaminated with sulphur and heavy metals resulting in a massive waste problem. Existing mines produce vast quantities of contaminated tailings and hydrocarbon sludge. So our new oil resources are far more environmentally damaging than oil wells. So our continued dependence on oil and our reluctance to switch to low-carbon technologies means that our global GHG emissions for oil will now continue to rise when it was thought that it would begin to shrink thereby stimulating research and investment in low-carbon technologies, as well as putting a ceiling on global CO_2e emissions from oil.

The key areas of production are the Orinoco Tar Sand in Venezuela with reserves of 513 billion barrels (8.16×10^{10} m^3), the Athabasca Oil Sands in Western Canada with 170 billion barrels (27×10^9 m^3) 1,400,000 km^2 and the Green River Formation, Utah in the USA with a massive reserve of 1.5 trillion barrels (38×10^{11} m^3) (Fig. 3.6). In theory, **these reserves represent more oil than has ever been extracted by drilling**!

The discovery of these new resources, although environmentally challenging and more polluting in terms of GHG emissions demonstrate that we still have to reach peak oil when all resources are considered. Production is linked to demand and demand is still not falling, and with these new discoveries it is unlikely now to decline. Complex production and waste treatment and disposal will mean more expensive oil and also higher associated GHG emissions. To some extent environmentalists were banking on oil running out and becoming very scarce and expensive in order to achieve a slowdown in GHG emissions. We can look at this in two ways. That we have to consider oil from these new resources as our long term supply of this useful product that we use in the manufacture of a broad range of materials, and so we should use them wisely. Or alternatively, we use these resources as a stop gap to help us develop better low-carbon energy technologies; perhaps both.

Table 3.1 Countries with a high dependency on coal for electricity generation in 2012

Country	%	Country	%
Mongolia	98	Israel	59
South Africa	94	Indonesia	44
Poland	86	USA	43
PR China	81	Germany	43
Australia	69	UK	29
India	68	Japan	27

Source: The World Coal Association http://www.worldcoal.org/resources/coal-statistics/. Reproduced with permission of the World Coal Association, London, UK. http://www.worldcoal.org/

What we cannot do is allow these resources to be used in the same way as we have squandered our previous oil reserves resulting in massive GHG emissions which have driven global warming. This would mean an end to our trying to control atmospheric CO_2 concentrations and lead to disaster.

> **Oil shales and sands have eliminated the problem of dwindling oil reserves in the coming century, but present an unparalleled environmental crisis in terms of surface pollution as well as out of control green house gas emissions.**

3.2.2 Coal/Lignite

Coal and lignite represent the most plentiful non-renewable energy resource on the planet and will eventually become the major source of fixed carbon energy in future. Coal is vitally important as a fuel source providing 41 % of global electricity (Table 3.1), 30 % of all primary global energy needs and is critical to many heavy industries, especially steel manufacture where it is used for 70 % of global production. Coal extraction is vast and continuing to rise, from 4677 Mt in 1990 to 7831 Mt in 2012, with an increase in output of 2.9 % between 2011 and 2012. China is by far the largest coal producer extracting about 45 % of the global annual output (Table 3.2) as well as being a major importer of both coal and lignite from countries such as Australia and Indonesia. In 2013 it imported 267 million tonnes, an increase of 14 % over the previous year. China has suffered from overproduction and unprofitable mines and as a result has closed many operations producing less than 90,000 tonnes per year. In the Autumn of 2014 they imposed an import tariff on all imported coal products of between 3 and 6 %, although its neighbour Indonesia was exempt. Under its 5 year energy plan (2011–2015) China estimated that it needed to create a further 860 million tonnes of new coal capacity to meet expected demand.

Table 3.2 Top ten producers of coal and lignite (brown coal) during 2012 in millions of tonnes (Mt)

Coal production per country		Lignite production per country	
Country	Mt	Country	Mt
PR China	3549	Germany	185
USA	935	Russia	78
India	595	Australia	73
Indonesia	433	USA	72
Australia	421	Turkey	66
Russia	359	Poland	64
South Africa	259	Greece	62
Germany	197	Czech Republic	43
Poland	144	India	43
Kazakhstan	126	Serbia	38

Source: The World Coal Association http://www.worldcoal.org/resources/coal-statistics/. Reproduced with permission of the World Coal Association, London, UK. http://www.worldcoal.org/

Globally reserves are vast, but there is some disagreement to exactly how large they are. There are two leading sources, the German Federal Institute for Geosciences and Natural Resources (BGR) estimates coal reserves to be 1038 billion tonnes equivalent to 132 years of global coal at the current rate of extraction (2012); while the World Energy Council puts reserves at 861 billion tonnes, equivalent to 109 years. In the USA alone, reserves at the current rate of use are estimated to last for at least a further 140 years.

Environmentally coal and lignite mining and their subsequent use as a fuel are very damaging. Both are now extracted primarily by strip-mining resulting in significant erosion, water pollution, landscape damage, and the reduction in biodiversity by reducing plant and animal habitats. The combustion of coal and lignite contributes to air pollution, toxic ash formation, black carbon as well as high GHG emissions. The picture at the start of this chapter shows open cast lignite mining in Hambach (Westphalia) Germany. The mine strip mine is currently 35 km^2 in area and reaches 293 m below sea level. Eventually it will expand to 85 km^2, which I think you will agree is a very large hole.

More information: http://www.worldcoal.org/

3.2.3 Natural Gas

Just like the other non-renewable energy sources, natural gas was also formed by decomposing plants trapped beneath rock millions of years ago, so once used it is not going to be replaced by natural processes any time soon. Its relative abundance and cheapness has seen a rapid increase in its use as an alternative to oil. It is transported from its point of abstraction via a network of thousands of miles of pipeline to where it is needed, for example, gas from Russia is being used throughout Europe including the UK. Also, like other fuels it is now transported via bulk tanker from more remote locations and offshore installations. It has the lowest GHG emissions of all the fossil fuels and so is considered to be a relatively "clean" energy source. So many of the reductions in GHG emissions associated in electricity production are due to a switch from coal and oil to natural gas (Sect. 8.1.1). Environmental impacts associated with drilling, extracting the gas and installing pipelines includes severe disruption to wildlife. Unlike coal and lignite it is much cleaner when burnt, but it can be a major source of the strong greenhouse gas methane if it escapes into the atmosphere. While gas fields are constantly being depleted new supplies to replace them are being found. However, at the current rate of use, known natural gas reserves in the US will be exhausted within 40 years, with best estimates indicating the total depletion of the Earth's supply of natural gas by 2075.

More information: http://www.naturalgas.org/

3.2.3.1 Shale Gas: Fracking or Hydrofracking

Oil and gas are also formed in thinner layers that are unable to form large reservoirs which are normally the source of these fuels. The name fracking comes from the process of hydraulic fracturing of rock at depth caused by the pressure from pressurized gas or oil in these layers. This creates veins that allow the oil or gas to migrate into reservoirs. However, much of this oil and gas is not released but trapped inside these veins. By injecting liquid at high pressure new channels are created which can gather this dispersed oil or gas which is then extracted.

Fracking is primarily used to recover natural gas and the fracking liquid or propellant that is used to open up these new channels and displace the gas is a mixture of 98 % water and a range of chemicals and additives including biocides, surfactants, benzene and methanol. Sand is also added to physically keep the small channels open and to prevent their collapse. Of course the key concern about fracking is safety, in particular minor earthquakes and groundwater pollution. Problems can occur in both ground and surface waters from contamination by the fracking liquid. The gas can also enter groundwater and move through the aquifer or can cause air pollution if it escapes to the surface. Hydrofrac zones are normally very deep and well below groundwater aquifers. For example, in counties Leitrim and Roscommon in Ireland the hydrofac zone is over 4 km deep. Working at this depth is extremely difficult and highly technical, with a possibility that frack liquid or gas can escape from the extraction wells into the aquifer above contaminating the groundwater. Fracking dates back to 1949 and since then over 350,000 wells have been fracked in the US alone with few pollution incidents reported (Fig. 3.7). In contrast, the socio-economic problems associated with the activity are more common. Fracking is compared by some commentators to the gold rush in the nineteenth century, leading to rapid uncontrolled expansion of rural and remote areas followed by a sudden economic downturn. The US Environmental Protection Agency has carried out

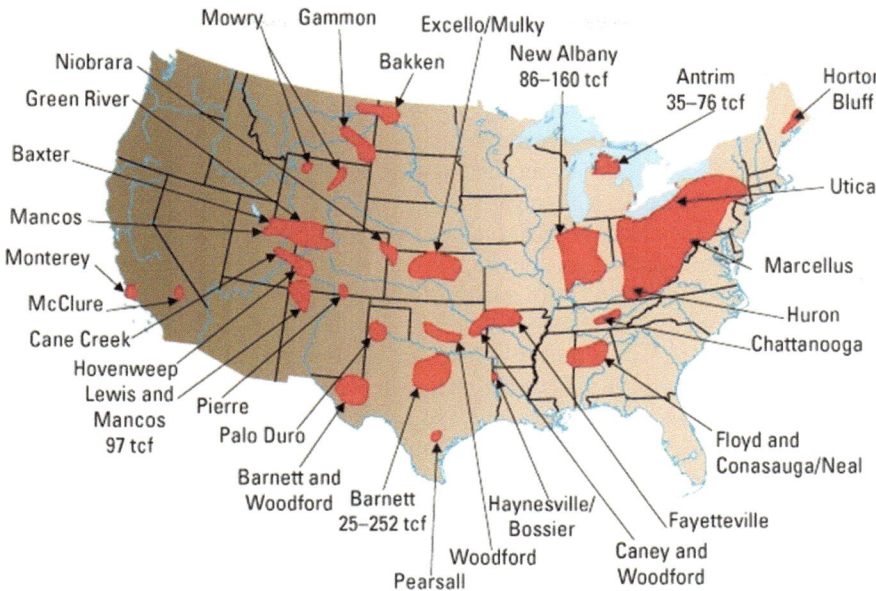

Fig. 3.7 Main shale gas basins in the US containing an estimated 500–1000 trillion cubic feet (tcf) of gas. One tcf is equivalent to 28.32 billion cubic metres. *Source*: USGS http://www.usgs.gov/. Reproduced with permission of the US Geological Survey, Reston, VA, USA

detailed studies on fracking since 2004 and has shown that the risk to groundwater is low, although there have been reported isolated problems due to poor operational management. The real problems occur with the uncontrolled migration of gas and frack liquid to the surface away from the extraction point, and with the improper disposal of the frack liquid. Although banned in France, it has been given the green light almost everywhere else including Ireland and the UK.

A study on the GHG emissions from the production of shale gas in the UK carried out for the Department of Environment and Climate Change in 2013 showed that emissions from shale gas would be very similar to locally sourced natural gas but lower than imported fossil fuels including gas. The average carbon footprint for shale gas extracted in the UK is estimated to be 200–253 g CO_2e per kWh equivalent as chemical energy compared to 199–207 g for UK sourced natural gas and 233–270 g for natural gas from outside the EU. This is similar to the footprint for liquefied petroleum gases (LPG) such as butane and propane. However, there is an important issue here, and that is when shale gas is used to generate electricity the carbon footprint rises to 423–535 g CO_2e per kWh. However, all these values are well below the carbon footprint for coal used for electricity generation at 837–1130 g CO_2e per kWh.

More information: http://pubs.usgs.gov/dds/dds-069/dds-069-z/

Fracking has often been referred to as the poor man's gas field, but in reality it will become the major source of natural gas by 2040. The deposits of shale gas are very large indeed with fracking already a very active industry in the US. Just as with oil shales and sands, the availability of these vast reserves has taken away the urgency to find alternative low-carbon energy resources that was driven by falling oil reserves, and has weakened our fight against climate change, especially in those countries such as the USA rich in gas shale resources.

Natural gas is the cleanest fossil fuel available and has already played an important part in our emissions reduction programme, especially as an alternative to coal, and **if used properly then shale gas may make the transition to a low-energy economy smoother**. The challenge is to use shale gas to help develop more low-energy technologies, so we must not lose our resolve to keep to carbon reduction targets and the increase in renewable energy sources. Another important factor is that fracking increases energy security, and in the UK with gas and oil production from the North Sea in rapid decline and dependence on gas from outside the EU rising rapidly, shale gas does allow increased security which could gradually be replaced by renewable without adversely affecting the economy (Sect. 8.1.2).

3.2.4 Helium

This unusual gas is the byproduct of radioactive decay and is occasionally a component of natural gas. It is primarily associated with a small number of gas fields in the US (mainly Texas and Kansas) and a handful of other areas in Algeria, Qatar, Poland and Russia. It is extracted by cooling natural gas to below 90 K when everything except helium liquefies. Helium is then distilled and compressed then finally cooled to a liquid. Supplies are critically low and no significant new reserves have been found in recent years, so it expected to run out within the next 20–25 years. Although the US government fixed the price of helium in 1996 in order to control its use, demand has continued to rise having doubled in the past 20 years. The gas has become critical for the manufacture and operation in a wide range of high technology areas including superconductors, space exploration, defence and in medicine which currently uses 28 % of all helium to cool the magnets in MRI scanners. It has been suggested that it may have significant roles in the development of new technologies, but in the meantime the vast majority is being wasted each year to fill party balloons! Once released helium is lost forever.

3.2.5 Uranium/Nuclear Energy

The nuclear industry is dependent on mined uranium, which is a radioactive metallic element. While it is a fairly common metal, only small reserves of uranium exist that are suitable for use in the nuclear power industry, which makes the notion that

Table 3.3 Metals thought	1981	2011
to be at critical levels for economic development change over time	• Cobalt	• Rhodium
	• Chromium	• Molybdenum
	• Manganese	• Platinum
	• Platinum	• Lithium
	• Titanium	• Rare Earth metals

nuclear power is a renewable energy quite erroneous. At the current consumption rates uranium will be fully depleted by 2090, but if fast reactor technology could be developed, which uses less uranium, these reserves could last for up to 2500 years.

Nuclear energy has small GHG emissions as it is not derived from fixed carbon, and there is little atmospheric pollution. Of course we are all aware of the downside to nuclear energy with incidents such as Chernobyl and Fukushima vivid reminders of the risks. Surprisingly, while accidents and leaks from nuclear power plants can have very serious effects, in terms of deaths per kilowatt produced then in theory it is the least dangerous of all energy resources. However, contamination to land and food, or to fisheries as seen recently in Japan, which can lead to long-term chronic effects are more serious as they are less quantifiable. It also produces routine radioactive waste which requires storage for thousands of years which makes the energy derived from such plants quite expensive. However, nuclear energy may well be suffering from poor research and development with physical scientists largely ignoring this area due to the problems and costs associated handling the material and disposing of waste, tending to focus on more high profile research areas which are less likely to have a negative press (Sect. 8.3).

3.2.6   Metals

The demand for metals changes with the needs of industry, although some metals such as iron, copper and zinc have been major drivers of industrial development for centuries and their ready supply continues to be essential. Rarer metals are also vitally important, especially in more modern technologies such as electronics, and because of their scarcity they are also very expensive. This scarcity drives a constant search for alternatives to scare and expensive metals that are both cheaper and more readily available. As different technologies are developed then different metals become critical in terms of their supply so the demand for certain metals seem to fluctuate (Table 3.3). However, new technologies are increasingly dependent on platinum and the rare Earth metals associated with it. For example, rhodium which is used in petrol is a secondary product of platinum mining, is now classed as a peak metal with supplies close to exhaustion. The supply of these rare Earth metals is unlikely to keep pace with the increasing demands so new technologies must be based on abundant materials if we are to achieve sustainable development.

Among the rare Earth metals which are vital for manufacture in new technology areas, including some low-carbon technologies, and their resource status (i.e. Critical, Near Critical and Not known) are:

Neodymium used in high performance magnets (**Critical**)
Erbium Optical fibres (**Not Known**)
Tellurium Solar panels (**Near Critical**)
Hafnium Computer chips (**Not Known**)
Tantalum All handheld electronics (touch screens) (**Not Known**)
Technetium Medical imaging (**Not Known**)
Indium Touch screen technology, solar cells (**Critical**)
Dysprosium High temperature magnets (**Critical**)
Lanthanum Batteries (hybrid) (**Critical**)
Cerium Batteries (hybrid) (**Critical**)
Europium Energy efficient lighting (**Critical**)
Terbium Energy efficient lighting (**Critical**)
Yttrium Energy efficient lighting (**Critical**)

Indium, gallium and hafnium are rare Earth metals heavily used in the electronics industry which are all thought to be close to or at peak. Predictions for exhaustion of global supplies of hafnium, which is an important part of computer chips, is as close as 2017. Indium, a critical component of liquid-crystal displays for flatscreen televisions and computer monitors, will be exhausted by 2020. This is not only posing a problem to the electronics industry, other technologies are also affected. This means that new designs for solar panels which are twice as efficient as most current panels may not go into production due to a lack of gallium and indium. The reality is that many of the advances in microtechnology and nanotechnology are probably not commercially viable due to lack of these rare metals unless new sources found.

Gallium became an important metal immediately after the second world war when it was used as critical component to stabilize plutonium in the form of an alloy which made the most effective and reliable atomic bombs. Today gallium is still in high demand in the manufacture of high speed electronic switches, solid state lasers and optoelectronic sensors and so is a critically important element both for industry and research. The metal is found in bauxite and produced as a secondary product of aluminium production. However, not all aluminium is produced from bauxite and not all bauxite contains enough gallium to be recoverable. Unlike aluminium itself gallium cannot be recovered from scrap due to the minute quantities used in manufactured products, a problem shared with all rare Earth metals. Currently it is at peak production and classified as a peak metal, so unless the electronics industry finds a substitute that is as effective, gallium will rapidly become exhausted.

Currently China has 97 % of the world's known reserves of rare Earth metals and they are so scarce that generally we have no idea where they are to be found. However, recent marine exploration has found that the areas of sea bed surrounding deep sea vents may be a possible source for such minerals in the future, although the environmental damage to these quite fragile marine habitats from mining is unknown. So once again our dependence on non-renewables leads us to take significant risks with our natural environment.

Some common metals are also close to peak with zinc supplies possibly becoming exhausted as early as 2037 at current demand levels, with the global demand for copper outstripping mineable supplies by the end of the century.

3.3 Land as a Finite Resource

Land is a key resource which has numerous life sustaining functions. Key to humankind is the production of food crops and the production of biomass that can be used as fodder for livestock, bioenergy, as well as building materials. Land is also important for recreation and in the creation of a healthy environment that promotes well-being. Next is its importance as a habitat and gene reserve for wildlife. But perhaps its most important role is the numerous environmental services it provides such as the removal of pollutants by intercepting and transforming ions from the atmosphere and land based sources, the prevention of runoff and erosion, and the role that plants and soils play in the global cycles of carbon, nitrogen and water, especially the sequestration of atmospheric carbon and other greenhouse gases.

With a population well in excess of 7 billion, land is under ever increasing pressure as more and more pristine land is taken for food production and development; and this demand seems insatiable. So it is only relatively recently that Governments have universally accepted land to be a critical resource that is finite, and which also plays an important role in both carbon emissions and carbon sequestration. The important of land was highlighted in June 2012 at the UN Conference on Sustainable Development held Rio de Janeiro with a call that all countries should adopt a system of land management based on sustainable principles where land is perceived as a non-renewable resource.

> *We find ourselves in the middle of a transition period, from a past time of "abundance" to a future time of "scarcity", which will affect the availability of several raw materials and natural resources such as water, soils and food.*
> Lester E. Brown, Director of the Earth Policy Institute, Washington (DC). EU Conference *Land as a resource*, Brussels, 19 June 2014

3.3.1 Soil and Processes

The basic component of land is soil which is often impoverished by bad management or contaminated from industrial activity and waste disposal. Soil is a major asset for every country on which fertility and other ecological services ultimately depend.

Soil evolves over considerable time scales and is a complex mixture of weathered inorganic and organic material which is colonized by a wide range of micro-organisms and larger mesofauna. Healthy soil provides the basis for terrestrial life and plays an important role in the global carbon cycle especially in sequestration of CO_2 and other greenhouse gases. Surprising soil holds 80 % of the world's terrestrial stock of carbon, far more than found in vegetation. While soil is one of the largest reservoirs of carbon on the planet, land use change and land degradation has resulted in huge losses of carbon in the form of greenhouse gases into the atmosphere. It is vitally important that in controlling global warming, soils and vegetation, especially natural forests retain their stocks of carbon and are managed in such a way as to be able to sequester even more CO_2 in the future.

Increasing food production through intensification of agriculture is leading to serious degradation of soil quality, and it is soil that is our key resource as oppose to simply area. Some land use changes, especially deforestation can cause local and even regional changes in weather pattern including extreme weather events.

3.3.2   Deforestation and Land Degradation

Throughout the world ancient forests are being destroyed to meet increasing demand for hard timbers, pulpwood, biomass and to provide land for farming. Forests are being destroyed not only for subsistence farming and traditional crops such as rice, but for highly managed plantations of oil palm, soy, rubber, coffee and tea. This isn't just confined to the rainforests, but temperate forests are also under increasing threat. Natural forests that have high biodiversity and conservation values as well as holding large reserves of sequestered CO_2, are being replaced with biofuel crops such as palm oil and pulpwood plantations. In a report prepared jointly by the Norwegian and UK Governments, '*Drivers of Deforestation and Forest Degradation*', it has been shown that 80 % of global deforestation is for agriculture. In Latin America it is mainly for commercial farming and forestry while in Africa and tropical Asia it is a mixture of both commercial and subsistence farming. Mining and urbanization are also important factors locally. Forests are also degraded though selective timber extraction in tropical and subtropical areas of South America and Asia, while in Africa forest degradation is caused by wood collection for household fuel and charcoal making. Grazing livestock in forests is also causing significant damage to the ecological quality of forests and their ability to store GHGs.

Report: Drivers of Deforestation and Forest Degradation: https://www.gov.uk/government/uploads/system/uploads/attachment_data/file/65505/6316-drivers-deforestation-report.pdf

Table 3.4 Countries with the largest net losses of forest cover and net gain of forests during the period 2000–2005 expressed in hectares lost or gained per year

Country	Net loss (hectares per year)	Country	Net gain (hectares per year)
Brazil	−3,103,000	China	4,058,000
Indonesia	−1,871,000	Spain	296,000
Sudan	−589,000	Viet Nam	241,000
Myanmar	−466,000	United States	159,000
Zambia	−445,000	Italy	106,000
United Republic of Tanzania	−412,000	Chile	57,000
Nigeria	−410,000	Cuba	56,000
Democratic Republic of the Congo	−319,000	Bulgaria	50,000
Zimbabwe	−313,000	France	41,000
Venezuela (Bolivarian Republic of)	−288,000	Portugal	40,000
Total	−8,216,000	Total	5,104,000

Source: FAO (2006). Reproduced with permission of the Food and Agriculture Organization of the United Nations, Rome, Italy

A staggering 27 million hectares of tropical forests were cleared for agriculture between the years 2000 and 2005. Over 50 % of cleared forest during that period was in Brazil with Indonesia next on the list at 12.8 %, with Brazil losing 3.6 % of its total forest cover while Indonesia lost 3.4 % (Table 3.4). Currently 13 million hectares of natural forest globally is being cleared each year and a further 12 million hectare become degraded releasing vast quantities of greenhouse gases as well as severely damaging biodiversity and ecosystem services. We have lost 75 % of the global primary forest and a further two billion hectares of land has been profoundly degraded since 1950. Poor farming and exploitative practices coupled with climate change has created over four billion hectares of man-made deserts with vulnerable areas found worldwide (Fig. 3.8). Today 1.5 billion people, mainly small-scale and subsistence farmers, are directly affected by land degradation. The United Nations set up a new initiative in 2008 along with the FAO and to monitor and protect natural forests for being felled. Named the 'initiative on Reducing Emissions from Deforestation and forest Degradation (REDD)' it works with developing countries by bringing together forest communities and other stakeholders to ensure forests are managed sustainability and carbon emissions are minimized (UN-REDD 2010).

More information on Land Degradation: http://www.fao.org/nr/land/degradation/en/
More information on UN-REDD programme: http://www.un-redd.org/Home/ tabid/565/Default.aspx

Approximately 40 % of all timber harvested globally is used for the manufacture of paper. In many countries this timber is coming from illegal logging or from unsustainable forests, leading to very high associated greenhouse gas emissions. Expansion of pulpwood and biomass forestry is leading not only to severe ecological damage

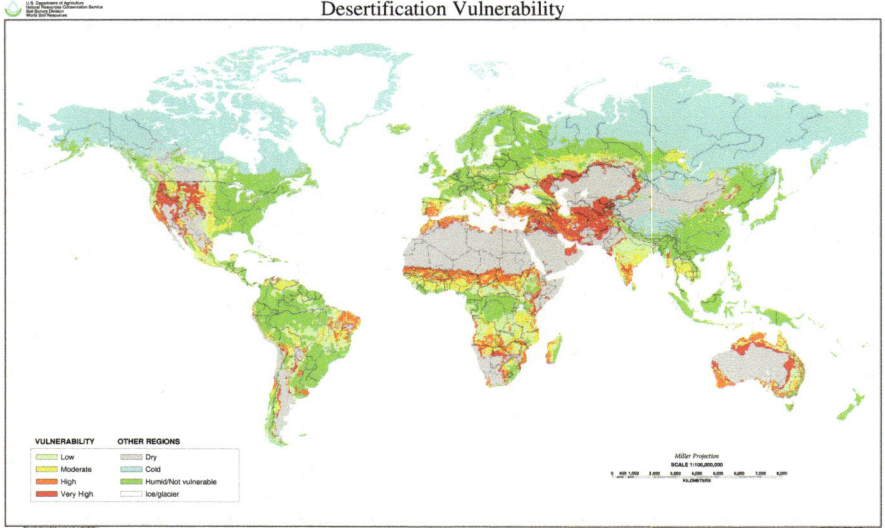

Fig. 3.8 Areas vulnerable to desertification. *Source*: USDA Natural Resources Conservation Service. http://www.nrcs.usda.gov/wps/portal/nrcs/detail/national/nedc/training/soil/?cid=nrcs142p2_054003. Reproduced with permission United States Department of Agriculture

and a huge release of GHGs, but also has significant social impacts, including exploitation of rural populations. Natural forests that are engendered are found in Southern Chile and Brazil, Sumatra, Boneo and New Guinea, and also Eastern Russia.

Much of the paper we use comes from well managed and sustainable plantations. However, this is not always the case. The World Wildlife Fund (WWF) has introduced a scheme called Check Your Paper which allows businesses and the public to identify paper that has been made from sustainable plantations. Working with both paper producers and retailers, they have produced environmental footprints for a huge number of paper products and rated them allowing customers to identify those products that have the small footprint and avoid those with the worst.

More information on the WWF Check Your Paper Scheme: http://checkyourpaper. panda.org/

Climate smart agriculture (CSA) is a system promoted by the FAO to reverse the trend in land degradation. It is aimed at increasing the fertility and quality of agricultural land to sustain food production and also optimize carbon sequestration. The sequestering carbon is increased through specific farming practices that also enhance fertility, such as reducing tillage, using intercropping and strict crop rotation, better crop residue management and better integration between crop and livestock management. So not only is food production enhanced and less prone to failure, soil quality and carbon content are also enhanced, ensuring farm incomes are increased and better sustained.

More information on climate smart agriculture: http://www.fao.org/climatechange/ climatesmart/en/

There has been a rapid expansion of building and paved areas in the past 50 years, especially of roads, which is continuing to grow. The demand for housing is creating urban sprawl with our towns and cities that has largely evolved rather than being planned with the idea protecting natural and productive land. Towns and cities need to be more compact and resilient, well served by public transport and sustainable. Those areas surrounding urban centres are especially at risk as are coastal areas. So ever growing areas of land are being built over, although the predominant uses of land remain for agriculture and forestry.

3.3.3   Action to Protect the Land Resource

How can we reduce the pressure on the demand for land? The starting point is simply better land-use planning, avoiding building on pristine land and fertile soils, stop destroying natural forest and lands, and ensure adequate and effective legislation to protect these areas. In terms of food production we can reduce demand for new land by cutting the amount of food that is wasted and by reducing the amount of animal protein, especially red meat, in our diets. Of course the ambitious biofuel and biomass targets is creating a new demand for land and this must be capped and targets met using land already in cultivation. Finally we have significant areas of degraded (i.e. brownfield and contaminated sites) and abandoned land that could and should be restored to increase our biocapacity. In general we are very bad at using land, often using it for a single purpose. The idea of land recycling is now gaining in popularity where areas are being used for multiple purposes and also smaller areas are being utilized for local food production or in water resource management. Above all we have to preserve and manage our soils to optimize carbon sequestering and maintain fertility.

Farmers are seen as the custodians of our land and soils yet they are not rewarded for the ecosystem services they help to sustain and so are generally unmotivated to managed their land to sustain such services. In developing countries millions of poor people rely heavily on land-based resources not only for subsistence but as their only source of income. Forests and other pristine areas including wetlands are increasingly being used for agriculture, putting ever more pressure on those areas remaining. So adequate and sustainable land management (SLM) is critical for the welfare of those who rely directly on these resources, and indirectly from a loss of environmental services. A major method of reducing poverty and food insecurity is through the promotion of more productive land use using an equitable and sustainable land management systems.

SLM can be defined as 'the use of land resources, including soils, water, animals and plants, for the production of goods to meet changing human needs, while simultaneously ensuring the long-term productive potential of these resources and the maintenance of their environmental functions'.

UN Earth Summit, 1992

More on sustainable land use management: http://www.fao.org/nr/land/sustainable-land-management/en/

Conflict and land use: http://www.un.org/en/land-natural-resources-conflict/pdfs/GN_Land%20and%20Conflict.pdf

More information: http://www.worldbank.org/en/topic/agriculture; http://www.fao.org/nr/land/lr-home/en/

3.4   What Are We Supposed to Do?

Well the way we use resources are critical and we can categorize this into four groupings:

- **Used once and destroyed** (e.g. fossil fuel for energy)
- **Used once but non-recoverable** (e.g. most rare earth metals)
- **Used once and recovered and used again but finally destroyed or non-recoverable** (e.g. zinc in pharmaceuticals; platinum, silver or copper in nanotechnology)
- **Used multiple of times and never destroyed** (e.g. majority of copper, iron, aluminium)

The way forward is for everyone to get involved in the use of resources. That includes scientists, manufacturers, we the consumers and of course waste operators; we all have a vital role to play in ensuring that resources last for as long as possible. It is inconceivable that we should squander critically limited resources denying future generations the use of these unique elements and compounds that still have incredible properties we have yet to discover. So we need to ensure the careful use of non-renewables in the design and manufacture of all our products. What we have failed to do is design products for better recovery of these non-renewables so that we can use them over and over again. We have become extremely good at marking plastic components for recovery after use, especially in cars and large household items, but we are very bad at taking this to the next level and including metals and even gases. I was once told that a refrigerator only lasts 5 years before it should be replaced, yet my mother had the same fridge for over 40 years; no doubt it was inefficient in energy terms, but 5 years seems a very short period of time. So we need to design manufactured goods, especially electronics, for a longer useful life and also make them repairable. One of my students looked at 40 electronic items thrown away at a recycling centre. Fourteen of the items (35 %) worked perfectly well and were in a safe and usable condition and had presumably been discarded for a more modern equivalent. Eighteen of the items could be easily repaired (45 %) and while nine of them were a bit shabby, they were all serviceable. Only eight of the items (20 %) were either past repair or had significant defects that were not economic to put right, or had casing damage or corrosion making them unsafe. I am not quite sure what this tells us. Certainly **we are not always discarding household electrical goods because they are defunct**. Also **we are more likely to replace an electrical appliance rather than have it repaired**. Unpopular as this may sound perhaps manufactured goods in general and electrical goods in particular need to be of higher quality, with a longer working life and designed to be easily repairable. This may make them slightly more expensive to purchase, but it will be cheaper in the long run as they will not have to be replaced so frequently. Designers should be designing for life, not profit. We also need to have an infrastructure of people who can repair and maintain these items at reasonable cost, but they will only survive if we support them. This goes beyond electrical goods of course, for example, did you know that **most sofas and armchairs are thrown out simply because people cannot get them reupholstered or new covers made**.

So the recovery of important elements and components is vital, and all this is down to the designer, and consumers demanding better quality. There is increasing interest in the replacement of non-renewables with biological materials, but that will not completely sort out our current problem of these valuable elements becoming exhausted. A significant increase in research and development, as well as investment at the local, regional and national levels in recovery and recycling is urgently needed.

> **Recovery/recycling is critical for resource sustainability**
> **What we have failed to do is design goods for longer working lives and for better recovery of non-renewables so that they can be used over and over again.**

3.5 Conclusions

- Natural resources vary from one country to another. This can result in either economic security or insecurity often leading to conflict (both political and military) over resources.
- Production is driven by demand; Price by scarcity.
- Without constraint on environmental damage then exploitation of increasingly inaccessible resources from increasingly fragile environments will continue putting even greater pressure on our biosphere.
- More complex and costly extraction, processing and transportation will result in higher GHG emissions per kg or kWh supplied (e.g. oil sands/shales)
- Apart from specialist metals then reserves of oil, coal, possibly gas and major metals will continue to be available to at least the end of the century with careful use. While fossil fuels are destroyed by combustion creating GHGs, other non-renewable resources have the potential to be recovered, recycled and re-used.
- Critical resources should not be used if they cannot be recovered … this should underlie the concept of a viable design.
- Protecting and improving soil quality is pivotal to sustaining our population as is ensuring that we preserve as much of our natural forests and other terrestrial ecosystems as possible.

- The list of non-renewable resources shouldn't be viewed in terms of how difficult they are currently to obtain or how expensive they are. It is more important to consider that non-renewable means that when they are gone there will be no more and so we should be more careful and selective in how and for what they are used.

> **The third step is accepting that all resources are potentially finite and that we must use them sensibly and sparingly, preserving and recycling them whenever possible.**
> **We can do this by developing simple strategies to maximize our use of non-renewables by careful initial product selection, maximizing product use and making an extra effort at the end of life of product to prevent the loss of non-renewables into the waste stream.**
> **Renewable resources must also be used respectfully and not wasted.**

Homework!

So how much service do you get from the resources used in the manufacture of your household items? Look around your home and make a list of things that you would like to replace or just things that you would like to have. I don't mean clothes, we will be coming to those later, I mean household items including white goods such as your cooker, fridge, washing machine, vacuum cleaner, or other items such as your TV, computer, garden mower or even your three piece suite and bed. Then I want you to have a good look at each item on your list and do two things. First analyse why you want to replace it or why you need it. Depending on what your reason is try and see if you can find an alternative solution to buying a new one, could you clean up the old one, redesign it, have it repaired or upgraded? For example, a friend of mine had a Dell laptop with only 254k of primary memory (RAM) and found it too slow. He was about to buy a new one when I suggested simply upgrading the RAM himself which is really easy to do. So he went online to a site called Mr Memory found out what extra memory he could install in his computer, ordered the new memory card and then watched the video of how to replace it (several times), and hey presto he increased the speed of his laptop to 2 GB RAM and it only cost him 12 euro. The real bonus here was that he was so absolutely delighted in doing it himself. The bonus for me was that he kept a perfectly good piece of equipment functioning and saved all those rare Earth metals in the processer and electronics which cannot be recovered and would have been lost. Seventy percent of PCs are replaced due to an operating (software) problem or memory capability, not because they mechanically fail, and failures are almost always due to the screen rather than the computer itself. Laptops are more likely to be physically damaged by dropping them or spilling coffee over them, but normally are reparable as long as the casing is intact.

Some of you will be heading for the screwdriver right now but most of you I suspect have glazed eyes. You don't have to do this yourself. Everything can be repaired even though modern designers appear to go out of their way to ensure that they can't, usually under the guise of safety. But let's imagine that your item has to go. Next I want you to carry out an investigation into the possible replacement item … this will involve looking at its energy rating, overall efficiency, possible carbon footprint, the materials it is made from, how it can be recycled, expected lifetime and how is this reflected in its warranty etc. You may have to contact their customer service people to find some of these answers and I would really encourage you to do that, because the more pressure and feedback they get from customers the more likely they are going to respond by producing better products in the future. So the outcome should be that you have prolonged the life of that object or replaced it with something that will last much longer and be more efficient in energy terms. In Chap. 12 we will be looking at what to do with those items you couldn't upgrade and so had to get rid of, where we explore the question *Recycling—does it matter?*

References and Further Reading

FAO. (2006). *Global forest resources assessment 2005, main report. Progress towards sustainable forest management* (FAO Forestry paper 147). Rome, Italy: Food and Agriculture Organization of the United Nations. Retrieved from http://www.fao.org/docrep/008/a0400e/a0400e00.htm

UN-REDD. (2010). *The UN-REDD Programme strategy 2011–2015*. Geneva, Switzerland: UN-REDD Programme Secretariat. Retrieved from http://www.unredd.net/index.php?option=com_docman&task=doc_download&gid=4598&Itemid=53

Oil Shales

http://fossil.energy.gov/programs/reserves/publications/Pubs-NPR/40010-373.pdf

Fracking

http://proamlib.blogspot.com/2011/01/strange-things-in-arkansas-dead-birds.html

Coal

http://www.worldcoal.org/

Natural Gas

http://www.naturalgas.org/

International Statistics

http://www.eia.gov/cfapps/ipdbproject/IEDIndex3.cfm

Part II
Greenhouse Gases and Global Warming

Chapter 4
Global Warming and CO$_2$

In this chapter we explore the science behind the processes that are driving global atmospheric carbon dioxide levels relentlessly upwards causing our planet to heat up. **"I don't believe in Global Warming" by Banksy.** Image by Duncan Hill. *Source*: https://www.flickr.com/photos/dullhunk/14205015878/in/photostream/. Reproduced with permission under the creative commons licence. https://creativecommons.org/licenses/by/2.0/

4.1 The Greenhouse Effect

In order to be able to tackle the problem of global warming we need to understand how and why it is happening. On the whole the mechanisms of global warming and climate change are fairly well understood and also fairly easy to understand. I suspect most of you know the basics already. As I have said, it is important that we fully appreciate what is happening to our global home before we can effectively start to mitigate it.

The greenhouse effect is simply *the trapping and build-up of heat in the atmosphere (troposphere) near the Earth's surface*. The atmosphere is made up of four layers. Closest to the earth is the troposphere which is about 15 km high. It is in this

© Springer International Publishing Switzerland 2015

N.F. Gray, *Facing Up to Global Warming*, DOI 10.1007/978-3-319-20146-7_4

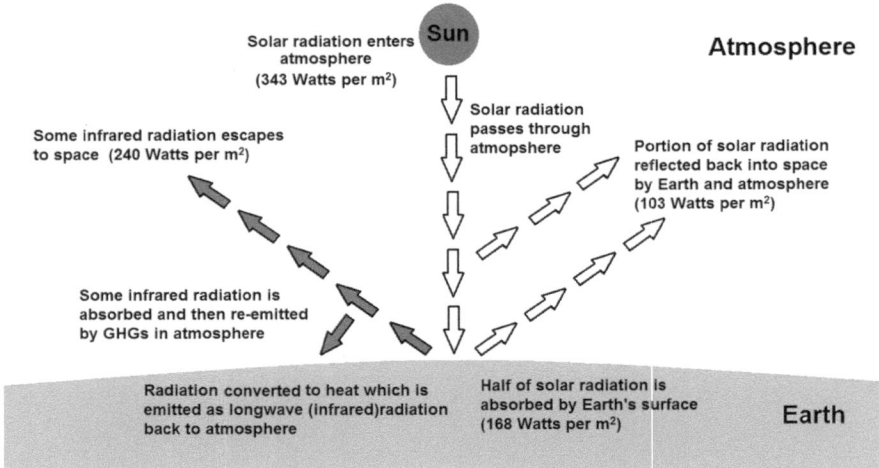

Fig. 4.1 The greenhouse effect

zone, which is closest to the planet's surface, that our weather takes place. Above this is the stratosphere which is 50 km thick. Here horizontal air movement occurs including the jet stream, which is caused by variable heating of both the land and primarily the oceans by solar activity. This in turn affects the weather in the troposphere below. So our weather is driven by solar energy. Sitting near the top of the stratosphere is the ozone layer which absorbs ultra violet radiation (UV). Above this is the mesosphere which is a further 90 km thick and above this is the ionosphere some 400 km wide and then space.

Solar radiation is energy emitted from the sun which passes through the atmosphere. Some of this radiation is reflected back into space but most reaches the surface of the Earth where it is absorbed and warms it. At the same time solar radiation is converted to heat which is radiated back into space as long-wave infrared radiation. Some of this infrared radiation escapes back through the atmosphere into space and is lost, but a portion is absorbed by the greenhouse gas molecules in the atmosphere heating it. The remainder is re-emitted in all directions with a certain amount of this reflected energy radiated back to the surface of the Earth (Fig. 4.1). Without greenhouse gases (GHGs) in the atmosphere the earth would be significantly cooler (approx. 16 °C less) so they are important in terms of temperature regulation. In fact we would be in a permanent ice age without having greenhouse gases present in the atmosphere. So GHGs act as an insulation jacket to the planet, similar to that on your hot water cylinder, trapping the heat inside the atmosphere. However, during the past century we have substantially added to the amount of GHGs in the atmosphere primarily through the use of fossil fuels (e.g. coal, natural gas, diesel, petrol etc.) to power industry, commerce, homes, utilities, cars, appliances etc. Since 1900 the average global temperature has risen from 13.7 °C to today's average of 14.6 °C due to the increased concentration of GHGs making that insulation jacket more effective.

> **Quite simply our planet is warmer than it should be due to the increased concentration greenhouse gases in the atmosphere. This is stopping solar energy escaping back into the atmosphere**

The three predominate GHGs are carbon dioxide (CO_2), methane (CH_4) and nitrous oxide (N_2O). The potential of each GHG to cause global warming varies significantly as does the amount of each gas that is emitted each year.

The majority of CO_2 comes from the use of fossil fuels (Fig. 4.2), while a significant portion of the CO_2 under the category associated with deforestation and decay of biomass is linked to agriculture. Other CO_2 sources would include cement production and natural gas flaring, and of course our own personal emissions.

The industrial revolution in Victorian times began the process of increasing emissions, but the current rapid rise began in the early 1950s (Fig. 4.3). Some countries such as the USA have always been relatively large emitters of GHGs, but emerging economies such as China are showing raid expansion while in contrast Europe is

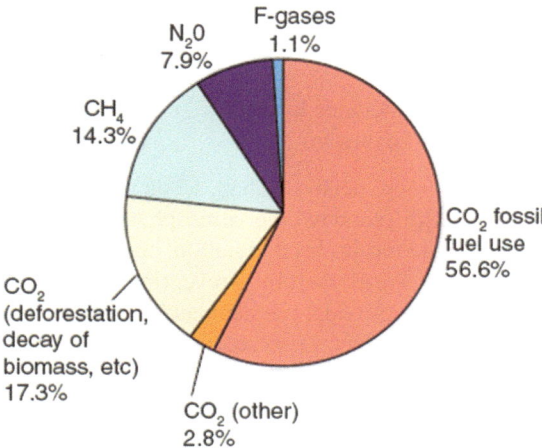

Fig. 4.2 The breakdown of global anthropogenic greenhouse gas emissions measured as CO_2-equivalents (CO_2e). *Source*: IPCC (2012). Reproduced with permission of the Intergovernmental Panel on Climate Change, Geneva

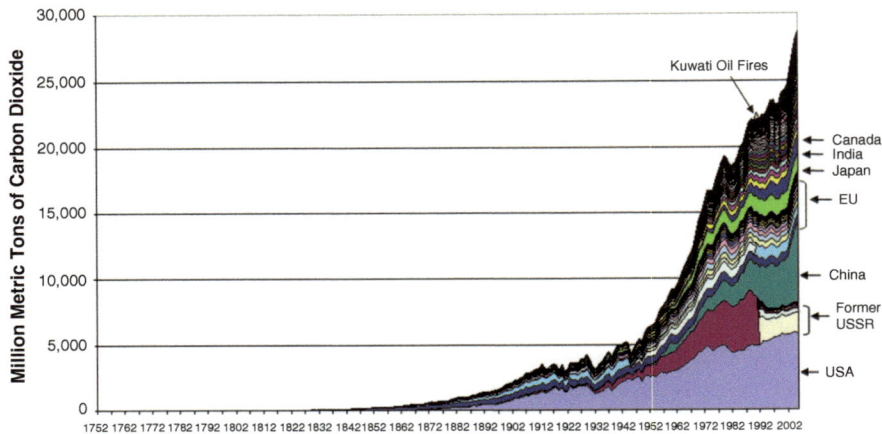

Fig. 4.3 Global CO_2 emissions from burning fossil fuels including cement manufacture and gas flaring by major emitting countries 1752–2006. *Source*: http://www.epa.gov/climatechange/. Reproduced with permission of the US Environmental protection Agency

showing a decline as its manufacturing base is moved to China and India. Greenhouse gases also come from other sources including:

- Volcanic activity (CO_2 and other gases)
- Natural respiration (the sea releases as well as absorbs and stores (sequesters) CO_2)
- Wildfires (CO_2 and soot)
- Decomposition (CO_2)
- Permafrost melt (CO_2 and methane hydrates)
- Cattle flatulence (CO_2 and methane)

These are considered in more detail in Chap. 13 on planet health.

So greenhouse gases are different to other atmospheric gases such as oxygen and nitrogen as they absorb and emit infrared radiation, controlling the speed at which heat can escape from the earth thereby preventing extremes of temperature (Fig. 4.1). Apart from carbon dioxide, methane and nitrous oxide other gases can also influence the greenhouse effect, these are classed as fluorinated gases. Details of each of these including where they originate are summarized below, but it is import to also realize that water vapour also acts as a GHG.

Major Greenhouse Gases

Carbon Dioxide (CO₂): The main sources are the burning of fossil fuels (e.g. oil, natural gas, and coal), solid waste, trees and wood products and chemical reactions (e.g. manufacture of cement). Carbon dioxide is also removed from the atmosphere (or "sequestered") when it is absorbed by plants as part of the biological carbon cycle.

Methane (CH₄): From the production and transport of coal, natural gas, and oil; livestock and other agricultural practices; decay of organic waste in municipal solid waste landfills and wastewater treatment.

Nitrous Oxide (N₂O): Generated from agricultural and industrial activities, including wastewater treatment; combustion of fossil fuels and solid waste.

Fluorinated Gases: Hydrofluorocarbons, perfluorocarbons, and sulphur hexafluoride are synthetic, powerful greenhouse gases that are emitted from a variety of industrial processes. Fluorinated gases are sometimes used as substitutes for ozone-depleting substances (i.e. CFCs, HCFCs, and halons). These gases are typically emitted in smaller quantities, but because they are potent greenhouse gases, they are sometimes referred to as High Global Warming Potential gases ("High GWP gases") (Sect. 4.3).

4.2   Radiative Forcing

The Earth's surface temperature is determined by the balance between incoming solar radiation and outgoing infrared radiation. *Radiative Forcing* (RF) is the measurement of the capacity of a gas or other forcing agents to affect that energy balance, thereby contributing to global warming and climate change. Factors include: ice albedo, tropospheric aerosols, deforestation albedo, and methane. The RF of a gas is defined as the difference between incoming solar radiation and outgoing infrared radiation caused by the increased concentration of that gas. Generally the balance of incoming to outgoing energy is construed at the surface of the troposphere—stratosphere boundary.

Positive radiative forcing causes an increase in the Earth's energy budget and ultimately leads to warming. Greenhouse gases absorb infrared radiation and re-emit it back to the Earth's surface thus increasing the Earth's energy balance, hence they all have positive RF values.

Negative radiative forcing results in a decrease in the energy budget, and ultimately leads to cooling. Aerosol particles in the atmosphere reflect solar radiation, leading to a net cooling, and therefore have negative RF values.

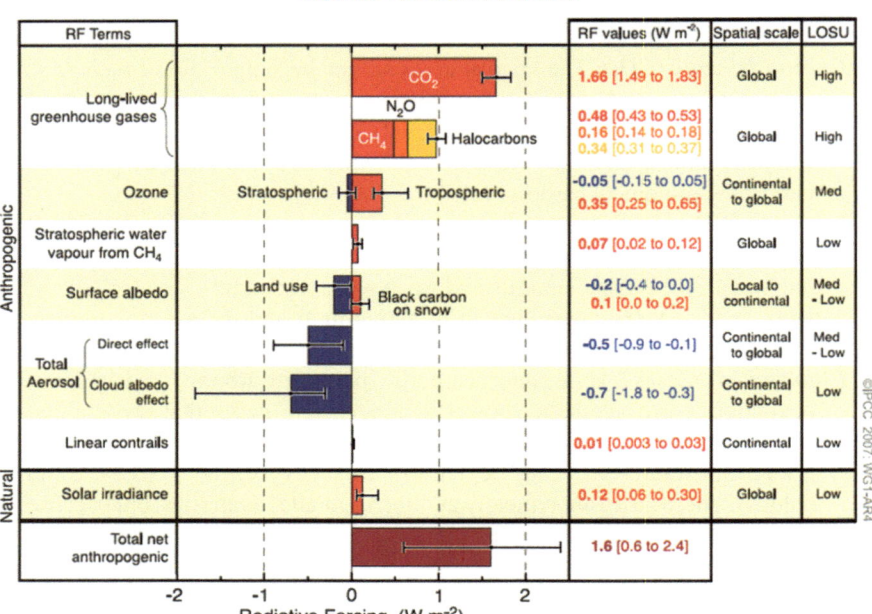

Fig. 4.4 Radiative forcing values for common natural and anthropogenic components involved in the control of energy reaching the Earth's surface. The LOSU refers to the Level Of Scientific Understanding of that particular relationship. *Source*: IPCC WG1-AR4 Report (IPCC 2007). Reproduced with permission of the Intergovernmental Panel on Climate Change, Geneva

Radiative forcing is also used as an index of the influence a factor has as a potential climate change mechanism. Forcing values are expressed in watts per square meter (W m^{-2}) (Fig. 4.4).

> *Positive RF increases the temperature at the Earth's surface.*
> *Negative RF decreases it.*

Other factors that control the effect of GHGs include their residence time in the atmosphere and also their spatial distribution globally. The **Residence Time** of GHGs refers to the time a GHG stays in the atmosphere which can be very variable. Some GHGs are short-lived while others remain in the atmosphere for hundreds or thousands of years. To properly asses the climate impacts of a combination of gases, the lifetime of each gas has to be taken into account. For example, the warming impacts of CO_2 persist for hundreds of years, whereas the warming impacts of aircraft contrails, which are made up of water vapour, or ozone last only days or months. **Spatial Distribution** refers to how far GHGs spread geographically.

Long-lived greenhouse gases spread across the entire global atmosphere (e.g. CO_2 and methane) therefore their warming impact is global in scale. Other gases which are short-lived result in warming effects that are local or regional (e.g. contrails from aircraft).

> *The residence time of GHGs control their spatial distribution.*
> *Globally-averaged radiative forcing calculations do not take into account these differences in spatial distributions.*

4.3 Global Warming Potential

The term Global Warming Potential (GWP) is used to total up the contribution of all the individual GHGs in the atmosphere and is also used as a tool to compare the potency of different greenhouse gases with that of CO_2. The GWP is calculated using the integrated RF and lifetime of each gas relative to that of carbon dioxide. Carbon dioxide has an assigned GWP of 1 and is used as the baseline unit (i.e. the reference gas) to which all other greenhouse gases are compared. Thus GWP is unitless, so a GWP of 2 is twice as potent as carbon dioxide. The GWP value can be used to convert various greenhouse gas emissions into comparable CO_2 equivalents when computing overall sources and sinks; greenhouse gases can thus be expressed in terms of **Carbon Dioxide Equivalent (CO_2e)**. So the GWP is used to express the impact of all GHGs using the same units (CO_2e).

All GWP values depend on the time span over which the potential is calculated. Therefore, short-lived GHGs initially have large effects that become less significant over time relative to CO_2, since the integrated RF of CO_2 increases over time. Methane, for example, has a GWP of approximately 25 over 100 years but 62 over 20 years. So in order to collectively calculate the effects of all GHGs, a similar time frame must be used, which is why the Kyoto Protocol uses the GWP time frame of 100 years which is the lifespan of CO_2 itself. If a climate policy is enacted to limit long-term temperature increase, effects of short-lived emissions may be overestimated if the time horizon chosen is too short. On the other hand, a time horizon of 100 years versus one of 20 years might underestimate the importance of short-lived emissions. This is shown in Table 4.1 which highlights the importance of preventing the release of some of these high GWP gases, many of which are vital for industrial processes. However, although they have high GWPs, they are released in very small amounts compared to CO_2.

All greenhouse gases are increasing in the atmosphere as shown in Fig. 4.5, with the exception of CFC's which have stabilized but have been replaced with fluorinated gases.

Table 4.1 Atmospheric lifetime and GWP relative to CO_2 at different time horizon for various greenhouse gases

Gas name	Chemical formula	Lifetime (years)	Global warming potential (GWP) for given time horizon		
			20-year	100-year	500-year
Carbon dioxide	CO_2	30–100	1	1	1
Methane	CH_4	12	72	25	7.6
Nitrous oxide	N_2O	114	289	298	153
CFC-12	CCl_2F_2	100	11,000	10,900	5200
HCFC-22	$CHClF_2$	12	5160	1810	549
Tetrafluoromethane	CF_4	50,000	5210	7390	11,200
Hexafluoroethane	C_2F_6	10,000	8630	12,200	18,200
Sulphur hexafluoride	SF_6	3200	16,300	22,800	32,600
Nitrogen trifluoride	NF_3	740	12,300	17,200	20,700

Source: Carbon Dioxide Information Analysis Center http://cdiac.ornl.gov/pns/current_ghg.html. Reproduced with permission US Department of Energy

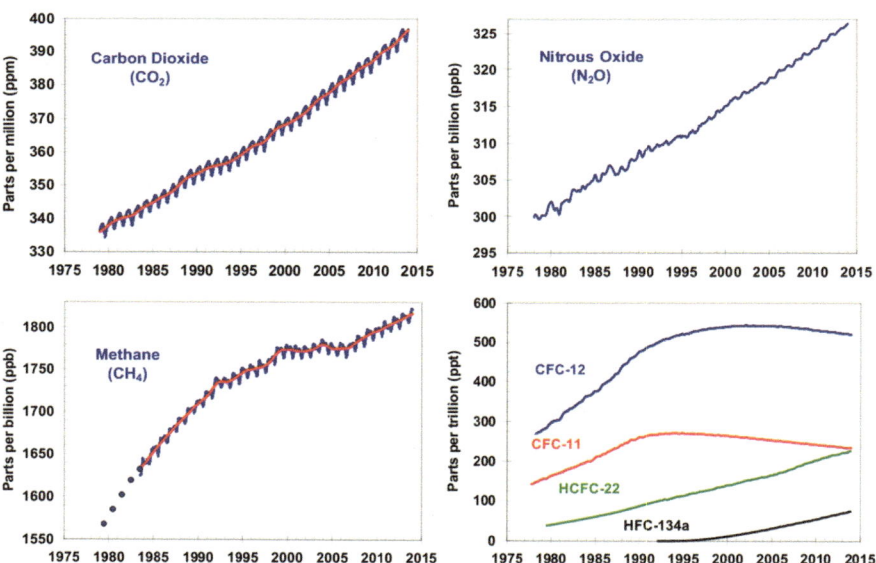

Fig. 4.5 Global GHG concentrations in the atmosphere 1978–2014. *Source*: National Oceanic and Atmospheric Administration http://www.esrl.noaa.gov/gmd/aggi/. Reproduced with permission of the National Oceanic and Atmospheric Administration, US Department of Commerce

Table 4.2 Details of the reflectivity of different surfaces including the albedo value

Surface	Details	Albedo
Soil	Dark and wet	0.05–0.40
	Light and dry	
Sand		0.15–0.45
Grass	Long	0.16–0.26
	Short	
Agricultural crops		0.18–0.25
Tundra		0.18–0.25
Forest	Deciduous	0.15–0.20
	Coniferous	0.05–0.15
Water	Small zenith angle	0.03–0.10
	Large zenith angle	0.10–1.00
Snow	Old	0.40–0.95
	Fresh	
Ice	Sea	0.30–0.45
	Glacier	0.20–0.40
Clouds	Thick	0.60–0.90
	Thin	0.30–0.50

4.3.1   Albedo

When the Sun's radiation hits the surface of the Earth a portion is reflected back into the atmosphere, this is known as the albedo. It is measured as the proportion (of 1), or percentage of solar radiation of all wavelengths reflected by a body or surface to the amount incident upon it. So an ideal white body will have an albedo of 1 with 100 % of the radiation reflected, while in contrast an ideal black body has an albedo of 0.0 with 0 % of the radiation reflected. In practice all surfaces reflect some radiation, for example, deserts and areas with snowfall have an albedo greater than 0.50 (>50 %). The values can however be quite variable. So for snow cover a value of 0.95 (95 %) could be obtained by an area of pristine white snow while a much lower value will be due to carbon black (soot) deposits which build up over time (Table 4.2). The effect of water on the albedo depends on the angle of the sun in relation to the surface of the globe (i.e. the solar zenith angle), which varies for time of year and location. The greater the angle then the greater the albedo effect. Albedo can have unexpected effects. For example global warming is leading to a loss of ice and snow cover, which results in less solar energy being reflected and so results in increased heating, while in contrast climate change is also leading to desertification that increases the reflection of solar energy and reduces the overall amount of heat trapped. Another is the effect of buildings (roofs), roads, car parks, railways and airports. All these tend to have relatively dark surfaces and so reduce the albedo causing greater heat absorption which is one of the reasons why cities and urban areas are always warmer than the surrounding countryside.

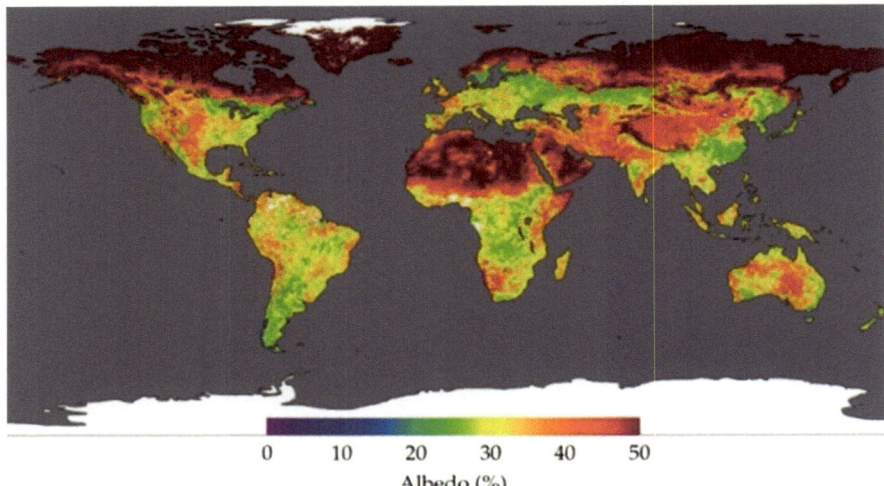

Fig. 4.6 Albedo of the Earth's terrestrial surface as measured by the TERRA satellite. Data collected from the period April 7–22, 2002. *Red areas* are the most reflective regions; *yellows* and *greens* intermediate; and *blues* and *violets* relatively dark surfaces. Areas where no data is available is shown in *white*. No data is given for ocean areas. *Source*: NASA (2004). Reproduced with permission of NASA, Washington

The albedo varies significantly in terms of location, and within quite small areas large variations will occur. Figure 4.6 summarizes the albedo at a regional scale.

4.4   Effects of GHG Emissions and Models

The concentration of GHGs in the atmosphere will continue to rise over the next century unless GHG emissions are significantly reduced from current levels. The direct effects of increased GHG concentrations in the atmosphere will be to:

- Raise average global temperatures
- Increase precipitation and severity of storms
- Increase sea level
- Increase ocean acidification

These are discussed in more detail in Sect. 1.3 and Chap. 13.

Climate models are not actually weather forecasts as we know them, rather they should be seen as providing us with more general indications of what the average conditions will be and of course we will see normal climate variability around these averages (Sect. 1.3). All major models agree on the above changes, although the magnitude and speed of these changes are uncertain and what we have at the moment are very sophisticated guesstimates. We do know however that the severity and speed of change depends primarily on GHG concentrations in the atmosphere and this is the only thing that we can actually control ourselves. **It is no longer a case of whether atmospheric CO$_2$ concentrations go up or down, we know for certain that they are going up and quickly. But it is still possible for us to control the rate of increase and perhaps one day stabilize the concentration and even perhaps over many generations actually see a reduction**. The things that we cannot control are natural processes such as volcanic activity and changes in suns activity. Neither are we able to control global changes in weather patterns caused by alterations in the circulation patterns in the atmosphere and oceans as a result of global warming. Our final problem is that we just don't know how quickly the key effects such as temperature, precipitation and sea level will respond to increasing GHG emissions. This is because the planet's response is extremely complicated and even unpredictable, as this is completely new territory for us. So this is what climate modelling is all about, trying to predict these key changes.

The one thing all these models tell us is we have to reduce GHG emissions by as much as possible and we have to do that right now!

4.4.1   Who Is the IPCC?

The IPCC is the acronym of the Intergovernmental Panel on Climate Change (http://www.ipcc.ch/index.htm) which is a scientific body which was set up jointly by the United Nations Environment Programme and the World Meteorological Organization in 1988. Its aim is to provide the world with a clear scientific view on the current state of knowledge on climate change and its potential environmental and socio-economic impacts. It reviews and assesses the most recent scientific, technical and socio-economic information produced worldwide relevant to the understanding of climate change. It does not conduct any research itself nor does it monitor climate related data or parameters, rather it presents unique overviews of our current state of knowledge on global warming and its effects . Since its formation it has produced a number of assessment reports with the first set published in 1990, and subsequent reports published in 1995, 2001 and 2007. The Fifth Assessment Report series (AR5) has been under preparation for a number of years and during 2013 underwent an extensive consultative process and replaced AR4 when finally published in 2014. Like the previous report series it is made up of three working group (WG) reports that contain the most up to date research, analysis and modelling of climate data. There is also a synthesis report written in a non-technical style primarily for policy makers (Fig. 4.7).

To download the IPCC AR5 reports: https://www.ipcc.ch/publications_and_data/publications_and_data_reports.shtml

Working Group I Report
The Physical Science Basis

Working Group II Report
Impacts, Adaptation and Vulnerability

Working Group III Report
Mitigation of Climate Change

The AR5 Synthesis Report

Fig. 4.7 The AR5 IPCC report series published in 2014. *Source*: IPCC http://www.ipcc.ch/. Reproduced with permission of the Intergovernmental Panel on Climate Change, Geneva

4.4.2   Predicting Emissions

Predicting the rate of global CO_2 concentrations has been underway for decades, under the umbrella of the IPCC. Models have become increasingly sophisticated and the most recent model (AR4-Fourth Assessment Report) is summarized below. The predications are made using different theoretical economic and social situations or scenarios. Four marker scenarios are used based on the best climate change models A1, known also as A1B, A2, B1, B2, plus two illustrative scenarios A1T, A1F1. Class A scenarios are economically focussed while the class B scenarios are more focused on the environment. **Capping the global population at nine billion is critical to all these model scenarios, as modelling with a population in excess of this not really possible if any sort of global stability is to be maintained**. A new process has now been completed and awaits publication (AR5).

The four main scenarios an which the models are based are summarized in Table 4.3 and depend on different economic scenarios and rates of population growth (IPCC 2007)

A1 Scenarios simulate an integrated world with the A1 family of scenarios characterized by:

- Rapid economic growth.
- A global population that reaches nine billion in 2050 and then gradually declines.
- The quick spread of new and efficient technologies.
- A convergent world—income and way of life converge between regions.
- Extensive social and cultural interactions worldwide.

Table 4.3 The four basic model scenarios on which the IPCC base many of their predictions

A1 (A1B) Emissions Scenarios
A future world of very rapid economic growth, low population growth and rapid introduction of new and more efficient technology. Major underlying themes are economic and cultural convergence and capacity building, with a substantial reduction in regional differences in per capita income. In this world, people pursue personal wealth rather than environmental quality
B1 Emissions Scenarios
A convergent world with the same global population as in the A1 storyline but with rapid changes in economic structures toward a service and information economy, with reductions in materials intensity, and the introduction of clean and resource-efficient technologies
A2 Emissions Scenarios
A very heterogeneous world. The underlying theme is that of strengthening regional cultural identities, with an emphasis on family values and local traditions, high population growth, and less concern for rapid economic development
B2 Emissions Scenarios
A world in which the emphasis is on local solutions to economic, social, and environmental sustainability. It is again a heterogeneous world with less rapid, and more diverse technological change but a strong emphasis on community initiative and social innovation to find local, rather than global solutions

There are subsets to the A1 family based on their technological emphasis:

- **A1FI**—An emphasis on fossil-fuels (Fossil Intensive)—current position
- **A1B**—A balanced emphasis on all energy sources.
- **A1T**—Emphasis on non-fossil energy sources.

The A2 scenarios are of a more divided world characterized by:

- A world of independently operating, self-reliant nations.
- Continuously increasing population.
- Regionally oriented economic development.

B1 scenarios simulate a world which is more integrated, and more ecologically friendly.

The **B1** family of scenarios are characterized by:

- Rapid economic growth as in A1, but with rapid changes towards a service and information economy.
- Population rising to nine billion in 2050 and then declining as in A1.
- Reductions in material intensity and the introduction of clean and resource efficient technologies.
- An emphasis on global solutions to economic, social and environmental stability.

The **B2** scenarios are characterized by:

- Continuously increasing population, but at a slower rate than in A2.
- Emphasis on local rather than global solutions to economic, social and environmental stability.
- Intermediate levels of economic development.
- Less rapid and more fragmented technological change than in A1 and B1.

So the A1 scenario is about rapid economic growth (groups: A1T; A1B; A1Fl) with a predicted rise in global temperatures by 1.4–6.4 °C by 2100. The A2 scenarios are based on regionally oriented economic development resulting in a predicted rise of 2.0–5.4 °C. While the B scenarios are B1 based on global environmental sustainability which is predicted to result in a 1.1–2.9 °C in global temperatures by 2100 and B2 focused on local environmental sustainability giving a rise of between 1.4 and 3.8 °C (Fig. 4.8).

Predictions are on target so far using the A1 (A1B) (red) scenario with the B2 (green) scenario now falling well below current CO$_2$ levels in the atmosphere (Fig. 4.9). The predictions are computed using two different models, the ISAM model (solid line) and Bern Carbon Cycle Model (dashed line), with the ISAM model using the A1 scenario being the most accurate. The rate of rise that we see most closely follows the A1F1 scenario which is one of high fossil fuel dependency, whereas we should be trying to reduce this rate of increase by adopting the B2 strategy of more environmental emphasis, local environmental sustainability and more regional action.

More information: http://www.ipcc-data.org/ddc_co2.html

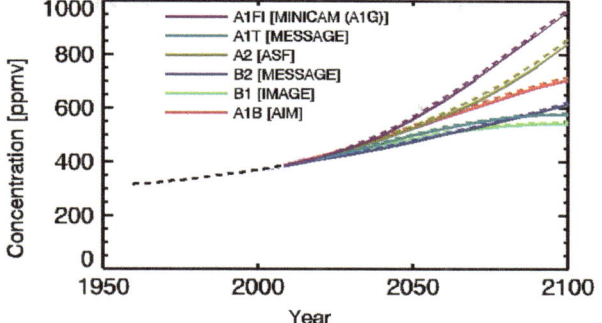

Fig. 4.8 Global prediction of atmospheric CO_2 concentrations based on the six modelling scenarios. *Source*: IPCC http://www.ipcc-data.org/ddc_co2.html. Reproduced with permission of the Intergovernmental Panel on Climate Change, Geneva

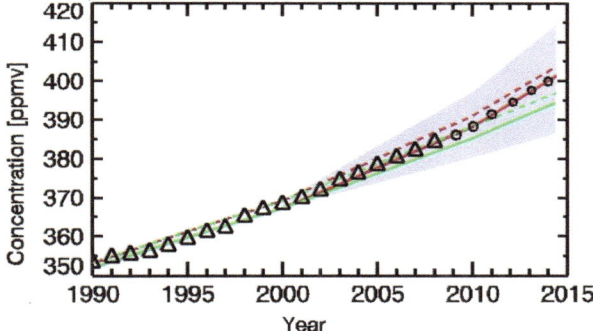

Fig. 4.9 IPCC predictions for global atmospheric CO_2 concentrations. These are on target so far for the A1 (A1B) (*red*) scenario with the B2 (*green*) scenario now falling well below current CO_2 levels in the atmosphere. *Source*: IPCC http://www.ipcc-data.org/ddc_co2.html. Reproduced with permission of the Intergovernmental Panel on Climate Change, Geneva

Using these predictions of global CO_2 concentrations climate modellers can begin to look at the effects of temperature increases on ocean currents which in turn affect atmospheric weather patterns.

4.5 Proposed Limits

The level of CO_2 in the atmosphere has been rising consistently (Fig. 4.10). Although there are minor seasonal fluctuations at any set location or region over the year it is best to use data from a single site. The data normally used is based on readings from

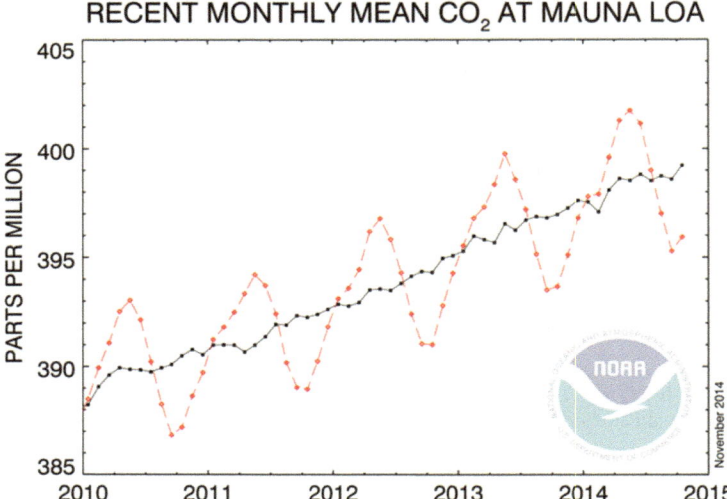

Fig. 4.10 The monthly averages of atmospheric carbon dioxide in ppm showing annual flux in concentration (*dotted line*) and the annual average as measured at Mauna Loa, Hawaii. *Source*: NOAA http://www.esrl.noaa.gov/gmd/webdata/ccgg/trends/co2_trend_mlo.png. With permission of the National Oceanic and Atmospheric Administration (NOAA), US Department of Commerce

the Mauna Loa Observatory in Hawaii which normally has its annual peak in atmospheric CO$_2$ concentration during May and its lowest value in October. The current level of CO$_2$ in the atmosphere exceeded 400 ppm for the first time in April 2014, with a peak for that year in May at 401.75 ppm. This is as predicted for the A1F1 scenario explored above.

More information: (http://co2now.org/)

So we exceeded the 400 ppm threshold for the first time during 2014, although the annual average will remain below this threshold until 2016. However, it has been a long term goal, since Kyoto, to stabilize global CO$_2$ concentrations to below or at 450 ppm in order to prevent average global temperature rise in excess of 2 °C. While many Governments maintain this goal, it is now evident that it is highly unlikely that this is possible given that the rate of increase is rapidly increasing. The annual rate of increase in CO$_2$ concentration in the atmosphere has steadily increased from 0.09 ppm per annum in the 1960s to 2.07 ppm per annum in the past decade (Table 4.4). This is due of course to increasing use of fossil fuels which continue to rise each year pushing emissions of GHG's ever upward rising from 29.8 Gt CO$_2$e per year in 2008, to 31.6 (2009), 33.5 (2010), and 34.8 Gt CO$_2$e in 2011. We have

Table 4.4 Rate of increase in global CO_2 concentrations by decade during the period 1963–2012

Decade	Total increase (ppm)	Annual rate of increase (ppm per year)
2003–2012	20.74	2.07
1993–2002	16.73	1.67
1983–1992	15.24	1.52
1973–1982	13.68	1.37
1963–1972	9.00	0.90

Source: NOAA http://www.esrl.noaa.gov/gmd/ccgg/trends/index.html
With permission of the National Oceanic and Atmospheric Administration (NOAA), US Department of Commerce

to peak at 44 Gt CO_2e per year by 2020 to best avoid >2 °C rise, although the best predictive models discussed at the IPCC meeting in Durban (2011) suggests emissions will be 55 Gt CO_2e per year by 2020!

GHG emissions rose by 3.2 % during 2011 which will mean a 6 °C rise in global temperatures by 2050 if that rate of increase is sustained. At the moment there is no sign that the rate of rise will fall. Of all fossil fuel derived CO_2 emissions, 45 % comes from burning coal, 35 % oil and 20 % natural gas. Emissions vary between countries quite significantly as we shall see. Of the four biggest emitters, China heads the list responsible for 29 % of global GHG emissions an increase of 9.3 % during 2011 alone. The US is responsible for 16 % and the EU_{27} 15 %, but both of those achieved on overall emissions reduction of 1.7 % and 1.9 % respectively during 2011. Next is India who is responsible for 6 % of global emissions which rose by 8.7 % during 2011. Even though China and India produce large weights of GHGs, their per capita emissions are only 63 % and 15 % respectively of the OECD average. This is a very worrying trend, for as these countries become wealthier so will the personal emissions.

Our problem is that we need a significant reduction in CO_2 emissions to initiate stabilization of GHGs in the atmosphere. What each of us emits today will affect people for at least 100 years, so we need to reduce emissions very quickly if we are going to stabilize atmospheric CO_2 concentrations. Industrial countries now need to cut emissions by 90 % by 2050 to achieve this. GHG emissions must start falling by 2015–2020 and continue to decline substantially if we are going to hope to stabilize at or below 450 ppm. So until global emissions fall there is no way atmospheric CO_2 will stabilize. Even if emissions are stabilized the CO_2 atmospheric concentration will continue to rise (Fig. 4.11).

Fig. 4.11 To stabilize atmospheric CO$_2$ concentrations (**a**) then we need a rapid reduction in emissions eventually stabilizing out at a significantly lower level than previously thought (**b**)

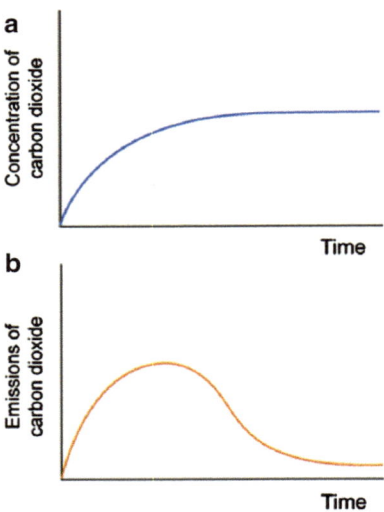

Most models currently use the IPCC's "moderate" A1B scenario which predicts atmospheric CO$_2$ concentrations of around 520 ppm by 2050 and 700 by 2100.
We are currently on the A1F1 pathway, which would take us to 1000 ppm by the century's end!

The Kyoto Protocol (1997) was ratified by most developed countries who accepted legally binding commitments to limit their emissions starting in 2005. The actual targets varied between countries with some allowed to increase emissions and other to stabilize or cut emissions. The average target was a reduction in CO$_2$e emissions of 5 % relative to 1990 levels to be achieved during the first reduction period of 2008–2012. The USA opted out on the basis that it excluded 80 % of the global population including China and India which gave them an unfair competitive advantage. Other countries have recently withdrawn (e.g. Canada in 2011) being unable to achieve their required reductions. During this time CO$_2$e has risen from 364 to 396 ppm, which has been a terrible lost opportunity.

More information on Kyoto targets: http://unfccc.int/ghg_data/ghg_data_unfccc/ items/4146.php;   http://www.eea.europa.eu/publications/ghg-trends-and-projections-2012

Of course there is an underlying problem which we will explore in depth later in the text and that is those living in the developed world use far more resources including fossil fuels per person than those in developing countries; and have been doing

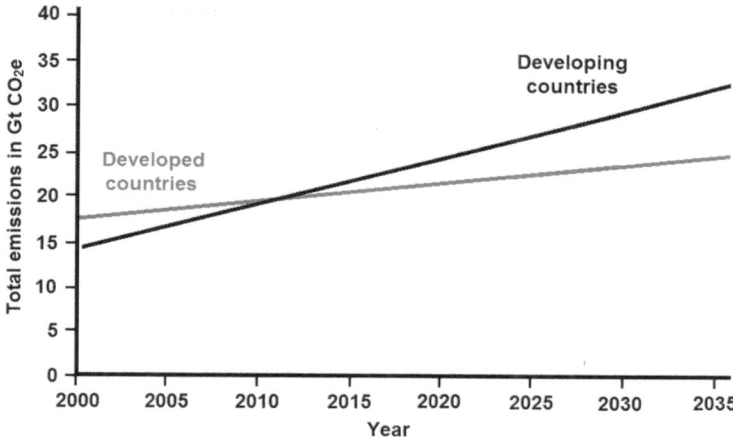

Fig. 4.12 Comparison of the increase in rate of total greenhouse gas emissions between developed and developing countries. Note one gigatonne (Gt) is equivalent to 1000 million tonnes

so since the mid eighteenth century. However, what we are seeing is a rapid growth in industrial development as well as consumerism in developing countries so that while the rate of increase in emissions in developed countries is marginally falling it is rapidly increasing in developing countries making their contribution for the first time greater than those in developed nations (Fig. 4.12). Another problem is that consumerism is also increasing in developed countries as well. However, the GHG emissions associated with the manufacture of those consumer products are not attached to the individual consumer or to their country, but rather to the country of manufacture (Chap. 5).

> **So until global emissions fall there is no way atmospheric CO_2 concentrations will stabilize. Even if emissions are stabilized at the current level the CO_2 atmospheric concentration will continue to rise.**

Global CO_2 emissions appear closely linked to global population. So reducing the global emissions by 50 % today without changing our current lifestyle or way of conducting business would require action equivalent to reducing the global population to 4.5 billion, a reduction of in excess of 2,500,000,000 people (Fig. 4.13). So the enormity in getting the required reductions in CO_2 emissions should not be underestimated in terms of the effects it will have on individuals and communities alike.

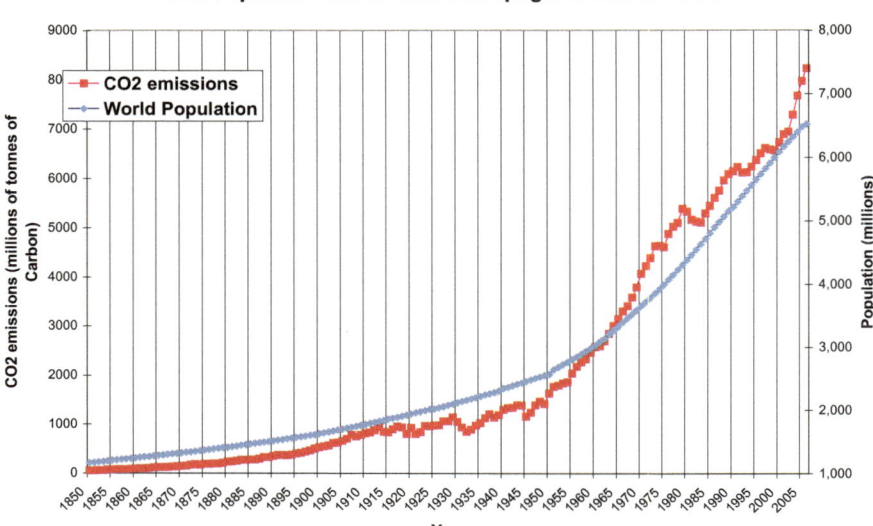

Fig. 4.13 The rise in population is closely linked to emissions. A 50 % reduction in CO_2 emissions on current levels is needed to meet emission targets. *Source*: The Carbon Dioxide Information Analysis Center http://cdiac.ornl.gov/. Reproduced with permission of the US Department of Energy, Oak Ridge, Tennessee, USA

4.6   Conclusions

- The greenhouse effect and global warming are real processes that are occurring right now.
- The energy balance is complex but is affected by man's activities in terms of burning fossil fuels, industrial and agricultural practice creating more greenhouse gases that reduce heat loss from the surface of the Earth.
- The rate of CO_2e emissions is continuing to rise annually and this is reflected by atmospheric carbon dioxide concentrations continuing to rise each year.
- What you emit today in terms of CO_2e will affect people for at least 100 years.
- We need to reduce GHG emissions significantly is we are to prevent atmospheric CO_2 concentration exceeding 450 ppm.

The third step is accepting that greenhouse gases actually control the surface temperature of the planet and that man's activities are releasing new GHGs, above those emitted by natural processes, at an ever increasing rate. Action is urgent because of the longevity of global warming potential (GWP) of the GHGs emitted, and that the targets set are realistic and necessary.

Homework

No homework for this section as we have a lot to cover in the next section which deals with measuring GHG emissions and the theory and practice of offsetting our emissions. However check this week's global atmospheric CO_2 concentration using the link http://co2now.org/.

References and Further Reading

IPCC. (2007). *IPCC Fourth Assessment Report. Working Group I Report: The physical science basis. WG1-AR4*. Geneva, Switzerland: Intergovernmental Panel on Climate Change. Published on behalf of the IPCC by Cambridge University Press, Cambridge, UK.

IPCC. (2012). *IPCC Fourth Assessment Report, Working Group I: Technical Summary (Revised)*. Geneva, Switzerland: Intergovernmental Panel on Climate Change. Published on behalf of the IPCC by Cambridge University Press, Cambridge, UK.

NASA. (2004). *Performance and Accountability Report: Part 2 Detailed Performance Data*. Retrieved from http://www.nasa.gov/pdf/83172main_PAR2.pdf

http://nca2014.globalchange.gov/highlights/report-findings/our-changing-climate#intro-section

National Research Council (NRC). (2002). *Abrupt climate change, inevitable surprises*. Washington, DC: National Academy Press. Retrieved from http://www.nap.edu/catalog/10136/abrupt-climate-change-inevitable-surprises

http://www.sierrafoot.org/global_warming/global_warming_greenland.html

The AR5 Working Group I Report. (2013). *The physical science basis*. Retrieved from http://www.climatechange2013.org/images/uploads/WGIAR5_WGI-12Doc2b_FinalDraft_All.pdf

Radiative Forcing

www.ipcc.ch/.../ar4/wg1/en/tssts-2-5.html

Greenhouse Gas Predictions

http://www.ipcc-data.org/ddc_co2.html

Chapter 5
Measuring CO$_2$ Emissions

In order to achieve carbon reduction targets we need to be able to accurately measure our green-house gas emissions. In this chapter we explore the concept of carbon footprints, how they are comprised and importantly how they can be reduced. **Oil filed at Prudhoe Bay in Alaska in the heart of the Arctic tundra.** Image by Peter Prokosch. Reproduced with permission of Peter Prokosch and GRID-Arendal http://www.grida.no/

5.1 Introduction

On a National basis we tend to measure carbon emissions by trying to carry out simple mass balances based on the amount of energy generated and consumer indices such as imports etc. More recently scientists have tried to identify all sources

© Springer International Publishing Switzerland 2015
N.F. Gray, *Facing Up to Global Warming*, DOI 10.1007/978-3-319-20146-7_5

that contribute to emissions, and the simplest way of doing that is by using a technique called **carbon footprinting**. The term emerged around the late 1990s and is defined as *a measure of the amount of carbon dioxide produced by a person, organization, or location over a given time*. So a carbon footprint describes the impact of carbon emissions, measured in units of carbon dioxide (CO_2) and other greenhouse gases (GHGs).

Although we all use the term carbon footprint the term is actually a bit misleading. Likewise the use of CO_2 can also cause confusion and possible inaccuracies because we are in fact dealing with all GHGs when we measure a carbon footprint. As was explained in (Sect. 4.3), each greenhouse gas which includes carbon dioxide, methane (CH_4), nitrous oxide (N_2O), fluorinated gases and so on, has a different atmospheric concentration as well as a different strength (or global warming potential) as a greenhouse gas. So a potent greenhouse gas at a very small atmospheric concentration can contribute to the overall greenhouse effect just as much as a weaker greenhouse gas at a much larger atmospheric concentration. Because of this variability, **carbon footprints are measured in grams (g), kilograms (kg) or tonnes (t) of carbon dioxide equivalent (CO_2e or CO_2-eq) which is the amount of CO_2 that would cause the same level of radiative forcing as the emissions of a given greenhouse gas** (Sect. 4.2).

> **The metric system is universally used for carbon accounting, but do watch out for tons which are sometimes used in the US compared to tonnes used elsewhere. To convert US tons into metric tonnes then multiply by 0.907.**

Footprints vary between countries and are largely a function of wealth, local and regional temperature (i.e. do you need air conditioning to keep cool or extra heating to keep warm) and infrastructural development. In 2002 wealthy countries had an average footprint of just over 12 tonnes per capita per year (t ca⁻¹ year⁻¹) compared to 0.8 t ca⁻¹ year⁻¹ for poorer countries. The global average remains close to 3.7 t ca⁻¹ year⁻¹ (Fig. 5.1).

> **The world or global average carbon footprint provides us with an interesting potential target for the individual. This allows poorer and developing countries to raise their standard of living while putting limits on the energy used by those living in the wealthier nations. If this was adopted it would maintain emissions at the current level and not reduce them which gives some indication of the problem of attaining global emission targets.**

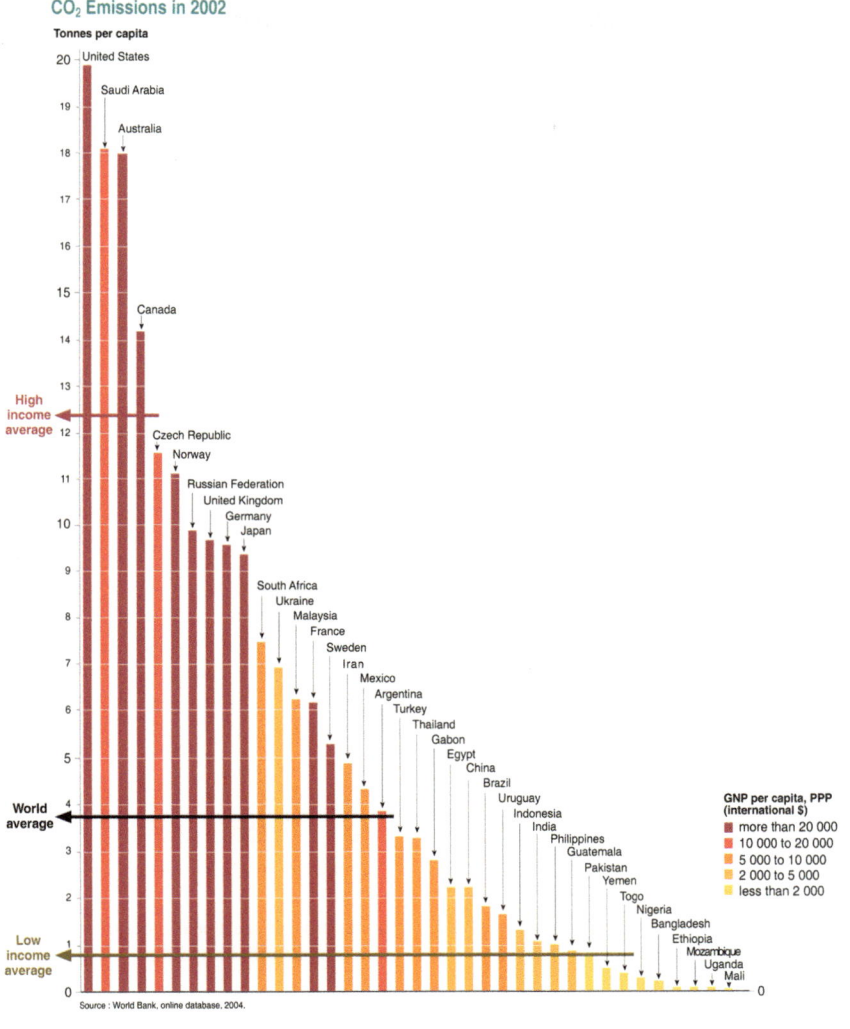

Fig. 5.1 Comparison of carbon dioxide emissions per country in 2002. *Source*: UNEP/GRID-Arendal http://www.grida.no/. Reproduced with permission GRID-Arendal and the United Nations Environment Programme

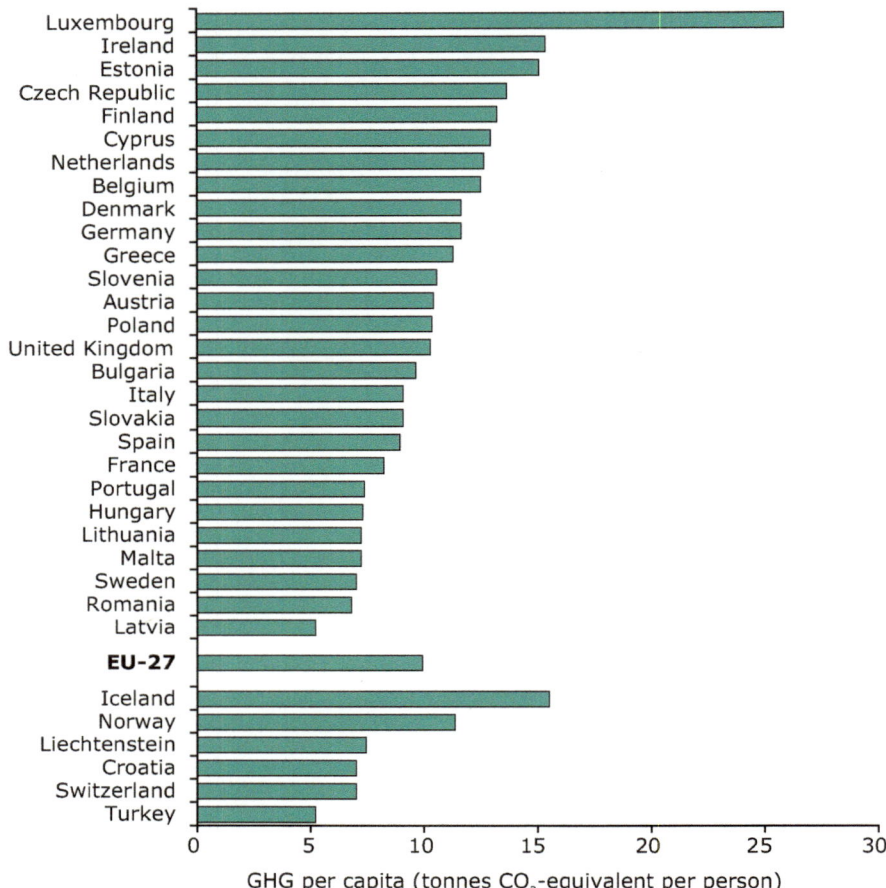

Fig. 5.2 Personal carbon footprints (tonnes CO₂e per capita) for European countries (2010). *Source*: European Environment Agency http://www.eea.europa.eu/themes/climate. Reproduced with permission of the European Environment Agency

It is sometimes difficult to reconcile the per capita emissions with those of the actual country as a whole. For example, in a European context both Luxemburg and Ireland are small contributors to overall European emissions. But as Fig. 5.2 clearly shows, both of these countries also have high per capita GHG emissions.

In 2010, Ireland had the second highest GHG emissions per capita within the EU at 13.6 tonnes of CO₂e compared to the EU average ≈10 tonnes CO₂e per capita (Fig. 5.2). Like many other European countries Ireland's per capita emissions have been slowly declining having peaked during 1990–2008 at 15.4 t CO₂e per capita. However, this is still nearly 10 tonnes over the average global per capita emissions! The Irish Environmental Protection Agency (EPA) has been responsible for producing annual estimates of GHG emissions since 1990. Each year in March a copy of

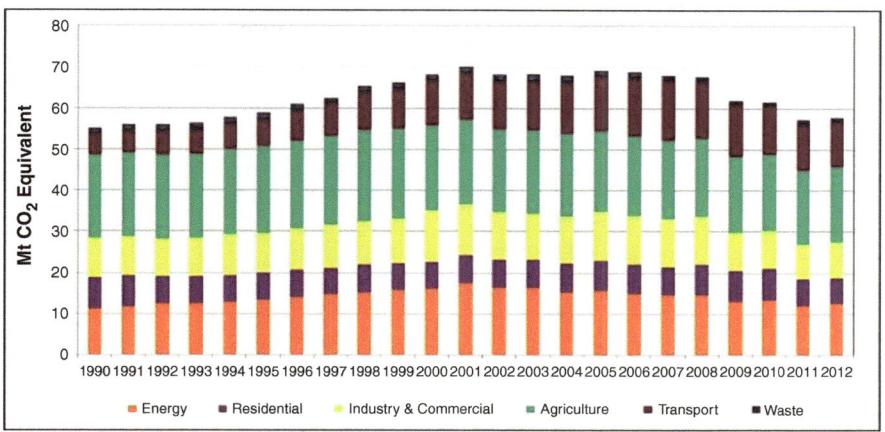

Fig. 5.3 Total greenhouse gas emissions in millions of tonnes of CO_2e for Ireland 1990–2012 broken down by major sources. *Source*: EPA (2014). Reproduced with permission of the Irish Environmental Protection Agency

Ireland's Greenhouse Gas Inventory is submitted to the European Commission and the UNFCCC (i.e. the United Nations Framework Convention on Climate Change). In their latest report, Ireland's GHG emissions rose by 1 % in 2012 to 57.92 million tonnes, reversing a 6 year downward trend. This was primarily due to a 5.9 % increase in emissions from power generation using coal and peat, and a 3 % increase from the agricultural sector due to increased livestock numbers. Industrial and commercial emissions rose by 1.8 % overall although the cement industry alone showed an 18 % increase in emissions as the economy began to recover. Transport, residential and waste emissions fell by 3.5 %, 5.9 % and 2.7 % respectively (Fig. 5.3). The reduction in residential emissions is thought to be due to the mild winter while the reduction in waste emissions is due to a 11.7 % reduction in methane emissions from landfill sites is due to organic waste being separated for composting. The EPA has blamed low-carbon prices for the increased use of peat and coal for electricity generation, but has also called for the decoupling of carbon emissions from economic growth.

More information: EPA Report—Ireland's Greenhouse Gas Inventory 1990–2012
http://www.epa.ie/pubs/reports/air/airemissions/irelandsgreenhousegasinventory1990-2012.html

The sources of GHGs in the US are similar to those in other developed countries and are summarized in Fig. 5.4, with trends shown in Fig. 5.5. Land use and forestry offset these emissions by 15 % overall by absorbing CO_2 from the atmosphere and storing it as plant material or as humus in the soil. To conform to the UN Framework Convention on Climate Change the US also carries out an annual review of its GHG emissions and possible sinks. This is carried out by the US Environmental protection Agency (US EPA) who use national data bases and reporting programmes to collect emissions data from energy use in the domestic,

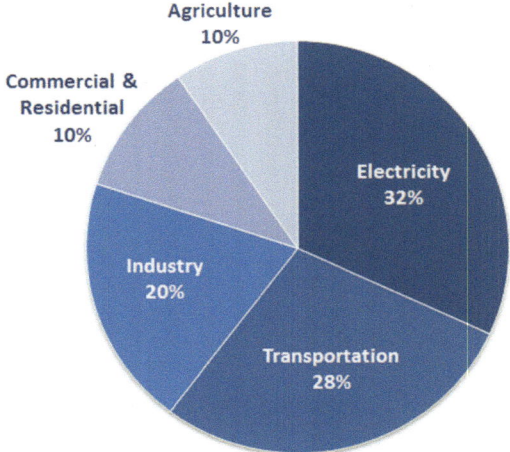

Fig. 5.4 Main sources of greenhouse gas emissions by sector in the US during 2012. *Source*: US EPA Inventory of US greenhouse gas emissions and sinks: 1990–2012. http://www.epa.gov/climatechange/ghgemissions/usinventoryreport.html. Reproduced with permission of the US Environmental Protection Agency

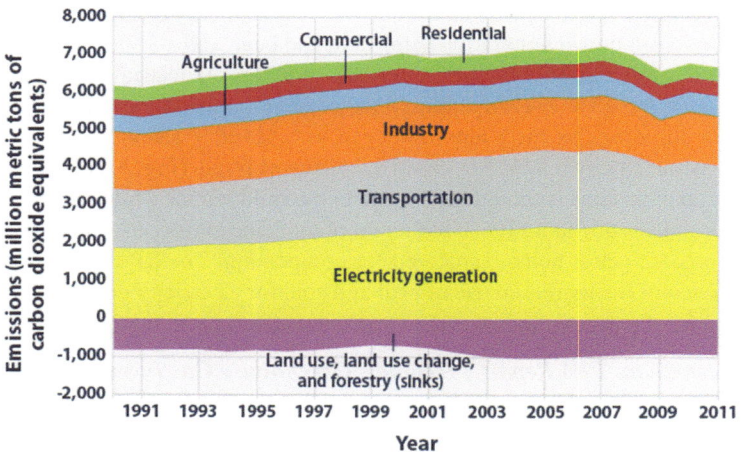

Fig. 5.5 US greenhouse gas and sinks by economic sector 1990–2011. *Source*: US EPA Inventory of US greenhouse gas emissions and sinks: 1990–2011. http://www.epa.gov/climatechange/ghgemissions/usinventoryreport.html. Reproduced with permission of the US Environmental Protection Agency

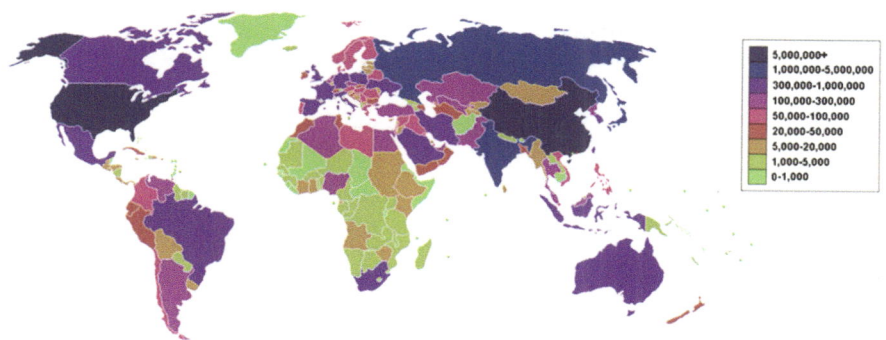

Fig. 5.6 Greenhouse gas (CO₂e) emissions in thousands of metric tonnes per annum (2006). Reproduced with permission under GNU Free Documentation License

agricultural and industrial sectors. Since reporting began GHG emissions in the US have increased by 5 % overall. The latest *Inventory of U.S. Greenhouse Gas Emissions and Sinks* was published in April 2014 and covers the period 1990–2012. This reported that GHG emissions for the US had declined by 3.4 % over the previous year to 6526 million tonnes of CO_2e during 2012, which is 10 % below 2005 levels. This reduction was attributed to improved fuel efficiency of vehicles, especially cars, less centralized electricity generation, Improved fuel mixes for power generation such as increase use of hydroelectricity and the reduction in the use of coal for electricity generation with more carbon efficient natural gas being used, and milder weather causing less demand for heating.

*More information: US EPA Report—**Inventory of U.S. Greenhouse Gas Emissions and Sinks: 1990–2012*** http://www.epa.gov/climatechange/Downloads/ghgemissions/US-GHG-Inventory-2014-Main-Text.pdf

Globally emissions per country are extremely variable. So who are the big emitters of GHGs? Figure 5.6 shows CO_2e emissions per country with the darker shades showing the large emitters while the lighter shades are the smallest. The difference between countries is vast. Those that are green have national emissions of less than 1 million tonnes CO_2e per annum while countries highlighted in dark purple have annual emissions above 5000 million tonnes CO_2e. As expected the poorest countries are the smallest emitters of GHGs.

5.2 Total Carbon Footprint

A carbon footprint is made up of the sum of two parts, the direct or primary carbon footprint and the indirect or secondary carbon footprint (Fig. 5.7). The primary footprint is considered to be those emissions that we have direct control over through the burning of fossil fuels such as domestic energy consumption and transportation.

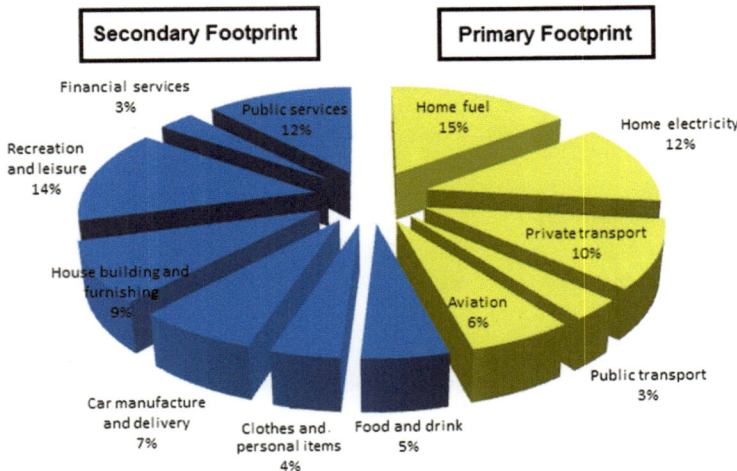

Fig. 5.7 Portioning of an individual's carbon footprint

In general terms our primary foot print represents approximately 46 % of the total. In contrast, the secondary footprint is much more difficult to define and includes emissions associated with the manufacture, distribution and disposal (calculated using life cycle analysis) of the products that we use, including hidden public services from the energy that powers street lights to the running of hospitals and schools. So all the activities we enjoy outside the home like going to a restaurant or bar and recreation in general, all fall into the secondary footprint even though we are the primary consumers of those services.

Carbon dioxide emission models and calculators are important for a number of reasons. In the first instance they are very powerful tools that enable individuals and businesses to quantify emissions and link these to specific activities or behaviour. So they allow us to understand where our emissions are being generated and to make informed choices about their management. A key element to all calculators, which some providers are good at exploiting, is educating us as to how CO_2e emissions are generated. Calculators can also be used for carbon taxation purposes, as well as helping us to develop mitigation strategies to reduce our emissions by comparing lifestyle choices.

At a personal level carbon footprint calculators fall into two broad categories, those that are freely available via the Internet which are operated by either government agencies, NGO's or individual companies, and those linked to offsetting mechanisms. There are also more complex models that are used for research and are very complex and detailed. Some of the more popular Internet calculators include:

Government, NGO or Company Examples

- *Ireland:* Combat Climate Change
- *UK:* Defra—ActonCO$_2$—Climate Trust
- *USA:* USEPA

Examples Linked to Offsetting Mechanisms

- Climate Care
- Carbon Neural Company UK
- Terra Pass

For carbon calculators to be useful then users must accept that at the outset that their individual behaviour or lifestyle is a source of global CO$_2$e emissions. This seems obvious but the critical point is that **we all contribute to global warming through CO$_2$e emissions, so in order to reduce them we need to act individually to address them**. Our personal actions affect the secondary emissions as well as the primary. So carbon calculators give us the means by which we can identify, manage and reduce our emissions most effectively. Calculators could also form the basis for future carbon taxation, allocation of carbon credits and personal carbon trading.

> *Carbon footprinting is a major mechanism for short to medium term CO$_2$e reductions in order to meet interim Kyoto targets; also if we are to reduce emissions by 80 % by 2050, this has to be one of the core mechanisms.*

More information: http://www.epa.ie/whatwedo/climate/calculators

Online carbon calculators are generally based on a limited number of inputs in order to make them easy to use. These input categories are normally broken down into **Home energy usage**, **Transport**, **Air travel** and **Waste**. The degree of resolution or accuracy varies between calculators and the more information given then the more accurate the output values will be. In order to make them simple then very few questions will be asked, or generic values will be generated from simple questions such as 'Do you recycle waste?'

Emission factors (real or generic) are established using a variety of methods and can be found in government and research literature. They use technical analytical methods such life cycle analysis of goods and services, as well as environmental input–output modelling to calculate these values.

Here are examples of input systems used by online calculators of increasing complexity and accuracy using just home energy usage and transportation as key input categories:

- Electricity usage (metered) only; Car usage (type of car or engine size not specified)

- Electricity, natural gas and oil usage; car usage by engine size using three bands (<1, 1.5, >2.0 L)
- Electricity, natural gas, oil and coal; Car usage (petrol/diesel), public transport (Bus/train)
- Electricity, natural gas, oil, coal, wood and bottled gas; Car usage (make/model/fuel type), motorbike/scooter usage, public transport (Bus/train/taxi)

Of course we live complex lives and use a wide range of heating options and travel methods, so the more simple the calculators the more will be excluded and the more difficult it becomes to get either a precise or useful measure of our real emissions. Calculators are becoming increasingly efficient and dynamic and also now normally include aviation, waste streams and some secondary allowances as well. While the degree of resolution is constantly improving the lack of consistency between calculations and models means that it is difficult to know if your emissions are truly accurate. There are in fact two key International Standards that improve our calculations for life-cycle assessment and for product carbon footprint analysis (i.e. ISO 14040 series, ISO, 14064 series and PAS 2050), as well as accepted protocols for their assessment (i.e. GHG Protocol) which are generally used by carbon foot printing companies and businesses alike (British Standards Institute 2006, 2011, 2012). However, in practice there is a lack of accuracy in many calculators and models.

Specific Protocols

ISO 14040   http://www.iso.org/iso/iso_catalogue/catalogue_tc/catalogue_detail. htm?csnumber=37456;
ISO 14064-1 http://www.bsigroup.com/en-GB/cfv-carbon-footprint-verification/;
PAS 2050 http://shop.bsigroup.com/en/forms/PASs/PAS-2050/;
GHG protocol http://www.ghgprotocol.org/standards/product-standard

More information: Greenhouse gas protocol organization: http://www. ghgprotocol.org/about-ghgp

Here is a summary of the minimum emission input categories that should be present in any calculators selected by you for carbon footprinting:

- Country selection
- Household details
- Food
- Waste
- Renewable heating
- Home energy
- Personal transport
- Air travel
- Recycling
- Secondary

To see just how reliable calculators are researchers from Trinity College Dublin compared six key calculators using basic data for a typical Irish household of three. They found that the annual household emissions varied between 12.1 and

27.2 tonnes of CO_2 per year (t CO_2 year^{-1}) giving an average of 18.1 t CO_2 year^{-1} which is equivalent to a personal footprint of 4.02–9.07 t CO_2 year^{-1} (mean of 4.6) a variation of over 5 t CO_2 year^{-1}. All these calculators excluded emissions from CH_4 and N_2O which underestimated emissions by 1.8 % overall, hence the use of CO_2 and not CO_2e.

More information: Kenny, T. and Gray, N.F. (2009) *Environmental Impact Review,* *29, 1–6.*

The study was extended to examine the top 20 carbon footprint calculators but this time using data for an Irish household of four (two adults and two children) Once again there was a large difference between the calculators with household emissions varying from 15.5 to 41.7 (mean 25.9) t CO_2 year^{-1}, or a mean personal footprint of 6.25 kg CO_2 ca^{-1} year^{-1}. That represents a massive variation between the calculators of 26.2 t CO_2 year^{-1} per household. The variation was seen in all the main input categories to the calculators: household energy 2.5–15.9 (mean 8.9) t CO_2 year^{-1}; transport 5.5–18.5 (mean 12.8) t CO_2 year^{-1} and air travel 1.3–14.3 (mean 5.2) t CO_2 year^{-1}. This shows that calculators are measuring different things and have different levels of accuracy. Of course it is imperative that we have access to the most extensive, accurate and reliable calculator as possible to be able to monitor changes over time using the original value as our base line. **The basis of successful household and personal emissions reduction is having access to high quality calculators**. Currently it is possible to achieve significant reductions in your personal or household emissions by simply switching calculators.

To try and identify the best calculators the same researchers created a very detailed carbon footprint using emission values calculated specifically for Ireland. They carried out a detailed assessment of 103 Irish households. The occupancy rates for the households varied between 1 and 6 with an average occupancy of 2.9. This is very close to the national household occupancy rate of 2.8, so the results give a good indication of an average household footprint. In terms of emissions 42.2 % was generated in the home, 35.1 % was from transport of different types including commuting to work and school as well as from recreational activities, 20.6 % was generated by private air travel and just 2.2 % from waste disposal. **This gave an overall mean Irish household emission 16.55 t CO_2e year^{-1} which equates to an average personal footprint of 5.70 t CO_2e year^{-1}, of which 1.15 t CO_2e year^{-1} comes from personal air travel.** The study showed that household energy consumption became increasing more efficient as the occupancy rate increased with the most efficient house type surveyed being terraced houses using natural gas and the least efficient being detached houses using oil fired heating.

More information: Kenny, T. and Gray, N.F. (2009) *Environment International,* 35, 259–272.

Such studies are still quite rare but are important in that they give some idea of the range of emissions per capita and per household. It also helps us understand what we are not measuring in our carbon footprints. In the context of primary and secondary (embedded) footprints we know that **in Ireland the average primary**

personal emission is 5.7 t CO_2e year^{-1}. Nationally derived values calculates **the total personal emissions per annum to be 13.6 t CO_2e year^{-1}**, so our average secondary or embedded personal emissions is the difference which is 7.9 t CO_2e year^{-1} or 58 % of the total. It is these secondary emissions that we appear to have very little control over, and to a certain extent this is true. Hospitals, schools, public buildings all have to be lit and heated for example. But also built into this is the energy attached to the products and services that we use. So what we should be trying to do is remove those embedded emissions to the personal footprint wherever possible, identify carbon avoidance and include this in our primary footprints, and of course put pressure on Government and sectors to reduce their emissions. Only in this way can we begin to really control where and how CO_2e is generated and released, and more importantly who is responsible for those emissions.

> **Using Fig. 5.7 as an example, we can see that only 15 % of our total footprint is in fact outside our direct control, so that as individuals we can manage up to 85 % of the carbon footprint.**

For carbon calculators to be successful they need to identify as many inputs as possible and to calculate the final footprint using specific conversion factors for that country or area. The study above developed a very specific calculator for Ireland which included specific conversion factors and fuel mixes which differ significantly between countries.

In their model they took into account a wide range of lifestyle options. For example, the average Irish household uses more than one fuel and this may include any of the following: Peat (hand cut turf or manufactured briquettes); butane, propane, wood (logs, chips, pellets); smokeless fuels; oil, natural gas and of course electricity. There is also a wide variety of Irish public transport options each with a specific emission per passenger mile such as the Luas (electric tram), DART (electric rail network), Irish Rail (diesel powered trains), Dublin Bus (double-decker city buses), Bus Eireann (intercity coaches), as well as taxis and even central cycle hire. They also included a range of waste and recycling options, a wide range of car engine sizes, petrol, hybrid or diesel, motorbikes and scooters. An interesting difference between taxis in Dublin and London is the increasing number of electric and hybrid cars now used by private taxi companies in the latter.

Examples of specific emission conversion factors for public transport in Ireland are:

- Irish Rail mainline train 54 g CO_2e per passenger km travelled
- Dublin bus 74 g CO_2e per passenger km travelled
- Luas 55 g CO_2e per passenger km travelled

Many sectors are missing from primary footprint calculators, including public transport, mainly due to the difficulty in allocating accurate emission values and also due to the problem of entering accurate information into the calculator.

So common categories such as food consumption, clothing, entertainment, purchase of manufactured goods, these are all things we all do and buy but are not included in our primary carbon footprint.

Some estimates put the US mean household footprint for food consumption alone at 8.1 t CO_2e year^{-1}. This is broken down into 83 % of emissions from food production (45 % CO_2; 31 % NO_x; 24 % CH_4), 11 % from transport from the farm to the processor then to the consumer, and 6 % from wholesale/retail such as refrigeration, lighting etc. Food is a major source of CO_2e with a significant portion arising as methane from cattle and manure. This GHG traps heat 21 times as effectively as CO_2 while remaining in atmosphere for between 9 and 15 years. Also nitrous oxide from fertilizers and manure is a major source of CO_2e and is 296 times more effective than CO_2 and lasts on average 114 years. So food in particular can often carry a hefty carbon tag. Even that quick shop for an evening mean can have quite an impact even if you are buying locally grown and manufactured produce. Here is a typical list for an evening meal for four excluding any alcohol or soft drinks.

300 g pack of Cheese	2600 g
300 g beef stewing steak	4800 g
1 kg potatoes	240 g
Half dozen eggs (organic)	1650 g
1 l carton milk	1050 g
1 kg Carrots	45 g
25 g packet crisps	55 g
4 pack apples	110 g
300 g punnet tomatoes (out of season)	2800 g
Total	**13.35 kg CO_2e**

That is an equivalent carbon footprint to driving a Toyota Yaris 112 km or an SUV 55 km!

Where food is included in calculators they usually allocate a generic or an average value which is frequently quoted as between 380 and 750 kg CO_2 ca^{-1}year^{-1}, with the lower value for vegans and the higher value for meat eaters. The actual allocation may be determined by asking you how many times a week you eat meat or a reference to frequency of using dairy products both of which have high emission values (Sect. 10.5). Out of season vegetables can also have very high emissions associated with them due to air miles or most likely because they have been grown in an artificially heated and lit greenhouse (Sect. 10.4). The reality is that emissions are generally much higher than this, although we know very little about emissions for specific foods grown in different locations. A word of warning. **These generic values exist for a wide range of products and are frequently assumed values. However, by repeating them so often, both on the Internet and in the press, they somehow become real.** If the generic value is correct then we should all be limiting ourselves to just to 1.36 kg CO_2 per day to achieve an annual food footprint of 500 kg CO_2, that is just 9.6 kg of associated CO_2 per person per week. This is considered further in Sect. 10.5.

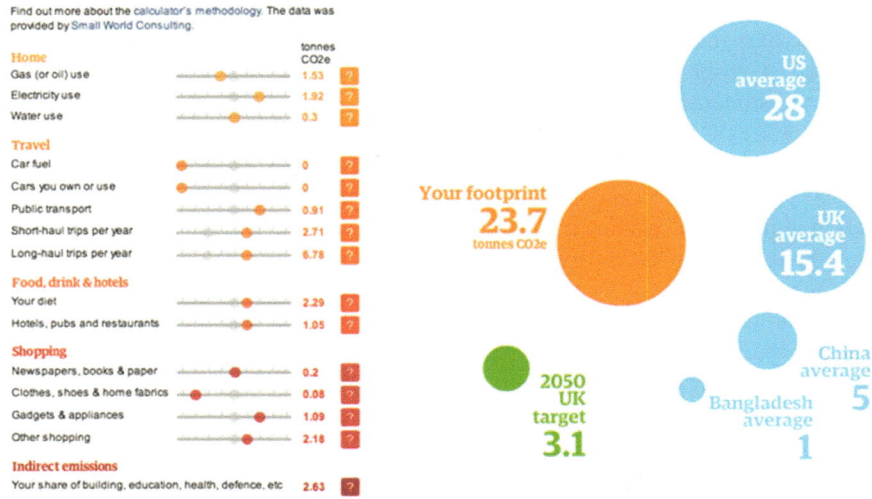

Fig. 5.8 The Guardian carbon calculator. *Source*: http://www.guardian.co.uk/environment/interactive/2009/oct/20/guardian-quick-carbon-calculator. Reproduced with permission The Guardian Newspaper Group

Some calculators have tried to take account of some of the more personal secondary components. The Guardian Newspaper has their own footprint calculator (Fig. 5.8) which includes food and shopping and then compares your footprint against the average UK value of 15.4 t CO$_2$ year^{-1}, the 2050 UK target of 3.1 and other countries.

Increasingly Apps for mobile phones are being developed to assess and manage our emissions. For example, many mobile devices now use global positioning satellites to work out automatically where you are and the distance you are travelling, while many apps are available that can detect whether you are walking, driving or flying and then calculate your CO$_2$e emissions

What are the problems and possible solutions of using carbon footprint models to help us manage and reduce our direct and indirect use of fossil fuels? First calculators should carry out appropriate calculations using flexible models of regional significance. They can be made more accurate by (i) ensuring as many of the secondary footprint categories end up in the primary footprint, (ii) using regional and locally accurate, updated and relevant conversion factors, (iii) not including offsets in the calculations, (iv) not allowing avoidance by omitting activities outside the home nor using generic values, (v) having a high degree flexibility in terms of inputs and finally (vi) ease of use.

It is imperative that we have access to the most extensive, accurate and reliable calculators as possible in able to monitor changes over time using the original value as our base line.

5.3   Embedded and Secondary Emissions

We have established that a significant portion of the emissions that we generate fall into the secondary part of our footprint which are not identified by most carbon calculators. Now a portion of these are not able to be allocated personally as we all benefit to a greater or lesser extent from them, for example for the provision and maintenance of the road network. Your business may depend totally on the use of the network, or you may just use it to get to work, school or the shops, but we all need the roads to provide food and other services. So there is no opting out of these types of emissions. However, there are a wide variety of emissions that should be in the primary footprint as they are used selectively by us for our sole benefit. In order to be able to really control of our emissions we need to transfer as many of these embedded or secondary emissions into the primary footprint itself, so that we can take full control. In order to establish what we are actually emitting at a personal level we need to recover these lost emissions by taking responsibility for all our emissions both inside and outside the home.

Context of Primary and Secondary (Embedded) Footprints

- Mean primary personal emission 5.7 t CO_2e year^{-1}
- Mean embedded personal emission 7.9 t CO_2e year^{-1}
- Mean total personal emission 13.6 t CO_2e year^{-1}

Key Responses

- Remove embedded emissions to personal footprint
- Identify carbon avoidance and include as many activities as possible
- Pressure on Government and sectors to reduce emissions
- Reduction of own emissions

5.3.1   Embedded Energy

Embedded energy is all the energy used in the manufacture, transport and disposal of a product and is often hidden completely from our calculation of our footprint. The majority of calculators take a simplistic view of energy usage, ignoring embedded energy costs and putting them into the secondary footprint. It may not even be taken into account during the calculation of the National Carbon Inventory when it is manufactured outside our own country, with the carbon footprint being allocated to the country of manufacture rather than consumption (e.g. China). However, our use of resources and consumables indirectly controls these embedded GHG emissions as well.

Our use of resources and consumables indirectly controls embedded (secondary) GHG emissions.

Table 5.1 Emissions from cars by engine capacity per km travelled

Fuel type	Engine size (L)	Engine category	Emission (g CO$_2$ per km)	Miles per gallon equivalent (mpg)
Petrol	<1.4	Small	159.2	40.8
	1.4–2.0	Medium	188.0	34.6
	>2.0	Large	257.7	25.2

Data from Defra (2008). Reproduced with permission the Department for Environment, Food and Rural Affairs, London

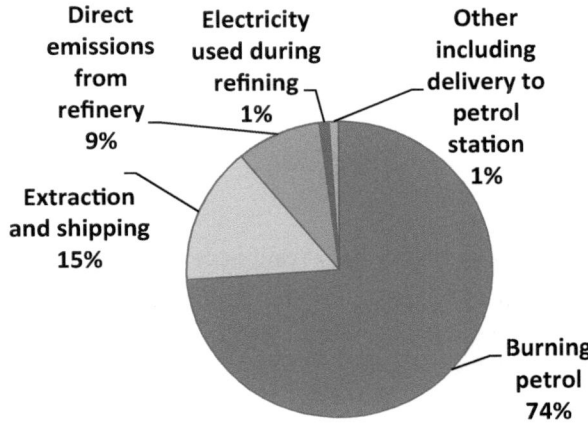

Fig. 5.9 The carbon footprint of transport fuel

For example, we tend to measure fuel used in car usage either in terms of kilometres or miles driven per size of engine, or the actual number of litres or gallons of fuel used. Generic values are derived from engine size and are commonly used in calculators as shown below (Table 5.1).

Yet when we look at the embedded energy in the production and supply of petrol the actual burning of the fuel in our cars only represents 74 % of the total carbon footprint of the product with 26 % associated with extraction, refining and transportation (Fig. 5.9).

Thus 26 % is lost into the secondary footprint both here and elsewhere. The extraction and shipping costs may be associated with one country, refining another and then be used in yet another. So this begs the question whose GHGs are they? Is it the company's responsibility, that of the individual country where the oil is extracted as exports form part of the GNP or the consumer of that product, that's you and me?

5.4 Examples of How We Use Energy

5.4.1 Driving

Of course using generic values does not reflect how much or how that product is used. Generic values for car emissions are measured under controlled conditions such as a constant speed, level road, just the driver plus a 25 kg payload, with a stripped down vehicle etc. These conditions have recently been improved under the New European Driving Cycle (NEDC) which is used in vehicle type approval. So these generic values or the emissions given by manufactures in their sales literature rarely reflect actual emissions which is partly due to how the vehicle will be driven (Fig. 9.7). There are a number of important factors that affect fuel efficiency including:

- Use of accessories (air conditioning, lights, heaters etc.)
- Payload (passengers, luggage equipment),
- Poor maintenance (underinflated tyres, poor tracking etc.)
- Gradients
- Variable weather
- Poor driving (aggressive, excessive acceleration/braking)

In order to approximate the actual emissions from cars, the Department for the Environment, Food and Rural Affairs (Defra) in the UK has added an extra 15 % onto the NEDC assessment to take into account real world effects, which means that Table 5.1 needs to be adjusted to give more realistic emission values for these three vehicle classes (Table 5.2). Of course these figures do not take into account emissions associated with the manufacture of the vehicle, replacement tyres, servicing, parts, repairs or the final disposal of the vehicle (Sect. 9.4.4).

Emissions are calculated most efficiently by simply measuring the amount of fuel used which automatically takes into account variability in both the vehicle and driving skill.

Table 5.2 Emissions from cars by engine capacity per km travelled with the Defra conversion for real driving conditions

Fuel type	Engine size (L)	Engine category	Emission (g CO_2 per km)	Miles per gallon equivalent (mpg)
Petrol	<1.4	Small	183.1	35.5
	1.4–2.0	Medium	216.2	30.1
	>2.0	Large	196.4	21.9

See Table 9.11 for an expanded version. Data from Defra (2008). Reproduced with permission the Department for Environment, Food and Rural Affairs, London

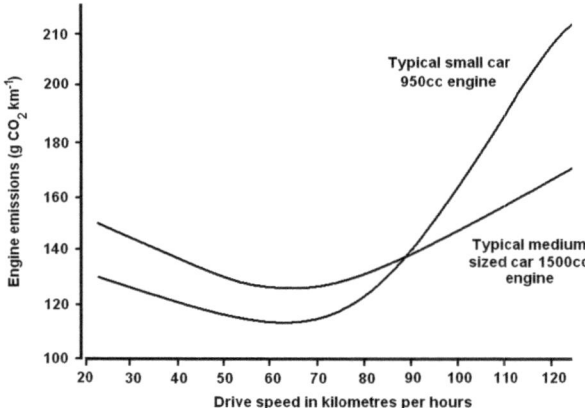

Fig. 5.10 Emissions vary at different speeds depending on engine size. Engines are often designed to give optimum performance at the speed at which emissions are tested

Car engines are also designed for specific driving conditions, so a car with a small engine is more efficient when driven relatively slowly and becomes increasingly inefficient at higher speeds. Conversely larger engine cars are far more efficient at higher speeds (Fig. 5.10).

Car engines are often designed to have their lowest emissions at or around 60–80 km per hour (km/h), so this is the optimum speed to minimize your GHG emissions. Small cars and SUVs driven above 100 km/h are usually very polluting in terms GHG emissions, so these vehicles need to be driven at or below 100 km/h. In urban areas the SUV will also be using a lot more fuel than the smaller car as well as releasing a lot of particulate pollutants known as particulate matter (PM) expressed as different particle sizes in micrograms (e.g. PM40, PM20 and PM10). SUVs are purpose built vehicles which should only really be used in rural areas where their capabilities can be fully exploited for off road and towing needs, so they should have low annual mileage. If you drive a lot on motorways or long distances you need a larger, probably a modern diesel, car. If you drive in congested areas and stop and start a lot you need a hybrid car, but these are less efficient if you are simply plying up and down the motorway as they have relatively large, usually petrol engines and carry a heavy battery. If you are driving in urban areas or making primarily short to medium distance trips, then you need a small car, preferably fully electric, petrol fuelled, or alternatively a car with a small engine with stop and go. Cars travel is discussed further in Sect. 9.4.

To really make a difference to your car emissions as well as reducing your fuel costs, use the fuel efficiency system and the rev counter in your car to help you drive more efficiently. I have a Toyota Auris and according to the sales material I should be using 4.2 L of diesel per 100 km (L per 100 km) driven. Over the years this has varied from 4.4 to 5.1. I now manage to keep it at around 4.4 L per 100 km by just being aware. Cars generally have two fuel efficiency indicators. The real-time indicator tells you how much fuel you are using per 100 km travelled and this changes every 15 s or so as it calculates the rate. So if you suddenly accelerate or go up a hill it will soar up to 10 or more L per 100 km in my car or when you are going downhill it drops close to zero. This indicator is useful as it allows you to see when fuel is being used and saved, it also allows you to see the difference in driving style and the effect of driving in too high or too low a gear when driving on the flat along a motorway for example. This is really a training tool and is well worth using. The second instrument is the integrated fuel efficiency gauge which gives you an average value (L per 100 km) over the period since you last reset it. So the longer it has been left without being reset the longer it takes for any changes to register. So my advice is to always reset you long-term gauge every time you refill the car with fuel and try and keep the value (L per 100 km) as low as possible. Driving at or below 100 km h^{-1}, or not exceeding a certain maximum rev count (2000 rpm in my own car) makes a huge difference in fuel efficiency. This depends on the size of your engine and the type of car you have, but with perseverance you can get much closer to the theoretical emissions level and you will also end up driving slightly slower, more smoothly and so more safely. I sometimes get into old habits and the long term fuel efficiency indicator starts to climb and then I have to start thinking about how improve fuel efficiency again. When I first started I couldn't get below 4.8 L per 100 km until I realized that my tyres were underinflated, and once at the correct pressure I was able to get to 4.5 L per 100 km immediately, so the condition of your vehicle is important as is regular servicing. Of course carrying extra passengers or weight in the car can increase this slightly again in my case by 0.2 or 0.3 L per 100 km, but again if you think about how the weight is displaced within the car then this can be minimized. Other factors also have an effect, the smoothness of the road surface, time spent at traffic lights or in traffic jams, inclines, bends, all these things add to your fuel efficiency. For example travelling on a flat motorway in Ireland compared to that in the UK in my car is the difference between 3.9 and 4.2 L per 100 km respectively, and that is due solely to the smoother surfaces found on UK motorways. So by optimizing my driving skill I can easily save 10–15 % on my fuel bills and my associated car GHG emissions. The savings are even greater in vans and SUVs. One friend of mine who drove his large van at high speed up and down the motorway found that he could save 25 % on his fuel bill when he reduced the amount of equipment he carried and drove more slowly and less erratically. It is quite easy to do and actually fun to have targets to try and beat. Of course extra features such as stop and go, which not only saves fuel but is so important in urban areas to reduce PM concentrations where traffic is stationary, eco drive indicators and an optimum gear change indicator, all add to that efficiency.

Compare your own car emissions at: http://www.car-emissions.com/

5.4.2 Lights

The campaign to replace standard incandescent light bulbs with low energy bulbs has been one of the great success stories in household energy reduction, helped along by various EU directives. The comparison is quite straight forward. If you use a 100 W incandescent bulb 24 h a day for a year the emissions are 500 kg CO_2e per bulb whereas the equivalent rated low-energy bulb is only 90 kg CO_2e for a year. The incandescent bulb is using 100 W of energy compared to just 18 W for the compact fluorescent bulb for the same light output, a saving of 82 %. You are making that saving no matter how long the bulb is on.

So household lighting is the simplest and most effective way of saving money and energy. Apart from using low-energy bulbs you can save energy by turning off lights when you are not in the room. It is widely believed that the life of the bulb is reduced by turning it on and off, and that when fluorescent bulbs are switched on they use extra energy. Both of these are true, but in reality the cost of the slightly reduced life of the bulb is negligible compared to energy saved, and the extra energy used to fire the fluorescent bulb is equivalent to just a few seconds of the energy used when the bulb is normally on. So I always turn off lights including low energy bulbs if I am not going to be in that room in next 15 min as this minimizes the damage to the bulb, ensuring it is completely cooled before switching it on again.

Always turn off lights including low energy bulbs if the area is not going to be used in next 15 min

Why do we need to turn off low-energy bulbs? Unfortunately, because low-energy bulbs are comparatively efficient people now feel that there is no need to turn them off when not needed. So whereas you would have seen a couple of lights on in a house a decade ago, with the advent of low-energy bulbs, you now see houses with all the internal lights on for most of the evening. Also, many new homes have incorporated in their design outdoor low energy soffit lights. There may be as many as 60 of these lights on a large bungalow creating a band of soft light around the house. These are normally light activated so they turn on at dusk and switch off again at dawn. The whole idea of using low-energy bulbs is to reduce emissions, so increasing the number of lights in your home or leaving more lights on reduces any advantage to the environment or your carbon footprint. Another problem is that with an incandescent light nearly all the energy used is released as heat into the room. A friend of mine who is a civil engineer built a remarkable house that was so well insulated that the excess heat from the lighting was sufficient to heat the room. We very often forget that those old bulbs were also heating the room, and once replaced an alternative heating source may be required.

One of the most wasteful uses of energy is the use of high powered outdoor spot lighting. These can be very powerful using up to 500 W. The angle of the light is critical and they should be facing downwards to where the light is needed, however, they are usually angled so they are at right angles to the house. I have a neighbour who lives on the other side of the valley and his outdoor light is so powerful and so badly adjusted that it lights up my yard, over 1.5 km away. I rarely use mine as I don't mind the dark and also strongly believe the countryside should be dark. So it is important that outdoor lights are less than 100 W, are positioned properly so as not to affect wildlife or dazzle motorists as they pass the house, and very importantly they should be turned off when not required. There are now low energy alternative bulbs that use just 25 or 40 W that are very effective.

More information: http://www.nef.org.uk/energysaving/lowenergylighting.htm

Let's look at an example of how energy is wasted. The wattage (W) of commercial fluorescent lamps varies between 30 and 80 W per metre length of the bulb. In Trinity College where I work we have an examination hall which has 40 individual 35 W lamps, so when they are all on the lighting is using 1.4 kW per hour or 33.6 kW each day. The problem is that the lights are often left on for days on end and as the hall is locked it is not easy to get them switched off. The conversion factor to calculate the CO_2 emissions for using electricity to kg CO_2e varies between countries depending on how it is generated, but in Ireland you multiply each kW by 0.607 (Sect. 8.4). So each day in the examination hall the lights are responsible for 20.3 kg CO_2e.

But what does this actually mean? Well an efficient electric kettle uses 76 g CO_2e to boil a litre of water, so leaving those lights on is equivalent to boiling 267 L of water enough water to make 1335 cups of tea every day. See in the box below how that figure was calculated.

How to Calculate the Emissions from Your Kettle
Look at energy rating on the bottom of your kettle. It is usually **3 kWh**
   Multiply this by 0.607 to convert kW to kg of CO_2e
   So $3 \times 0.607 =$ **1.821 kg CO_2e per hour used**
   or 1.821/60 = **0.304 kg CO_2e per minute**
   Time how long it takes to boil exactly 1 L of water, in my case 2 and a half minutes, so
   $0.304 \times 2.5 = 0.0758$ kg or **76 g CO_2e**

In the home a 100 W light bulb produces 60.7 g CO_2e per hour or 0.728 kg CO_2e over 12 h compared to just 10.9 g CO_2e per hour or 0.131 kg CO_2e over 12 h for a 18 W low energy light. My office has two double-fluorescent lamps equivalent to 85 g CO_2e hour (4×35) or 0.68 kg CO_2e per 8 h working day. A standard seminar room has eight double fluorescent tubes (16×35) which produces 340 g CO_2e per hour or 4.08 kg CO_2e per 12 h. So suddenly you can see that you really can save a lot of energy by turning off lights both in the home and in the work place. Of course only turn those lights off which are not needed, one of my colleagues is always turning off the lights on the stairs which is a bit dangerous, so be sensible.

Replacing lighting, but more importantly turning lights off, are simple but very effective ways to reduce CO_2e emissions.

5.4.3   The Internet

We tend to think that a search on the Internet via Google or another search engine is emissions free. In reality energy is being used not only to carry out the search as well as generate the results but to operate your electronic device and display the results. Searches have become very fast with a typical Internet search taking an average of 0.2 of a second. According to Google itself the average search will generate GHG emissions equivalent to about 0.2 g CO_2e although this depends on the type of server employed. So the emissions from driving 1 km in that Toyota Yaris are equivalent to 600 Internet searches. In 2011 Google alone handled 1,722,071,000,000 searches which is equivalent to 350,000 tonnes of CO_2e. This does not include the emissions generated by using your laptop, pad or smart phone. Now while the emissions from using your device will be your emissions, those from the Internet belong to Google or the other search engine. Google offsets these emissions, but they are really part of our personal emissions as we are using the device.

A study at Harvard University has shown that a typical search using a standard PC could generate up to 7 g CO_2 which includes the emissions both from the search and the device. That means there is a possibility that in 2011 those Google searches could have generated as much as 12 million tonnes of CO_2 but with the search engine responsible for less than 2.9 % of the emissions. So while we have made impressive saving in energy efficiency and use in many areas, we are continuously producing new sources of emissions.

More information: http://googleblog.blogspot.ie/2009/01/powering-google-search.html

So Surfing the Net does have a definable impact in terms of GHG emissions, so how can these be minimized. Well we can try and make our devices last longer and not replace them quite so often. Do you really need to upgrade? We can try and reduce the amount of searches that we make. Another option is perhaps to try and use a renewable energy charger. If our electronic devices are powered by sunlight then we have already made a significant reduction in those associated GHG emissions.

5.4.4   Mobile Communication

Some activities have a very small footprint or global impact, or do they? Our mobile and smart phone usage is rapidly growing each year, but should this be a cause for concern in terms of global warming? Each text or tweet is equivalent to just 0.014 g CO_2e with the largest portion generated while typing in the message. However, the number of texts sent everyday is enormous. It has been estimated that 25 % of all global texts are sent in China, equivalent to 2,500,000,000,000 text messages per annum. So globally in that year it was estimated that texting generated 32,000 tonnes of CO_2e, which is about 5 g CO_2 e per person on the planet whether they text or not. In the UK, OFCOM (the Independent regulator and competition authority for the UK communications industries) reported that on average 200 text messages are sent each month by each registered phone customer and as there are currently 87.2 million registered mobiles in the country that means that 209,280,000,000 (209 billion) messages are sent annually which is equivalent to GHG emissions of 2930 tonnes of CO_2e per year! That is just 46 g CO_2e per person per year in the UK (the population of the UK being 63,705,000 in April 2013), which is equivalent to making a cup of tea.

In 2009 there were 2.7 billion mobiles in use globally and this has risen to 6.8 billion in 2013. In China there are 1.150 billion phones with 85 % of the population using one. In 2008, 94 % of adults in the UK had a phone, with 75,750,000 million subscribers, which is 1.23 phones for every person at that time. According to the Central Statistical Office, in 2011 there were 5.5 million mobile subscribers in Ireland and with a population of 4.59 million that is equivalent to 1.20 phones per person almost identical to the UK.

Although a mobile phone will last at least 10 years or more, devices are replaced on average every 18 months with a billion phones sold annually between 2004 and 2008. In the past 24 months there has been an enormous surge in the replacement of mobiles with smart devices which in turn are being replaced every 22 months on average in the UK and USA. To manufacture a mobile phone emits between 16 and 60 kg CO_2e and charging it is equivalent to 6 kg CO_2e per year, the latter being the only emissions allocated to Ireland. The energy used to support the network system works out at 36 kg CO_2e per phone per year, so add the emissions from charging, allow for the manufacture and transport of the phone spread over 18 months then the global impact of each mobile phone is approximately 47 kg CO_2e per year. The 47 kg CO_2e value is based on a year's typical usage of just under 2 min per day per phone which is equivalent to 128 g CO_2e per day. Using the base figure this means that currently global mobile phone use is generating at least 320 million tonnes CO_2e per year and possibly far more. However, the footprint of your mobile phone is largely determined by how often you use it. What are the alternatives? Well a 3 min mobile call is equivalent to sending a letter by post using recycled paper in terms of emissions, while landlines only use a third of the power to transmit calls compared to mobile to mobile calls.

Apple is one of the world's leading manufacturers of smart phones, pads, laptops and computers. They have made a significant attempt to calculate the GHG emissions generated from their products. For example the iPhone 2, 3G and 3GS each generate, according to Apple themselves, 55 kg of CO_2e while an iPhone 4 generates 45 kg CO_2e during their entire life cycle. The new iPhone 5S has a footprint of 70 kg CO_2e. What is interesting that the majority of this footprint is generated during production so for the iPhone 5S, 81 % of that 70 kg CO_2e is from production with the remainder associated with it use by the owner.

Compare Apple computers: http://www.apple.com/environment/reports/

Such values and comparisons are really useful, but the real problem is trying to actually measure the usage footprint especially given the change in use from mobile phone voice and texting to email and Internet searches via smart phones. The main energy usage for the owner of a smart phone is the frequency it needs to be recharged and this not only depends on usage but also on what apps are being used. For example, if you have a large number of active apps on your smart phone such as those with GPS functions, apps that are collecting data from your phone in relation to optimizing advertising, continuous Internet connection such as 3G, Bluetooth etc., then all these functions are continuously active and draining your battery. So if apps are not in use then turn them off. When these apps are all active you may find yourself having to recharge your battery 4–8 times more frequently and this will have an effect on your footprint. The phone itself uses very little power, it is all those active links. So get into your settings and start to manage the apps more actively and reduce the interval between charging the phones battery. Overall, compared to other activities, the use of mobile phones is a very small portion of your daily carbon footprint.

Every action has a carbon footprint and your personal decision can affect how much that action is responsible for adding to global CO₂ emissions. Something as mundane as drying your hands is a good example. It is a compromise between cost, hygiene and footprint. There are generally four options when it comes to drying your hands, although I doubt you will in practice be given a choice. These are the standard electric dryer, drip dry (that's shaking your hand and wiping them on the back of your jeans), paper towels (recycled low-grade), or an airblade. Which is most carbon friendly? Well no marks for guessing that drip dry is the most climate friendly option (0 g CO_2e), with the airblade next at just 3 g CO_2e per use. This compares very favourably over the other more traditional options of 10 g CO_2e per paper towel (recycled low-grade) or 20 g CO_2e using a standard electric dryer. The difference between the two dryers is due to the airblade using unheated air. By using a powerful fan (1.6 kW) to push the air through small slits the water is physically removed from the hands rather than evaporating it off, and it does this in just 10 s. In comparison a standard electric dryer takes longer and heats the air using 6 kW combined heater and fan.

5.5 Making the Right Choice

As every action has the potential to create GHGs, we all have numerous opportunities every day to reduce our emissions by making small changes to our lifestyle. You will be surprised if you start adding up these small changes in terms of grams of CO_2e saved each day.

> **Everyday numerous opportunities arise when you can make simple choices that will reduce your carbon footprint.**
> **Try recording carbon saved each day and actively look for more ingenious ways of avoiding GHG emissions.**

Of course you can mislead or be misled by carbon footprinting and this can often lead us into dangerous or at least contentious territory, especially when talking about green or organic options, so common sense is required.

An example of a contentious comparison would be comparing car and bike use. Which produces less CO_2e, two people individually cycling 1 mile or driving together in a 990 cc Toyota Yaris? Well cycling uses more energy than simply sitting in a car, so on average a cyclist uses 50 extra calories per mile depending on fitness, their weight, how fast they go etc. So emissions from cycling depend on the fuel you use. So here are some options to get those extra 50 cal and the emissions generated from the production and supply of that extra food:

65 g CO_2e Banana
90 g CO_2e Cereal with milk
200 g CO_2e Bacon
260 g CO_2e Cheese burger

So if the car is emitting 193 g CO_2e per mile this is equal to 97 g CO_2e for each passenger per mile. So cycling appears to be more or equally as efficient in terms of GHG emissions as the car if you rely on a banana or cereal for that extra energy; but with emissions of 400 g CO_2e per mile for the two fueled by bacon then suddenly cycling produces more emissions! I could have been even more unfair and included air freighted asparagus at 2800 g CO_2e. But this is clearly a ludicrous comparison as cycling has many other benefits such as improving fitness, health and stamina, it produces less pollutants and is far better for personal wellbeing. The comparison also doesn't work under every condition, but it does demonstrate that we must be sensible when using carbon footprinting. It also reminds us that every action uses energy and that energy can be expensive sometimes in terms of GHG emissions.

Making the right choice is not always obvious, often because there are more reasons for choosing a particular option apart from its potential emissions.

5.5.1 Plastic vs. Paper Bags

Globally between 500 and 1000 billion standard supermarket high density polyethylene (HDPE) bags are used each year with six billion handed out free to shoppers just in UK during 2012. The problem of these bags is really the major litter problem that they pose as well as being a major threat to wildlife. In terms of their carbon footprint this is significant due to the large numbers used. The transparent ultra thin lightweight HDPE bags (just 0.025 mm thickness) that you use to put loose vegetables in have a footprint of 3 g CO_2e each. The standard printed lightweight supermarket HDPE bag is equivalent to 10 g CO_2e each. The thicker reusable supermarket HDPE bag can vary between 100 and 280 g CO_2e in their manufactured footprint. So for those worried about their footprint, what are the alternatives to the standard supermarket HDPE bag?

Well interestingly the standard printed lightweight supermarket HDPE bag contains 35 % less CO$_2$e than the equivalent sized paper bag. When paper bags are used for groceries then you are often given two bags one inside the other to give the bag sufficient strength, so an equivalent paper bag may have six times more embedded CO$_2$e than the plastic version to give it the same strength. A standard reusable HDPE bag contains 10–28 times more CO$_2$e than the standard printed lightweight supermarket HDPE bag, while cotton or canvas bags contain on average 171 times more CO$_2$e. This means that you would need to reuse your plastic or canvas bag in excess of 20 or 170 times respectively in order to gain an emissions advantage over using the thinner bags just once. A study by the Environment Agency in the UK found that canvas bags are used on average only 51 times with composting the best final disposal option, making them potentially far less climate friendly than the plastic bags in terms of overall carbon emissions. To put the standard supermarket HDPE bag into perspective, then it contains only one thousandth of the CO$_2$e than food inside the bag.

What to do with plastic bags? Interesting there are three options. Burning or incineration releases toxins and 100 % of the embedded CO$_2$e is released, although some heat is generated. If we recycle the bags then up to 50 % of the embedded CO$_2$e is lost in the process. In contrast, burying the bag in a landfill ensures the embedded CO$_2$e is stored permanently with virtually no CO$_2$e released to the atmosphere. Difficult decisions eh?

In Ireland the standard supermarket HDPE bags have a levy of 15 cent each on them so that they are virtually never used by customers. Before the levy was introduced the countryside was littered with the bags, and now such bags are virtually never seen as litter items, except unfortunately containing dog poo, but that is another story. Transparent ultra thin bags are still given out free in supermarkets. In contrast, China banned both the standard and ultra thin HDPE bags and called people to use baskets or cloth sacks instead to reduce environmental pollution, including marine litter. This ban, introduced in 2008 saves 35 million barrels of oil a year in their manufacture. Ironically China has been the largest manufacture and exporters of such bags to the rest of the world.

So perhaps in terms of emissions the best option if you need a plastic shopping bag is to use a reusable HDPE bag and make sure you use it at least 20 times for a thick plastic bag or 35 times for a woven plastic bag otherwise you will have emitted more CO$_2$e than if you had used disposable bags. I called the standard lightweight supermarket HDPE bag disposable because we tend to consider them as a single use item. In fact they can be used again and again if you are careful, which is putting the investment in reusable bags under pressure. Lightweight supermarket HDPE bags have loads of different uses so it is important that they are reused as many times as possible, finally using the bag to wrap contaminated unrecyclable waste for disposal to landfill.

> **We all have cupboards full of reusable bags … the trick is to actually use them carefully to ensure that they have long lives.**
> **The challenge is how many times can you use them beyond their break-even point?**

More   information: http://www.independent.co.uk/environment/green-living/
plastic-fantastic-carrier-bags-not-ecovillains-after-all-2220129.html

What about paper bags? Using them certainly feels that you are caring for the environment being a natural product. The paper industry is very energy intensive so that the carbon footprint for all paper is quite high ranging from 12 g CO_2e for a lightweight recycled bag to in excess of 80 g CO_2e for a virgin paper (e.g. clothing) bag excluding the non-paper handles. Virgin paper releases between 2.5 and 3.0 kg CO_2e per kg of paper manufactured compared to 1.25–1.50 kg CO_2e per kg of recycled paper. Of course the embedded CO_2e in the paper is released when paper is either burnt, landfilled or composted which is equivalent to 500 g CO_2e per kg paper on top of the emissions from its manufacture. So only use paper bags when necessary and ensure that they are full.

**Paper is hard to reuse so
ALWAYS RECYCLE
to save CO_2e emissions**

Plastic bags, paper bags, and recyclable bags … what do we do? The golden rule is to avoid plastic and paper bags whenever possible. If you buy reusable bags put a date on them and ensure that you use them well over the breakeven point in terms of emissions. I know it looks awful but rips and holes can be repaired with sticky tap. I would recommend that you invest in long life alternatives (e.g. rucksack, stacking trays for the car). I brought a high quality canvas bag from Oxfam 10 years ago which I have used everyday not only for shopping but for work. Of course neither my wife nor my daughters will be seen out with me when I am carrying it, but it has proven exceptionally climate friendly. So you need to find a compromise.

Having a strategy in place for carrying goods without disposable bags is a surprisingly effective way of minimizing CO_2e emissions over a lifetime

5.6   Rebound Effect

Rebound effects are common in the science of climate change and carbon reduction technologies. Expected reductions in emissions due to a mitigation action can unexpectedly result in an increase in emissions due to an unforeseen knock on effect. For example, offsetting flights increases aviation travel miles while installing low energy light bulbs increases the number and time low energy bulbs are left on resulting in no emissions reduction or even an increase in emissions. Another example is that many business sustainability strategies are based on stimulating growth by reducing product cost through energy efficiency. Effects can be categorized as direct, indirect or economy wide.

Direct effect: For example, increased fuel efficiency lowers the cost of consumption, and hence increases the consumption of that good.
Indirect effect: For example, decreased cost of a specific item enables increased household consumption of other goods and services, increasing the consumption of the resource embodied in those goods and services.
Economy wide effects: New technology creates new production possibilities and increases economic growth.

So offsetting and emission trading projects, as we shall see later, all have the potential to have a rebound effect and so need to be managed very carefully (Sect. 6.2).

So every action we take uses energy and so we need to make sure that we not only get value for money but value for CO$_2$e emissions. So what could you get instead of driving your medium sized petrol car 1 km? Well as the box below shows quite a lot or very little.

What does 160 g CO$_2$e give you:

- 11,429 texts
- 879 min of 18 W lighting
- 158 min of 100 W bulb lighting
- 16 recycled paper towels
- 16 standard lightweight supermarket HDPE bags
- 3 min mobile to mobile talk
- 2 high street paper carrier bags
- 1 km travel in a medium sized petrol car
- 0.5 km travel in a diesel SUV
- 1.6 m of travel in a Jumbo jet going to US

The challenge is how can we reduce our current emissions in all aspects of our life by 80% by 2050?

5.7   Conclusions

- Average individual emissions per annum range from 0.8 to 12.3 t CO$_2$e year^{-1} for low income and high income countries respectively.
- In Ireland per capita emissions are around 13.6 t CO$_2$e year^{-1} of which 5.7 t is our current primary footprint. In terms of global justice, we should in theory all strive to reduce our total emissions to the current global average of 3.7 tonnes per capita per annum. Which would stabilize our global CO$_2$ emissions at the current rate while allowing those in developing countries to increase their personal energy usage to the same level.

- Footprint calculators can allow us to effectively measure and manage our emissions. However, there is huge variability between carbon calculators in terms of what data and information is required to be entered and calculated footprints that are then produced. Also there is overlap between primary and secondary footprints in many models. Therefore it is important to select one that gives you detailed values which you can then use for a long period in order to measure the effect of your personal plan to reduce emissions.
- Calculators do not take into account embedded CO_2e. Embedded CO_2e can be part of the manufacturing, supply or disposal cycle.
- Every action uses energy and has a measurable CO_2e footprint so personal choice controls the energy used and CO_2e emitted.
- Choices have other benefits apart from purely emissions including environmental as well as personal wellbeing. Therefore choices are often difficult.
- Everyday numerous opportunities arise when you can make simple choices that will reduce your carbon footprint.

The fifth step is accepting that an individual's everyday behaviour or life style is a source of global CO_2e emissions and that carbon footprinting enables individuals, households and companies to measure and manage their emissions.

This also involves you agreeing to carryout regular analysis of your CO_2e emissions and agreeing to set yourself personal emissions targets, goals, or limits

Homework!

There are a lot of tasks this time and they are going to give you the basic information that you need to tackle your own personal emissions and make a positive difference to global warming.

Carry out an inventory of the lighting in your home and evaluate where you can intervene to reduce unnecessary lighting. Start by asking the following questions:

- How many lights are in your home?
- What is their wattage?
- How many are low-energy?
- How many lights do you have on at any one time?
- Do you turn them off when you leave the room?
- Do you leave your outside lights on unnecessarily when you are inside?
- How much CO_2 would you emit if all your lights were on for 1 h?

Take advantage of wasted light to reduce your own footprint.
Adopt the 15 min rule

Table 5.3 Cork vs. Down GAA Match September, 2010 Croke Park, Dublin: travel statistics

Point of departure	Numbers attending	Car[a] (%)	Train (%)	Bus (%)	Air (%)	Mean distance *one way* (km)	Car kg CO_2e	Train kg CO_2e	Bus kg CO_2e	Air kg CO_2e	Total kg CO_2e
Cork	19,000	60	20	15	5	260					
Donegal	6000	65	0	35	0	222					
Derry	5500	60	0	35	5	235					
Belfast	6500	50	30	10	10	167					
Dublin	18,500	40	35[b]	25	0	15					
Kerry	1500	75	0	20	5	320					
Athlone	1500	50	20	30	0	188					
Sligo	2000	50	20	25	5	210					
Galway	2000	40	40	20	0	225					
Kilkenny	1500	50	30	20	0	120					
Monaghan	1500	75	0	25	0	130					
Portlaoise	1500	70	10	15	0	84					
Wicklow	1000	75	25	5	0	50					
Navan	2000	65	20	15	0	55					
Wexford	1000	60	25	15	0	142					
Kildare	2500	70	10	20	0	58					
UK[c]	3000	5[d]	10	5	80	438					
Europe[e]	3500	0	0	0	100	1307					
Totals											

[a]Mean car occupancy rate is 2.37
[b]5 % of which used Luas. **Emission conversion factors for Ireland:** Car assume 1.4–2.0 L engine 216 g CO_2e/km; Train 54 g CO_2e/passenger km; Bus 74 g CO_2e/passenger km; Luas 55 g CO_2e/passenger km; Air internal 158 g CO_2e/passenger km, London 130 g CO_2e/passenger km; Berlin 120 g CO_2e/passenger km. Does not include RF (1.9)
[c]Ignores travel to stadium to and from airports etc. Assumes London as mean distance
[d]Ferry 20.35 kg CO_2e/passenger one way Dublin Holyhead High Speed Ferry
[e]Ignores travel to stadium to and from airports etc. Assumes Berlin as mean distance

Next we are going to measure the CO_2 emissions for your household or just yourself. Select a minimum of four different on-line carbon calculators and input the data required. To do this you will need to find your gas, oil and electricity bills to calculate the amount of energy used. There will be other questions asked so this may take some investigation by you, such as the energy usage of some of your larger appliances. Section 8.4 may help you with this.

If you like mathematical puzzles the you may like to try and calculate the travel carbon footprint for all the spectators who went to see the Cork v Down GAA Match in September, 2010 at Croke Park, Dublin (Table 5.3).

References and Further Reading

British Standards Institute. (2006). *Environmental management—Life cycle assessment—Principles and framework ISO 14040:2006*. London, England: British Standards Institute. Retrieved from http://www.iso.org/iso/iso_catalogue/catalogue_tc/catalogue_detail.htm?csnumber=37456

British Standards Institute. (2011). *Specification for the assessment of the life cycle greenhouse gas emissions of goods and services 2050:2011*. London, England: British Standards Institute. Retrieved from http://shop.bsigroup.com/en/forms/PASs/PAS-2050/

British Standards Institute. (2012). *Greenhouse gases. Specification with guidance at the organization level for quantification and reporting of greenhouse gas emissions and removals* BS *EN ISO 14064-1:2012*. London, England: British Standards Institute. Retrieved from http://www.bsigroup.com/en-GB/cfv-carbon-footprint-verification/

Defra. (2008). *2008 guidelines to Defra's GHG conversion factors: Methodology paper for transport emission factors*. London, England: Department for Environment, Food and Rural Affairs. Retrieved from http://archive.defra.gov.uk/environment/business/reporting/pdf/passenger-transport.pdf

EPA. (2014). *Ireland's Greenhouse Gas Inventory 1990–2012*. Wexford, Ireland: Environmental Protection Agency. Retrieved from http://www.epa.ie/pubs/reports/air/airemissions/irelandsgreenhousegasinventory1990-2012.html

World Resources Institute. (2014). *Product life cycle accounting and reporting standard*. Washington, DC: World Resources Institute. Retrieved from http://www.ghgprotocol.org/about-ghgp; http://www.ghgprotocol.org/standards/product-standard

Per Capita Emissions

http://unstats.un.org/unsd/environment/air_co2_emissions.htm
http://www.epa.gov/climatechange/ghgemissions/

Emissions Map

http://unstats.un.org/unsd/environment/qindicators

Calculator Examples

http://carboncalculator.direct.gov.uk/index.html
http://www.carbonfootprint.com/index.html
http://www.carbonfund.org/holiday?gclid=CJnAicTj7a0CFcFO4QodownE4w
http://cmt.epa.ie/en/calculator/
http://www.carbontrust.co.uk/
http://www.epa.gov/climatechange/emissions/ind_calculator.html

Chapter 6
The Real Cost of Carbon

The major mechanism to control emissions is to set a price on greenhouse gases that both encourages energy efficiency and investment in low-energy systems, but also stimulates the development of low-carbon technologies. In this chapter we explore how the price is fixed and how this is implemented

6.1 How Do Government's Tackle Climate Change?

It often feels that nothing is being done about global warming, but in fact behind the scenes a lot has been achieved already. There are no Governments that are now unaware that we have entered a very unsafe period in the history of humankind. Likewise, there are no Governments that are unaware of the enormous challenges that lie ahead and that do not accept that we have to act now. What sometimes gets in the way are vested financial interests, pressure from certain industrial and retail sectors such as energy and aviation, political concerns about re-election, the

© Springer International Publishing Switzerland 2015
N.F. Gray, *Facing Up to Global Warming*, DOI 10.1007/978-3-319-20146-7_6

economy and generally responding to concerns about taxation and sustaining current standards of living. As an environmentalist I would like to have seen a lot more done much earlier, because we have known about the problems of population growth and carbon emissions for many decades and if we had acted earlier ... but that is the past and we didn't. However, things are now happening behind the scenes at an increasingly rapid rate. So whatever an individual politician may say in public, the reality is that collectively countries are now acting very positively to tackle the problems. Just like a large ocean-going oil tanker, its takes a long time to deviate from a particular course and it is going to take time to get to grips with dealing with global warming.

So at a Governmental level, carbon emissions can be reduced through a number of policy or economic mechanisms. These include **emission limits**—where Governments set targets or actual limits in some cases; **cap and trade**—which is trading in carbon credits or units of carbon dioxide equivalent (CO_2e); and **direct taxation**—making consumers or manufacturers pay for the CO_2e emitted. Trading schemes such as cap and trade, provide a direct incentive to businesses to reduce and limit GHG emissions and are amongst the most effective tools in driving international carbon reduction policies, or at least are supposed to be.

The **United Nation Framework Convention on Climate Change** (UNFCCC) was set up to deal with the problems of global warming in the mid 1990s. It has so far been adopted by 188 countries who all acknowledge that "*the change in the Earth's climate and its adverse effects are a common concern of humankind.*" The Convention was extended by the International community by the adoption of the Kyoto Protocol in 1997. This recognized that economic instruments play a key role for the effective implementation of the Convention's objectives. Kyoto was vitally important in our journey to reduce GHG emissions because for the first time emission limits were agreed and economic incentives to meet these targets were created by adopting the concept of flexible trading mechanisms for CO_2e which was eventually launched in January 2005.

> *'Clearly, reaching the Kyoto targets involves costs. We cannot achieve climate mitigation for nothing. However, the flexible mechanisms incorporated into the Kyoto Protocol represent the most cost-effective measures to achieve the Kyoto targets.'*
> United Nation Framework Convention on Climate Change

6.2 Background to Emissions Trading

The Kyoto Protocol is pivotal in emissions trading as it sets a limit on the total CO_2e that can be emitted by any country thereby creating a trading base. Gradual reductions in the limit creates a constant demand for CO_2e credits that can only be created by implementing CO_2e reduction programmes, thereby releasing credits to be traded. Trading creates income for industries which can be used to pay for low CO_2e emission technologies. Emissions trading is not new and had been around previously in other guises, but this is unique as it is truly a global enterprise.

> **Emissions trading** is a market-based system that reduces GHG emissions through using economic incentives. Known also as **cap and trade**, a limit or cap is set on the amount of CO_2e a company can emit. Companies are allocated a set volume of carbon credits each year to cover their emissions. Companies can then trade excess credits created by investing in low-emission technologies or energy efficiency. Alternatively they can buy extra credits to cover emissions in excess of their cap.

Kyoto is managed through a series of annual meetings of the various parties involved in trading and is known simply as the *Conference of the Parties or **COP***. During this main gathering there is a separate meeting of the key players, *the Conference of the Parties serving as the meeting of the Parties to the Kyoto Protocol or **CMP***. The first meeting of the Parties to the Kyoto Protocol was held in Montreal, Canada in December 2005, in conjunction with the eleventh session of the Conference of the Parties (i.e. COP 11/CMP 1). More recent conferences on climate change are summarized in Table 6.1. While details of the future meetings as well as links to all pass meeting of UNFCCC can be access at:

List of meetings: http://unfccc.int/meetings/unfccc_calendar/items/2655.php

Under the Protocol, countries agreed to limit or reduce GHG emissions. This is done primarily through the setting and achieving of National (i.e. country specific) targets and secondly through the development of a carbon market. The protocol introduced three key (flexible) mechanisms to achieve this:

- Emissions trading
- The clean development mechanism (CDM)
- Joint Implementation (JI)

Table 6.1 Details of the most recent conferences on climate change

Date	Meeting	Place	Link
December, 2011	COP17/CMP7	Durban (South Africa)	http://www.cop17-cmp7durban.com/
November, 2012	COP18/CMP 8	Doha (Qatar)	http://www.cop18.qa/
October, 2013	COP19/CMP 9	Warsaw (Poland)	http://www.cop19.org/ http://www.cop19.gov.pl/
December, 2014	COP20/CMP 10	Lima (Peru)	http://www.cop20lima.org/
November, 2015	COP21/CMP11	Paris (France)	http://www.cop21paris.org/

The UN Trading System first allocates carbon units to countries, with each carbon unit equivalent to 1 tonne CO_2e. As industry is seen as the largest emitter of GHG emissions they are expected to make the largest reduction which it does by the use of these flexible mechanisms.

6.2.1   Emissions Trading Scheme (ETS)

Originally the Emissions Trading Scheme (ETS) covered only stationary installations and across the EU some 11,000 heavy energy consuming installations involved in power generation and manufacturing were covered. In Ireland alone over 100 such installations are covered by the scheme. Each company is allocated a generous carbon unit allowance (credits) against which they are able to offset their existing emissions. They can then save credits by investing in and implementing sustainable low emission technologies which then allows them to sell off any carbon credits they have saved to those who are unable to meet emissions targets within their own allocated carbon budget. This is the basis of emissions trading. By reducing the allocation of credits over time, this creates a scarcity of credits which then drives up the unit price making the investment in low emission technologies increasingly more attractive financially.

Rather than buy carbon credits, companies can invest in CO_2 reduction technologies outside their own country, normally in developing countries, for which they are allocated carbon units in return. This is problematic because in some cases it is actually creating new credits whereas simple emission trading is based on the concept of a limited and over time a reducing number of credits. In practice companies have opted to use a combination of all three mechanisms.

6.2.2   Joint Implementation (JI) and the Clean Development Mechanism (CDM)

Industries can earn **emission reduction units** (ERUs) from an emission-reduction or emission-removal project in another country. Each ERU is equivalent to one tonne of CO_2e which can be counted towards meeting the industry's Kyoto target. Therefore, these two project-based mechanisms are an alternative to reducing emissions domestically. In theory this is a fantastic idea. Companies transfer new low-emission technologies to other developed countries or invest in emissions reduction in developing countries and in return they get paid in carbon credits. The problem is of course that in some circumstances this creates new carbon credits which instead of feeding and stimulating the carbon market, tends to flood the market making trading both unstable and unpredictable.

A Joint Implementation or JI project enables an industrialized country to carry out emissions reduction programme with another developed country usually involving technology transfer. In this case project-specific credits are converted from existing credits from another country and so should not create new credits. This is in contrast to clean development mechanism or CDM projects which promote investment in sustainable development projects that reduce emissions in developing countries. For example, this may be a rural electrification project using solar panels or the installation of more energy-efficient boilers. All these projects result in the creation of new carbon credits and many companies have made significant profits through this mechanism. In some cases overheads are excessively high making the actual investment on the ground quite small resulting in limited success. Sometimes the long term results of these projects are quite poor with little or no long term management, support or assessment taking place. Projects are often one-offs, with little co-ordination between projects. One example is reforestation in some areas of Africa. Trees that have been planted under the scheme have subsequently died due to a lack of maintenance or have been rapidly cut down for firewood within a couple of years, often less. In contrast there have also been many successful projects with long lasting positive effects.

More information on CDM: http://cdm.unfccc.int/about/dev_ben/ABC_2012.pdf

So more than just emissions credits, known as European Union Allowances (**EUA**) can be traded and sold under the Kyoto Protocol's emissions trading scheme. There are three other credits or units that may be transferred under the scheme, each equal to 1 tonne of CO_2e:

- A removal unit (**RMU**). This is a sequestration measure based on *land use, land-use change and forestry* (LULUCF) activities such as reforestation and agriculture.
- An emission reduction unit (**ERU**) generated by a joint implementation (JI) project.
- A certified emission reduction unit (**CER**) generated from a clean development mechanism (CDM) project activity.

All of these individual credits or units can be traded and all have a different value on the market.

Clearly the management of the scheme is going to be complex, as all the transfers and acquisitions of these units have to be tracked and recorded through registry systems under the Kyoto Protocol. The UN Framework Convention on Climate Change provides the core management system. The **Emissions Trading Registry** is managed primarily by the UN and is rather like a bank that keeps track of all transactions in the trading and investment market (i.e. similar to a stock exchange). There are a number of these registries around the world with the EU having its own. The **International Transaction Log** ensures the secure transfer of emission reduction units between countries.

These mechanisms are strictly monitored and overseen by three groups. **The clean Development Mechanism (CDM) Executive Board** supervises the CDM under the Kyoto Protocol and prepares decisions for the CMP. It undertakes a variety of tasks relating to the day-to-day operation of the CDM, including the accreditation of operational entities. **The Joint Implementation Supervisory Committee (JISC)** operates under the authority and guidance of the CMP and supervises the verification of emission reduction units (ERUs) generated by JI projects following the verification procedure under the JISC. Finally there is a compliance regime which consists of a **Compliance Committee** made up of two branches: a Facilitative Branch and an Enforcement Branch. I told you it's complicated.

More information: http://unfccc.int/essential_background/items/6031.php

6.2.3   The Cap and Trade Mechanism

The limit or 'cap' on the total number of allowances creates the scarcity needed for trading.

Companies that keep their emissions below the level of their allocated allowances can sell their excess allowances at a price determined by supply and demand at that time. Those facing difficulty in remaining within their allowance limit have a choice between several options: (i) They can take measures to reduce their emissions such as investing in more efficient technology or using a less carbon intensive energy source; (ii) They can buy extra allowances and/or CDM and JI generated credits on the market, or (iii) they can use a combination of the two. This flexibility ensures that emissions are reduced in the most cost-effective way.

Prices on the carbon market vary with supply and demand. As the price of CO_2e increases polluting becomes more expensive. The CDM and JI mechanisms drive the carbon market and at the same time reduce global emissions through investment in clean technologies in developing countries and the transfer of low-emission technologies to other developed countries. Individuals are not part of the trading system, although some countries (e.g. Norway) have introduced a voluntary offset system where carbon credits are purchased and deleted from the UN Trading System. To be fully successful the Emissions Trading Scheme needs to be expanded. Since 2012 the ETS has also included emissions from air flights to and from European airports.

More information on climate markets: http://www.cdcclimat.com/Carbon-markets.
   html

6.3   Emissions Trading

Within the EU, carbon is traded under the European Union Emissions Trading Scheme (EU-ETS) which is a cap and trade system. The 27 member states of the EU have all agreed that the ETS will be the key mechanism in the collective actions to reduce emissions of GHGs. It was brought into force on the 25 October 2003 by EU Directive 2003/87/EC.

More information: Directive 2003/87/EC.

The EU-ETS was implemented in distinct phases or trading periods:

- Phase 1 (1 January 2005 to 31 December 2007): **Pilot phase**
- Phase 2 (1 January 2008 to 31 December 2012): **First trading period**: coincides with the 5-year period during which the EU and its Member States must comply with their emission targets under the Kyoto Protocol.
- Phase 3 (1 January 2013 to 31 December 2020) **Extended trading period**: to create market stability and encourage long-term investment in emission reductions

More information: http://www.eea.europa.eu/multimedia/how-does-the-carbon-offset-scheme-work/view

At the outset all EU governments were required to set an emission limit for all installations covered by the scheme. The common trading 'currency' of the EU-ETS is an **emission allowance**. One allowance equates to 1 tonne of CO_2e.

> **Emission allowance**: *Permission to emit to the atmosphere, 1 tonne of carbon dioxide equivalent, during a specified trading period. The allowance is only valid for the purpose of the Directive and is only transferable in accordance with the Directive.*

Each installation is allocated emission allowances for the particular commitment period, with each Member State required to draw up a **National Allocation Plan** for each trading period setting out how many allowances each relevant installation will receive each year.

The EU Emissions Trading Scheme creates a price for carbon and thereby offers the most cost-effective way to achieve the planned reductions in greenhouse gas emissions. Examples of National Allocation Plans can be downloaded from the Internet. For example, Ireland's National Allocation Plan (2013–2020) can be accessed at:

http://www.epa.ie/pubs/advice/air/climatechange/phase/euets2013-2020preliminar yfreeallocationtoghgpermitholders.html#.VHcDmdKsVGs

More information: http://ec.europa.eu/clima/policies/ets/pre2013/nap/ documentation_en.htm

Carbon credits have created a whole new business sector including carbon brokerages, consultancies and commodities desks at banks. The economic and environmental effectiveness of a cap and trade system crucially relies on the size of the cap that has to be fixed before implementing the trading scheme. Allocating emission allowances must aim at emission reduction significantly below business-as-usual levels. Scarcity must be initially created so that subsequently a functioning market can develop. However, some analysts feared at the beginning that the generous allocation of emission allowances in many EU member states could result in carbon dioxide prices that were so low that they would have little if any effect at all in reducing emissions or stimulating low-emission technology development and adoption. It is now evident that the cap on the size of those allowances was far too generous at the outset with far too many credits allocated resulting in a large surplus creating a relatively small demand. This has been the subject of much heated discussion in some countries where overly generous allocations were assigned. The justification for this over allocation was to give industry time to react to the new scheme and not to put companies at a disadvantage with global competitors. However, it was potentially a massive cash bonus to those industries large enough to qualify for the scheme, and did little to encourage investment in low-emission technologies, with the extra cash often being converted into profits rather than inward investment. The problem of excess credits in the trading system has also been exacerbated by the creation of new credits.

Carbon is traded via two key exchanges, **the European Climate Exchange** (**ECX**), which is the leading market for trading CO_2e emissions in Europe, and the **Chicago Climate Exchange** (**CCX**) in the US, although there are four other exchanges at the moment such as NASDAQ OMX which covers trading within Scandinavia. The value of carbon fell significantly with the recession both in Europe and the USA. By December 2011 the value of carbon as traded on the ECX was EUA €8.29 (i.e. European Union Allowance—the main cap and trade unit), CER €4.14 (i.e. Certified Emission Reduction the carbon unit from CDM projects) and ERU €3.99 (i.e. Emission Reduction Unit from JI actions). A year later in December 2012 these prices had collapsed further EUA €4.20, CER €0.34, and ERU €0.18 falls of 49 %, 92 % and 96 % respectively in just 12 months, leaving the whole trading system in crisis (Figs. 6.1 and 6.2). Trading was very low during the recession, but as the price fell then trading picked up (Fig. 6.2). There has been a slow recovery with the EUA trading at EUA €7.11 at the end of November, 2014, which is half of the initial price set at the start of trading.

More information on current EU carbon market: https://www.eex.com/en/

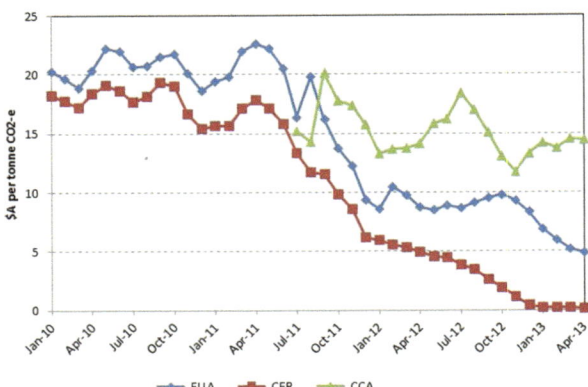

Fig. 6.1 Trend in carbon prices per tonne for EU (EUA and CER) and California (CCA) ETS (January, 2010 to April, 2013) in Australian dollars ($A) showing global variation in the trade price for carbon. *Source*: http://www.aph.gov.au/About_Parliament/Parliamentary_Departments/Parliamentary_Library/pubs/BN/2012-2013/EmissionsTradingSchemes

Fig. 6.2 Trend in carbon prices on the European Climate Exchange (ECX), compared to volume of credits traded showing the rapid decline in value (January 2010 to June 2012). *Source*: European Energy Exchange AG. http://www.eex.com/en#/en. Reproduced with permission of the European Energy Exchange AG, Leipzig, Germany

6.4   The Cost of Sequestration

So how do we set a realistic cost for greenhouse gas emissions? Currently the impact of GHG emissions can only be permanently reduced by either using less fixed carbon energy or actually removing CO_2 from the atmosphere. Removing carbon from the atmosphere is very difficult and in theory it should always be cheaper to capture and remove it at the point of release rather than trying to remove diffuse sources of CO_2 and other GHGs.

There has been much interest in sequestering CO_2 by capture and storage of gaseous emissions from power stations and major manufacturing installations. While some success has been achieved in the oil extraction and refining industries, especially by Statoil in the North Sea, it is still very much in its developmental stage with the long term safety of the underground storage of the captured CO_2 uncertain. Currently the predicted cost of capture and storage in the US is $0.04 per kWh of electricity generated (*Science* Dec 2009). In the US some 0.630 kg of CO_2e is produced per kWh of electrical energy produced, which is equivalent to 1587 kWh of energy per tonne of CO_2e emitted. So this gives a carbon price of $63.50 per tonne CO_2e. However, this does not include all costs or site specific costs, such as storage facility development and security or the development costs, but it does give us some idea of what the cost of permanently removing CO_2 might be in practice, although the actual cost may well be in excess of $70–$100 per tonne CO_2e, especially if operated by a third party, we just don't know for certain yet.

It is of course already possible to safely and permanently sequester (i.e. permanently remove) CO_2 through the production and storage of wood charcoal and biochar. Charcoal has a number of key advantages over other sequestering technologies. It is a low cost solution to sequestration because it requires less energy than its potential alternatives. It is affordably stored with no danger to the environment because it is an extremely stable and non-polluting material. It can also be reused as a clean fuel when more efficient carbon sequestration technologies are developed. The most modern production methods use the less valuable timber fractions, are less polluting and require only small amounts of energy in its production with the potential to even recover this waste heat for combined heat and power generation. The challenge is to produce charcoal in a sustainable manner in the volume required and **so while it may not be the idea solution it is feasible, and just like carbon capture and storage, it gives us a real sequestration cost on which to base carbon costs**.

Commercial charcoal prices vary around the world as do potential production rates, manufacturing methods, and scale of production. Commercial bulk charcoal prices in the UK range from €310 to €425 per tonne, giving an average price of €345 (2009 values). If we allow €35 per tonne for other costs like forest and production development, storage, and security, €380 appears a realistic mean estimate for the production and long-term storage cost per metric ton of charcoal. As carbon dioxide comprises only 27.3 % of carbon by weight, this is equivalent to an offset cost of approximately €104 per tonne of CO_2 produced, similar to the cost by capture and storage.

Charcoal and capture and storage provide a fixed carbon price that is both economically stable and high enough to act as a real incentive to encourage us to meet our carbon-reduction targets.

So what does this mean in practice? A round-trip flight from New York to London produces 2.5 tonnes of CO_2e (or 6.7 tonnes when radiative forcing (RF) is included). Emissions can be offset when purchasing a ticket for approximately **€8–€47** (based on December 2011 values) depending on the online offsetting company selected. If we use the European Climate Exchange (ECX), then the offset would be using ERU credits just **€0.45** (or €1.20 if RF is included) (based on December 2011 values). Most countries that introduced carbon taxation of fuels, such as Ireland, used the initial ECX value of €15 per tonne of CO_2e (Table 6.2). Using this value would give an offset cost per passenger for the return flight to New York of **€37.50** (or €101 including RF). So the offset values being offered online seem justifiable, and indeed are quite expensive when compared to the current trading prices for CO_2e. However, if we use the charcoal-derived cost, which represents a real cost for removing the CO_2e emitted, then the offset would increase to **€260** (or €697 including RF) or **€121** (or €323 including RF) based on the estimated cost of capture and storage. There is a huge difference between the offset cost derived from our current carbon

Table 6.2 Comparison of carbon taxes levied per tonne of CO_2e based on the exchange rate at 12th August 2013

Country		Cost per tonne of CO_2e	Notes
Australia		€16.29	Introduced July 2012 will link to EU-ETS in 2015
Canada	Alberta	€10.95	Introduced cap and trade scheme in 2007
	British Columbia	€21.90	Introduced 2007
	Quebec	€2.56	Introduced 2007
Ireland		€20.00	Introduced 2010 for all fossil fuels, since 2013 coal and peat
France		€17.00	Introduced 2010 for businesses and householders
Finland		€18.05	First to introduce carbon taxes in 1990
Norway		€15.81	Introduced in 1991
New Zealand		€9.05	Proposed in 2005 but never implemented
Sweden		€79.06	Introduced 1991, industry only pays around 25 % of this

tax value on which most offsetting companies base their estimates, compared to the very low values at which carbon is being traded on the carbon exchanges. All these estimates are however way below the real cost of capturing and storing that CO_2e. These higher values are the reality of our high energy lifestyles and the actual cost of negating the associated emissions.

More information: Gray, N.F. (2009) *Sustainability: Science, Policy and Practice,* 5, (2), 1–3.

6.5 Carbon Taxation

Although carbon taxation was first introduced in Finland in 1990, it is still not very widespread and where it has been introduced there are different rules on applicability and charges per tonne of CO_2e emitted. Most countries apply carbon taxation to vehicle fuels only, while others cover all fossil fuels, including electricity, natural and bottled gas, coal and other fuels. No two countries are identical in their approach with a wide range of taxation bands from €2.56 per tonne of CO_2e in Quebec to € 79.06 in Sweden. However, while actual carbon taxes rarely reflect a realistic cost for dealing with emissions evidence from nearly all the countries who have introduced carbon taxation shows that they do seem to encourage us to use less fuel, invest in renewable energy and reduce overall carbon emissions. While some carbon taxes are aimed at all sectors including the domestic sector as in Ireland, others are aimed at specific high energy industries such as in Australia or all business sectors which is the situation in the UK. However, costs of carbon taxation in all cases are eventually passed down to the consumer in the form of price rises. To offset this many countries use part of their revenue from carbon taxation to subsidize those on lower incomes.

Ireland introduced a carbon tax on some household and vehicle fuels in 2010 at a rate of €15 per tonne of CO_2e emitted. Household fuels included oil, liquid petroleum gas (LPG) and natural gas. This was increased to €20 per tonne of CO_2e emitted in late 2011. In the 2013 budget the carbon tax was extended to also include solid fuels such as coal and peat at a rate of €10 per tonne increasing to €20 per tonne on the 1st May, 2014. This will increase the cost of a tonne of either coal or peat briquettes by €26.33 and €18.33 respectively in 2013 and doubling the following year to €52.67 and €36.67 respectively. This will have a significant effect on householders using these fuels and bring them into line with other fuel types used for heating. So while it is been expensive for consumers, it has reduced carbon emissions by over 15 % in Ireland since 2008.

More information: www.revenue.ie/en/**tax**/excise/leaflets/solid-fuel-**carbon-tax**-guidance.pdf

The approach in the UK has been different, with the concept of a *climate change levy* introduced in 2001. This levy is based on energy usage by all businesses from Industry to agriculture with domestic users exempt. It is set at 0.524p per kWh for electricity, 0.182p for gas, 1.172p for LPG and 1.429p per kWh from coal and other solid fossil fuels (levies at 1 April, 2013). Energy from renewable or modern combined heat and power systems are exempt. In 2013 the UK Government introduced a system (Climate Change Agreements—CCAs) of 65 % discounts on the levy to energy intensive industries who meet specified targets for reducing their carbon emissions or improving their energy efficiency.

More information: http://customs.hmrc.gov.uk/channelsPortalWebApp/channelsPortalWebApp.portal?_nfpb=true&_pageLabel=pageExcise_InfoGuides&propertyType=document&id=HMCE_CL_001174

Of the three provinces that have introduced carbon taxation in Canada it is in British Columbia that the tax has been most successful. First introduced in July 2008 at Can $10 per tonne of CO_2 emitted, it has risen by Can $5 per tonne annually until its current level of Can $30 (equivalent €21.90), well in excess of carbon trading price on the EU-ETS. In practice this equates to an extra 7.24 cents on a litre of petrol and 8.29 cents per litre of diesel. Although the tax is eight times higher than that paid in Quebec, it has wide support with 54 % of voters supporting the tax. Contrary to expectations it has had no negative effect on the overall economy of the province with the extra tax revenue used to fund tax cuts. The results have been impressive with overall GHG emissions declining by 9.9 % over the period of 2008–2010 compared to just 5 % in the rest of Canada, which represents a 15.1 % reduction in the use of vehicle fuel in the Province.

Australia is the latest country to introduce a carbon tax. Set at Au$23 per tonne CO_2e at the outset in July 2012, it is currently at Au$24.15 (2013). It is levied at the largest emitters in the country with about 500 high energy installations covered including electricity generation, mining, and other industries, **and also business travel**. Based on the EU-ETS system the price is set to rise to Au$25.40 (€17.40) per tonne CO_2e in 2014. In July 2015 it will then be linked to the EU-ETS cap and trade system which will probably see the unit cost fall significantly. The tax is very controversial in Australia which has one of the world's highest CO_2e per capita emission rates. But interestingly it has had significant effects, especially in how electricity is generated. The use of lignite, a poor quality and highly polluting type of coal which is plentiful in Australia, has fallen by 14 % since the introduction of the tax, with coal-fire power generation also falling by 5 %, with a massive investment in renewable energy and a 9.5 % increase in the use of natural gas. While some of the tax is invested in renewable energy and related projects, much is being used to support lower income earners to compensate for any rise in prices caused by the tax. Examples of carbon tax levels in other countries are compared in Table 6.2.

> *The ability to offset greenhouse gases at these extremely low prices is a major disincentive for the adoption of real CO_2 reduction policies and actions, which would be far more expensive and inconvenient.*
>
> *Offsetting does not permanently or even temporarily remove GHG emissions from the atmosphere.*
>
> *The impact of GHG emissions can currently only be reduced by either using less fixed carbon energy or actually removing CO_2 from the atmosphere.*

6.6   The Real Cost of Carbon Offsetting

Carbon neutral is a widely used term, applied primarily to energy efficient homes. In Germany, a 'low energy' house is defined as one which requires less than 42 kWh of heat energy per square meter of floor area per annum ($kWh\ m^{-2}\ year^{-1}$). In contrast a house compliant with the 2002 Building Regulations in Ireland would be in the 55–70 $kWh\ m^{-2}\ year^{-1}$ range. An ultra-low energy or PassivHaus is defined as one which requires less than 15 $kWh\ m^{-2}\ year^{-1}$. These figures refer only to the energy required for space heating and do not cover energy required for domestic hot water supply and other uses. So while houses have the potential to have a very small carbon footprint, they are not carbon neutral and neither is it possible for a family to live in one without using carbon-based fuels either directly or indirectly.

So, while the aspiration to become either carbon neutral or have zero CO_2 emissions is very laudable it is of course impossible to live a zero carbon existence. Almost everything we do or purchase requires energy derived primarily from fossil fuels that release CO_2 and other greenhouse gases (GHGs). So it appears perverse that increasing numbers of companies, as well as individuals, do claim to be carbon neutral yet continue to travel, heat buildings, produce and purchase manufactured goods much in the same way as before this miraculous transformation took place. A major mechanism in achieving apparent carbon neutrality is through **offsetting**, a process whereby **someone else is paid to eliminate the CO_2 you have emitted** by investing, in theory, in carbon reduction technologies and investment projects.

> *Everyone uses energy either directly or indirectly and therefore emits CO2e emissions.*
> *No one expects you not to emit some GHGs ... it is impossible.*

According to offsetting ideology, air and car travel, electricity, gas and oil use, in fact whole business and household carbon footprints can be neutralized by a quick visit to an offsetting company website and the payment of an *appropriate* fee. In other words if you can afford to pay you don't need to reduce your emissions. It's a win–win situation; you have the moral high ground while continuing to do exactly what you want. To me this seems ludicrous ... it is just like paying for indulgences in medieval times. An indulgence was a full or partial remission of the punishment for a sin granted by the Church in return for, you've guessed it, an appropriate fee. So we now see large companies claiming to be carbon neutral on business air travel, for example, due to offsetting, while in reality the amount of air travel done by its employees during this period may well have increased significantly.

More information: http://www.monbiot.com/2006/10/19/selling-indulgences/

> *A carbon offset is a reduction in emissions of carbon dioxide or greenhouse gases made in order to compensate for or to offset an emission made elsewhere.*

6.6.1 How Does Offsetting Actually Function and Does It Work?

The clean development mechanism (CDM) is the EU sanctioned version of offsetting that operates under the Kyoto Protocol. The impact of climate change is based on the total emissions of long-lived GHGs over a period of 100 years, so in terms of impact we have to look at the total sum of emissions over that period. Does offsetting also deal with this timescale? There is no point in reducing emissions by 1 tonne now if the knock on effect in 10 years is 1.5 tonnes. The emissions from your flight will have an effect on global warming for 100 years. So investing in an offset project is clearly not eliminating your CO_2 contribution to global atmospheric GHGs, at best it is a contribution to reduce the rate of emissions in the future. That is why many consider that radiative forcing should also be considered in offsetting.

There are a number of offset options:

* Renewable energy
* Energy efficiency
* Land use, land use change and forestry (LULUCF)
* Emission trading schemes
* Investment in carbon international reduction projects (CDM and JI projects)
* Methane collection and combustion (primarily at landfills and wastewater treatment plants)
* Destruction of industrial pollutants

The most popular offsetting option is carbon sequestration using agriculture and forestry, which means simply storing carbon as biomass either as timber or as organic matter in the soil. Tree planting is the most popular offsetting option for businesses and individuals alike as this appears to actually turn your CO_2 into a fixed carbon product.

6.6.2   Carbon Sequestration in Agriculture and Forestry

Carbon sequestration and reductions of methane (CH_4) and nitrous oxide (N_2O) emissions can occur through a variety of agricultural and forestry practices. Conversely, carbon can also be lost and CH_4 and N_2O emitted to the atmosphere through a number of land-use changes and practices. So within these industries sequestration is a dynamic process where carbon can be stored or lost and released back into the atmosphere.

Much of our forests have been in existence for hundreds and sometimes even thousands of years and so are actually part of our stored carbon and when we cut down these virgin woodlands and forests we are essentially adding to our atmospheric CO_2e emissions with what was essentially permanently stored carbon (Fig. 6.3). **Extra carbon sequestration can only be achieved through the planting of new forests**.

Terrestrial carbon sequestration is the process through which atmospheric carbon dioxide (CO_2) is taken up through tiny openings in leaves and incorporated as carbon into the woody biomass of trees and agricultural crops (Fig. 6.4). Roughly half of this biomass is carbon. Some of this carbon makes its way into soils where vegetation, leaf litter and roots decay. Carbon in forests and soils can return to the atmosphere as CO_2 when agricultural tillage practices stir up soils or when biomass decays or is burt. Forests and agricultural soils can therefore act as either a net carbon sink or source. The term "sink" is also used to refer to forests, croplands, and grazing lands, and their ability to sequester carbon. As agriculture and forestry activities can also release CO_2 to the atmosphere, a carbon sink occurs when carbon sequestration is greater than carbon release over the same time period. The movement of carbon in and out of trees is part of the Earth's global carbon cycle. The advantage of forestry is that sequestration activities can be carried out immediately

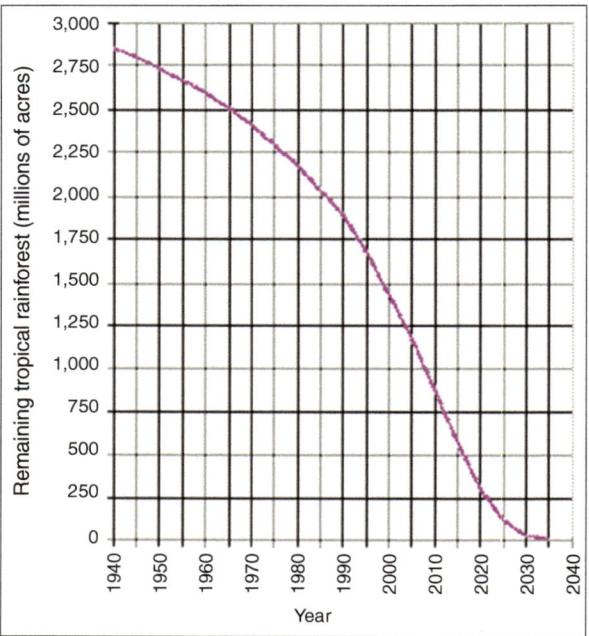

Fig. 6.3 Loss of tropical rainforest in millions of acres 1940–2035. Tropical deforestation is responsible for about 20 % of the world's annual CO_2 emissions, so it makes sense to control the loss of forests to prevent further emissions. Similarly it is important to ensure better control of wildfires to prevent the release of GHGs and lose all the carbon bound up in the trees and soil below. *Source*: IPCC Special Report on LULUCF (2000)

and is seen as a very cost-effective method of emission reduction. Of course there can often also be large environmental and landscape benefits to planting trees.

Tree carbon sequestration rates vary according to tree species, soil type, regional climate, topography and management practice, although well-established values for carbon sequestration rates are now available for most tree species grown in different parts of the world. Likewise, soil carbon sequestration rates vary by soil type and cropping practice. So for example, pine plantations in the Southeast US accumulate 100 tonnes of carbon per acre after 90 years (equivalent to 1.0 tonne of carbon per acre per year or 3.67 tonnes CO_2 per acre each year). An acre is equivalent to 0.405 ha. Changes in forest management, such as lengthening the harvest-regeneration cycle, generally results in less carbon sequestration on a per acre basis. So as a general guideline, a single broad leaf tree planted in Ireland or the UK would absorb 1 tonne of CO_2 over a period of 80–100 years.

Carbon accumulation in forests and soils eventually reaches a saturation point, beyond which additional sequestration is no longer possible. This occurs when trees reach maturity and when the organic matter in soils has accumulated to original levels before losses occurred. In agricultural soils, carbon can be sequestered for 15 years or longer, depending not only on the type of soil but also on the continuity and length of management.

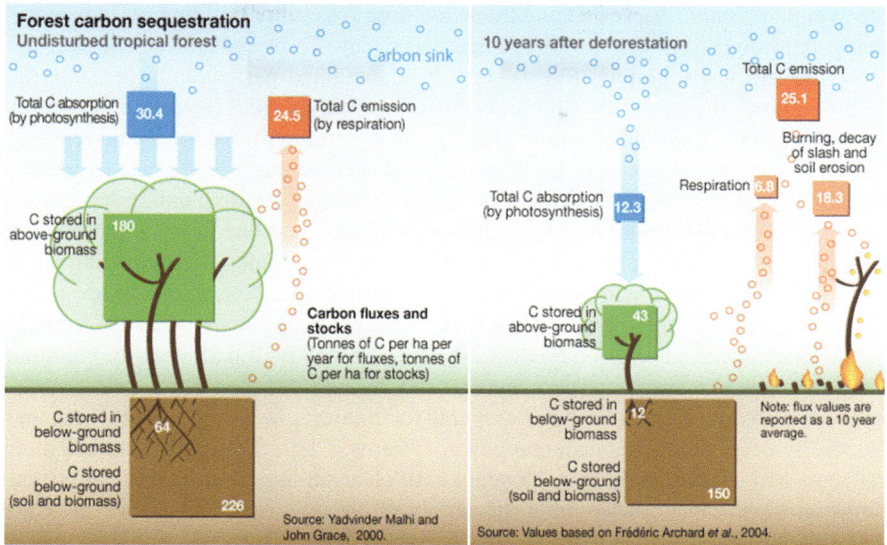

Fig. 6.4 How carbon dioxide is cycled through established and plantation forests. Much greater sequestration is achieved in established natural forests. Image by Riccardo Pravettoni, UNEP/GRID-Arendal. Based on the work of Yadvinder Malhi and John Grace, with values from Federic Archard and his team details of which are found in the *source*: http://www.grida.no/publications/vg/biofuels. Reproduced with permission Riccardo Pravettoni, UNEP/GRID-Arendal

Natural decay and disturbances such as harvesting or fire releases nearly all the carbon back into the atmosphere as CO_2.

U.S. forests and croplands currently sequester over 600 million tonnes CO_2e after accounting for both gains and losses in carbon. This current amount of sequestration in forests and croplands offsets approximately 12 % of total U.S. CO_2e emissions from the energy, transportation and industrial sectors. However, this is not permanent and does not reflect new planting or agriculture. This level of carbon storage already existed before climate change was understood and so is already part of the equation as reflected by the carbon cycle. **So only the creation of new forests is relevant to offsetting**. For tree planting to be eligible for inclusion in carbon accounting under the Kyoto Protocol, then only forests planted since 1990 can be included. Therefore the potential of carbon sequestration by tree planting is dependent on the rate of afforestation.

The problem is that CO_2 is removed from the atmosphere very slowly by absorption into the oceans, which in turns drives acidification (Sect. 13.8). This process can take up to 200 years and removes between 65–80 % of the CO_2, the remainder being removed through chemical processes such as the weathering of rocks which takes significantly longer. So carbon sequestration by forestry cannot reverse the effects of CO_2 in the atmosphere unless the forest is permanently retained. In practice trees are eventually either harvested and the wood used for fuel or building, or they eventually fall down from disease or old age and decay. To permanently sequester the CO_2 then all newly planted forests must be maintained permanently just like the carbon stored in coal or as carbonates in sedimentary rock. However in all cases almost all of the available carbon taken up by the tree will eventually be released back into the atmosphere as CO_2 when the wood is either burnt or allowed to decompose usually within 80–100 years of planting. Currently we are planting more trees than we are harvesting globally which is good news. The bad news is that we are continuing to cut down ancient virgin forests and woodlands and replacing them with plantations which will be largely used for biomass in the future. The economic maturity for coniferous forests in Ireland and elsewhere in Western Europe, such as Sitka Spruce, planted on good quality soil could be well within 35 years of planting. This increases to 40 years for poorer soils or more exposed sites, although this varies between species. There is a huge demand for biomass to generate both heat and electricity using combined heat and power systems (CHP). Forests are thinned after 20 years and subsequently every 5 years until clear felling takes place. These thinnings are important for sustaining the supply of biomass along with other biomass crops such as willow or *Miscanthus* (elephant grass). Short rotation coppicing (SRC) is a high production system to produce biomass using traditional forestry techniques. Trees are planted and grown as a single stem until it reaches a diameter of 10–20 cm at breast height when they are cut back to the base to promote the growth of multiple stems. Trees take between 8 and 20 years, depending on species and site conditions before they can first be harvested. This can be repeated every 2–4 years promoting fast growth and higher yields overall. Common SRC species in the UK and Ireland are Eucalyptus, Popular, Sycamore, Ash and the fast growing *Nothofagus* or the Southern Beech.

More information on SRC: http://www.forestry.gov.uk/pdf/FRMG002_Short_rotation_forestry.pdf/$FILE/FRMG002_Short_rotation_forestry.pdf

There is a finite amount of carbon that can ultimately be sequestered by forestry. Eventually this mechanism of sequestering would result in the entire land mass becoming engulfed by forests which would have to be maintained indefinitely. Also, as the area of land suitable or available for tree planting is limited, it is clearly neither a realistic or sustainable option for neutralizing CO_2 emissions. Forestry is also taking land from food production as does land for biomass, ethanol and oil crops. Planting trees in developing countries is frequently used as an offsetting option or to create new carbon credits under CDM projects. However, they are frequently cut for firewood shortly after planting or die due to poor management, lack of water or unsuitable location so may not reach maturity.

Accounting for how much carbon is sequestered by forestry is also complicated, with increasing amounts of biomass, especially timber, being used for fuel. The new biomass power stations in the UK source timber from around the world with much coming from forests that have already been used for offsetting or already included in global carbon accounting, so it's getting complicated. So planting trees as a mechanism of carbon sequestration simply results in a temporary storage of the CO_2 and is of limited capacity.

Is planting trees a good thing? Of course it is. It is not possible to eliminate all CO_2 emissions immediately, so in the interim the impact of those emissions that are unavoidable can be temporarily mitigated by absorbing some of the excess CO_2 in the atmosphere by planting trees. At the same time this creates habitat for a wide range of species, thereby preserving biodiversity. Tree planting also creates a future resource of sustainable building materials and fuel. So the concept of planting trees creates a positive attitude towards the problem of global warming and helps to educate and inform others. But does it address the problem of rising global CO_2 levels, not really.

> **It is crucial that offsetting is combined with effective action from each of us to reduce our total CO_2 emissions.**
> **Tree planting is a temporary storage system for CO_2 and is of limited use in terms of reducing atmospheric GHG concentrations. It does however help to sustain wildlife and create a better environment for everyone.**

6.6.3   Offsetting as a Mechanism of Controlling Emissions

In practice GHG emissions continue to be well in excess of the required reduction targets and many consider that this may in part be due to offsetting. The attraction of offsetting is that it prevents CO_2 becoming a limiting factor either in business or personal life. Offsetting gives you the moral high ground without any of the pain and this has been exploited to the full in business GHG footprinting.

The ethos of the offsetting industry can be summarized as: *'The atmosphere does not care where GHGs are emitted, nor does it care where they are prevented. What is essential from the point of view of climate change is reducing the total amount of emissions.'* I couldn't agree more, but offsetting must not allow one sector of society to use more than its fair share. Can offsetting achieve the 80 % reduction in total global GHG emissions required by 2050? The answer is a resounding no, as offsetting is not reducing existing emissions.

Those supplying offsetting tend to accept its limitations but believe they play an important role in supporting low-energy technologies while both educating and encouraging the public to reduce their CO_2 output. They also see it as a mechanism to prepare for a future where carbon is likely to be high cost and highly regulated. However, offsetting is primarily seen as another form of carbon tax allowing companies to appear to be climate friendly which is good for business. For the individual the voluntary offsetting of flights, for example, is generally a way of giving travel legitimacy. In most cases selling offsets is a company's only source of income making the offsetting market very competitive resulting in a wide variety of charges and charging mechanisms, and anyone can do it. So the amount actually ending up being invested by companies can be shockingly small.

For me **offsetting means business as usual**, a simple and cheap mechanism to avoid real GHG emission reduction. It is perceived as a voluntary carbon tax, and so by paying the tax you are somehow making good your emissions. But the tax in this case is not being used to reduce the GHG emissions of those paying the tax. Can paying €20 to offset a flight from London to New York really mitigate the effects of your flight CO_2e emissions for the next 100 years? Of course not.

> **Can offsetting achieve the 80 % reduction in total global GHG emissions required by 2050? The answer is a resounding NO, as offsetting is not reducing existing emissions.**

More information: http://www.carbonneutral.com/

6.7   So Where Do We Stand on Carbon Pricing?

Regardless of whether direct carbon taxation or cap and trade schemes are adopted neither will succeed unless a credible price for carbon is set. Making carbon expensive is the simplest way to incentivize both carbon reduction and the innovation of low-carbon technologies. Likewise those who are developing new low-carbon technologies need to profit from their inventions to stimulate further research and development, and this can only be done if **carbon prices are both high and stable over the long term**, which will also stimulate the level of investment in low-carbon technologies that are urgently needed. The current European trading price of CO_2 is fluctuating around €7 per tonne (2014–2015). This low market cost is a major disincentive for both innovation and reduction activities and will need to be much higher and credible for global warming to be taken seriously within the market place. So how do we set the price? What is required is something similar to the gold standard. That is why it is important to adopt a fixed sequestration process such as carbon capture and storage or charcoal production on which to fix carbon pricing.

There is hope, however, and some governments have made bold stands through carbon taxation. In 2011 the UK coalition Government agreed the importance of creating a so called '*carbon price floor*' or a minimum value for carbon for the power industry which they set in 2012 at £16 per tonne of CO_2 emitted. This will increase incrementally towards an interim target of £30 per tonne by 2020 and eventually £70 tonne by 2030 creating an anticipated £30–40 billion for new investment in low-carbon energy generation. The idea is one way of ensuring that not only is inward investment made available for the power supply industry to restructure itself into a more sustainable low-carbon electricity supplier, but it also puts a more realistic value on energy that will encourage more careful usage and also reduce emissions.

Personally I feel that offsetting is being misused as a mechanism. For example, many organizations and businesses have made massive cuts in their emissions by simply pressing a few buttons on their computer. For example, in 2009 there were two suppliers of electricity in a particular European country, the state owned generating company, which at that time generated 95 % of the country's needs using a wide range of non-renewable and renewable methods; and a renewable energy company which at that time was generating a very small percentage of the country's electricity using wind turbines, and of course this was dependant on wind strengths. The emissions from the two suppliers was 0.538 g of CO_2 per kWh generated by the state company and just 0.142 g of CO_2 per kWh generated by the renewable energy company. Changes in European competition laws allowed customers to switch from the state owned company to another supplier. So customers quickly realized that by switching from the State company to the other supplier they were able to claim a 73.6 % reduction in CO_2 emissions, even though the energy they were using was still being generated largely by the state company. To be fair to the renewable energy supplier, although they could not meet this demand by generating electricity in that particular country, they invested money in renewable energy projects elsewhere, and at the same time were able to develop their company. So who is the winner and loser here? I am not sure, but what is clear is that companies did not reduce their emissions by 70 % because they did not reduce their energy consumption but simply changed supplier. The winner seems to be the new supplier and the loser, well the environment, and that means all of us. There is absolutely nothing wrong with this and we now have more renewables in Europe thanks to this initiative; but companies taking advantage of this lower conversion factor are not necessarily reducing their use of fossil fuels and so are not directly helping to reduce the ever growing global CO_2 level.

There is clear evidence that offsetting is increasing emissions in other ways. Increasing economic prosperity in developing countries due to the many EU CDM projects and offsetting initiatives is leading to a rapid rise in emissions in some countries. We need to control the readjustment of wealth and emissions more carefully. So while developed countries need to significantly reduce emissions, the developed countries must also restrict the rise in their emissions and set targets.

As I said earlier, there is absolutely nothing wrong in using energy and emitting CO_2, we all do it, but we can also manage and reduce our emissions more effectively. So stop worrying about being carbon neutral. It is simply not possible. What we need to think about is how we can really minimize our emissions on a daily basis at work, at home, when we are out enjoying ourselves. This especially applies to consumer purchases from small things like food and drink to larger things such as computers, cars and even houses.

6.8 Conclusions

- Emissions trading is a key mechanism in making the transformation to a low-carbon economy without destabilizing society. However, it must be equitable and carbon credits must be allocated in such a way as to genuinely reduce carbon emissions overall.

- Neither direct carbon taxation nor a cap-and-trade scheme will succeed unless a credible price for carbon is set.
- Making carbon expensive would be an incentive for both carbon reduction and the innovation of low-carbon technologies.
- The economy sustains society but it cannot achieve the required GHG emissions alone. Personal reduction in GHG emissions is key to achieving Kyoto, although we must be careful not to be pulled into unrealistic offsetting options. Carbon taxes, personal cap and trade ideology are all potential mechanisms.
- Promoting the concepts of 'carbon offsetting' and 'carbon neutral' runs the risk of providing an apparent justification for continuing with a fossil-fuel intensive lifestyle and culture, whereas the only certain way to reduce CO_2 emissions is to use less fixed carbon fuels and fewer products that employ them either in their manufacture or production.
- We can't buy ourselves out of our personal moral responsibility to act for all those people and other species we share our planet with, especially those who have very small personal footprints or in the case of the vast majority of species … no footprint at all.
- Offsetting can and does invest in future emission reduction technology and helps those in developing countries to improve their lifestyle through the more efficient use of energy, but can also drive extra growth, prosperity and so result in increased emissions.
- Offsetting is not a substitute for the reduction in our use of fossil fuels through our high energy lifestyles. At best it is investing in a low-carbon society but does not address current escalating CO_2e emissions or our ability to meet our Kyoto commitments.

> The sixth step is accepting that the price of carbon should be set at a credible value in order to be incentive for both carbon reduction and the innovation of low-carbon technologies.
>
> Step sixth is also accepting that offsetting can be a positive action against climate change but not in reducing existing GHG emissions. As individuals we can't buy ourselves out of our personal moral responsibility to act for all those people and other species we share our planet with by offsetting. So the only certain way to reduce CO_2 emissions is to use less fixed carbon.
>
> So use money or time equivalent to offset carbon activities by investing in the reduction of your own or community GHG emissions.

Homework!

If you have completed all carbon footprinting exercises in Chap. 5 then well done. What I want you to examine is how much it will cost to offset your own or family footprint in terms of trading against the EUA price of €7.11 and using the charcoal standard of €104?

How many trees would that be equivalent to each year? As a guide one broad leaf tree grown in Ireland or the UK would adsorb 1 tonne of CO_2 over 80–100 years. So if you are emitting 5 tonnes of CO_2 per year then you will need to plant five broad leaf trees and maintain it for 100 years just to cover what you emitted this year. In fact you need to multiply 5 by 1.5 to take into account the long term effects of the GHGs over a 100 year period. So to be precise you need to plant 5 times 1.5 which makes 7.5 trees this year. So after 10 years you will have to be managing 75 trees for between 90 and 100 years, after 20 years that will be 150 trees for between 80 and 100 years and that is just for your personal primary emissions!

References and Further Reading

UN Framework Convention on Climate Change

http://unfccc.int/2860.php

Trading in Carbon Units

http://www.ecx.eu/

Kyoto Limits

http://www.guardian.co.uk/environment/blog/2012/nov/26/kyoto-protocol-carbon-emissions

CDM Projects

http://cdm.unfccc.int/about/dev_ben/ABC_2012.pdf
http://reforestation.elti.org/resource/718/

Offsetting

http://www.carbontradewatch.org/pubs/carbon_neutral_myth.pdf
http://www.nature.com/news/the-inconvenient-truth-of-carbon-offsets-1.10373
http://www.co2offsetresearch.org/policy/Australia.html
http://www.epa.ie/downloads/pubs/research/climate/CCRP_6_web.pdf

Offsetting Examples

http://www.boghill.com/eco-stuff/carbon-offsetting/
http://www.gco2.ie/index.aspx
http://www.climatecare.org/index.htm?redirected=true

Forestry

IPCC Special Report on LULUCF. (2000). Retrieved from http://www.grida.no/publications/
    other/ipcc_sr/?src=/climate/ipcc/land_use/019.htm
McKay, H. (Ed.). (2011). *Short Rotation Forestry: Review of growth and environmental impacts*.
    Forest Research Monograph 2. Surrey, BC: Forest Research, 212pp. Retrieved from http://
    www.forestry.gov.uk/pdf/FRMG002_Short_rotation_forestry.pdf/$FILE/FRMG002_Short_
    rotation_forestry.pdf

Deforestation

http://rainforests.mongabay.com/deforestation.html

Chapter 7
Ecological Footprint

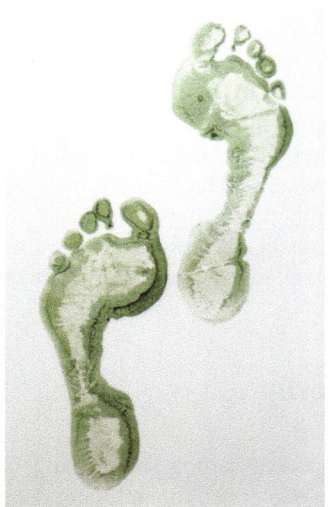

To tackle the problem of global warming and climate change effectively we need reliable measures of both the impact of man's activities as well as the ability of planet Earth to sustain us. Here we are going to explore ecological and sustainability measures and indicators

7.1 Action and Reaction

The most important thing when thinking about the problem of global warming is not to panic and simply turn your back on the situation. **This is a big challenge, as big as any war or life challenge any previous generation has ever faced, although**

I suppose it doesn't feel like it … yet. It is insidious, rather like cancer perhaps, it is creeping up on us very slowly and soon it will be terminal rather than treatable. The brave ones amongst us **are those who are going to tackle this problem**. Whether it is as a leader or a mere foot soldier, there is work for everyone, and it is vital that everyone takes an active and responsible part. As time goes on those who don't will feel increasingly marginalized.

So remember, DON'T PANIC, because Governments are slowly getting their acts together through a range of actions. Regardless of what sceptics may saying in public, and this includes a lot of politician's who have to balance a large number of issues. Every nation is putting together serious contingency plans to tackle climate change and slowly these are coming into place; whether it is introducing carbon taxes, encouraging energy suppliers to optimize their fuel mixes to reduce emissions from electricity generation, giving grants for home insulation and other energy saving initiatives, incentivizing renewable energy projects, even building nuclear power stations. Each country is having to make serious climate commitments under the various protocols they have signed. However, I suspect that these plans are just too slow to deal with the current rate of rise of GHG emissions? Plans have certainly been side tracked in recent years by other political issues such as the recession, but the reality is such issues are short term and while they may be very harmful at a personal level, they are largely insignificant to the larger problems of planet health and global warming in particular. That is why we must all become proactive in reducing GHG emissions right now.

So how are these problems being tackled and monitored? Here we are going to explore the world of ecological and sustainability indicators.

7.2   Ecological Footprint

We are familiar with the idea of carbon footprinting, but this does not take into account all the pressures on planet health, including the potential carrying capacity of our planet to support the ever growing dominant species … us. The ecological footprint was a concept developed by a Ph.D. student Mathis Wackernagel and his supervisor William Rees at the University of British Columbia (Vancouver) and first published in 1992. The idea was originally called *Appropriated Carrying Capacity* with the name ecological footprint being adopted soon after. Apparently the name came from a computer technician who used it to describe the small size of his latest computer, and the term just stuck. The concept took off after Wackernagel and Rees published the book *Our Ecological Footprint*: *Reducing Human Impact on the Earth* in 1996 and today it is acknowledge as a major tool in measuring human impact on the planet, although there are some scientists who find the technique flawed.

More information: http://blog.nature.org/science/2013/11/05/science-ecological-footprint-plos-biology-critique-blomqvist/

Image reproduced with permission of the Global Footprint Network http://www. footprintnetwork.org

The ecological footprint measures human demand on the Earth's ecosystems as a whole. It represents the amount of biologically productive land and sea area necessary to supply the resources a human population consumes, and to assimilate associated waste. Using this assessment, it is possible to estimate how much of the Earth (or how many planet Earths) it would take to support a global population. For 2008, humanity's total ecological footprint was estimated at 1.52 planet Earths or in other words, humanity uses ecological services 1.52 times as fast as Earth can renew them. Every year, this number is recalculated with a 3 year lag due to the time it takes to collect and publish all the underlying statistics.

> The Global Footprint Network was founded by Mathis Wackernagel in 2003 as a non-profit organization to promote sustainable living with the core aim to accelerate the use of ecological footprinting as a resource accounting tool. Ecological footprints tell us: how much natural capacity we have; how much we use, and who uses what.

More information: http://www.footprintnetwork.org/en/index.php/GFN/

The 2007 statistics published in 2011 indicated that we would reach 2.0 planet Earths per person by 2038, but revised figures for 2008 released in 2012 shows that we are currently now at 1.52 Earths which will rise to 2.8 by 2050, reaching the 2.0 earths per person barrier by 2025 (Fig. 7.1). This rate of growth is totally unsustainable as sustainability is when the individual, family, community, or country lives within the bio-capacity of a single Earth which has to be our personal and community goals. However, the predictions are based on a moderate rate of growth in emissions and the reality is that these emissions are increasing much quicker than anticipated, so urgent action is needed. What ecological footprinting allows us to do is to look more holistically at our impacts as well as our mitigation strategies not only at the national level but more importantly at an individual and community level.

Ecological footprinting measures how quickly we use resources and emit pollutants that damage our environment compared to how fast nature can generate

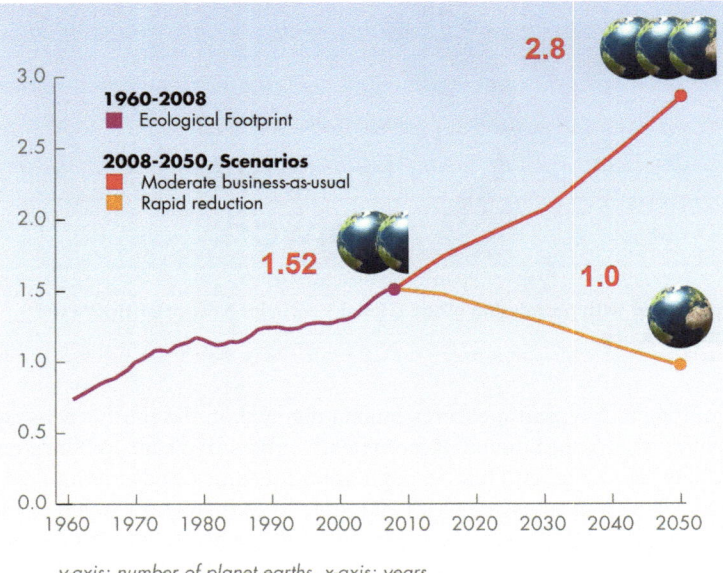

y-axis: number of planet earths, x-axis: years

Fig. 7.1 Increase in the global average ecological footprint in planet Earths. In 2012 the footprint was 1.52 Earths which will rise to 2.8 by 2050. *Source*: http://www.footprintnetwork.org. Reproduced with permission of the Global Footprint Network, Oakland, CA, USA

new resources, assimilate our waste and repair itself. It does this by assessing the biologically productive land and marine area required to produce the resources a population consumes and to absorb the corresponding waste, using prevailing technology. Footprint values are calculated individually for **Carbon**, **Food**, **Housing**, and **Goods and Services** as well as the total footprint number of Earths needed to sustain the world's population at that level of consumption. This approach can also be applied to an activity such as running a business, a university, driving of a car or having a pet. This method of resource accounting is similar to life cycle analysis where the consumption of energy, biomass (food and fibre), building materials, water and other resources are converted into a normalized measure of land area called '**global hectares**' (**gha**).

In 2006, the average biologically productive area per person worldwide was approximately 1.8 global hectares (gha) per capita. However when we compare this value to that of individual countries then the USA had a footprint per capita of 9.0 gha, and that of Switzerland was 5.6 gha per person, while China's was at the global average of 1.8 gha per person (Fig. 7.2). The World Wildlife Fund (WWF) claims that the human footprint had exceeded the bio-capacity (the available supply of natural resources) of the planet by 30 % by 2008, and this has continued to rise. Wackernagel and Rees originally estimated that the available biological capacity for the six billion people on Earth at that time was about 1.3 ha per person. This is smaller than the 1.8 global hectares published for 2006, because the initial studies

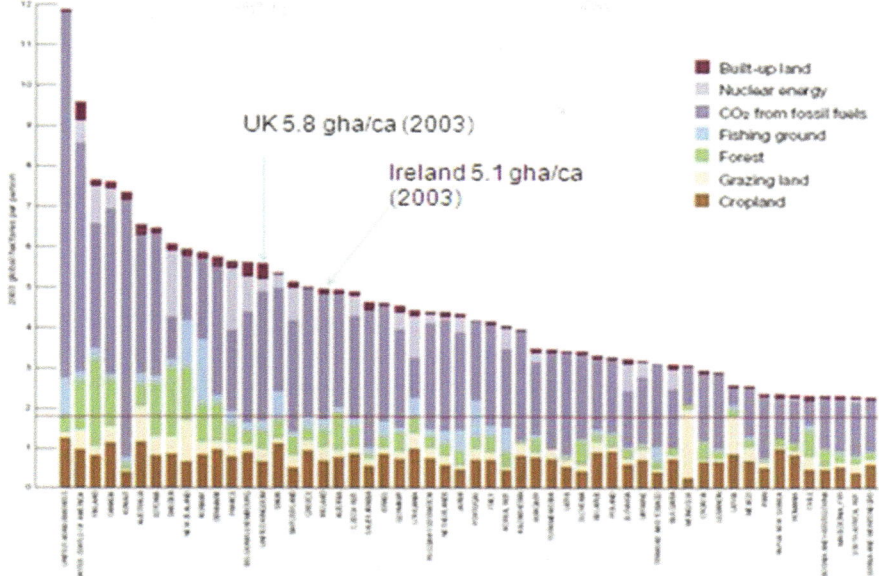

Fig. 7.2 Ecological footprint per person by country expressed in global hectares (gha) per capita. *Source*: http://www.footprintnetwork.org. Reproduced with permission of the Global Footprint Network, Oakland, CA, USA

neither used global hectares nor included bio-productive marine areas. We convert global hectares into planets using the equation:

$$\textbf{Planets} = \textbf{gha per capita} \div \textbf{1.8}$$

> **The basis of sustainable living is doing just that, living on a small piece of land. This is of course not possible for everyone, but each of us has the equivalent of 1.8 global hectares to sustain us. This area of land is constantly being reduced by increased consumerism, increased greenhouse gas emissions and continued population increase.**

7.2.1   Calculation of Ecological Footprint

The first step in calculating an ecological footprint (EF) is to determine the yields of primary products (i.e. cropland, forest, grazing land and fisheries) and the area necessary to support a given activity (i.e. output per unit area). **Bio-capacity** is measured by calculating the amount of biologically productive land and sea area available which is multiplied by yield data. Also included is the **assimilation and**

breakdown of wastes using current technologies and management practices. A nation's consumption is calculated by adding imports to, and subtracting exports from, its national production.

The National Footprint Account identifies whether or not a country's ecological footprint exceeds its bio-capacity. The national footprint account varies between countries. So **ecological creditors** have an ecological footprint which is smaller than its bio-capacity (i.e. has a reserve bio-capacity); whereas **ecological debtors** have an ecological footprint which is greater than bio-capacity (i.e. operating with an ecological deficit). In practice there are always going to be ecological debtors and creditors, but ideally we should all be ecological creditors.

The Global Footprint Network have developed a range of standards for calculating ecological footprints which is also available under license as software applications for countries, businesses etc. A new set of standards for their calculation was published in 2009 with a major revision in 2012 which has resulted in recalculations of earlier footprints.

http://www.footprintnetwork.org/images/uploads/Ecological_Footprint_
    Standards_2009.pdf

A new publication *Calculation Methodology for the National Footprint Accounts, 2010* describes the fundamental calculations and principles used.

http://www.footprintnetwork.org/images/uploads/National_Footprint_Accounts_
    Method_Paper_2010.pdf

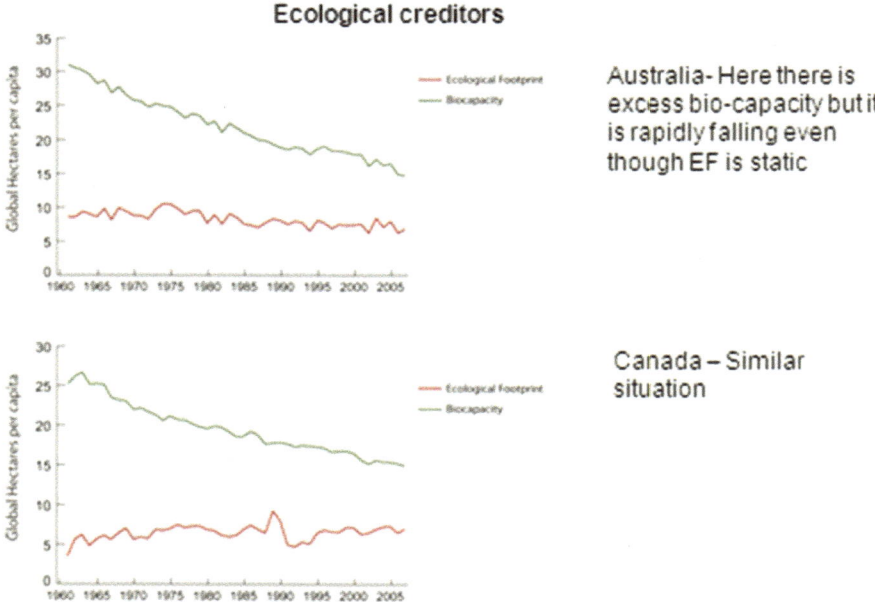

Fig. 7.3 Examples of countries that are ecological creditors. *Source*: http://www.footprintnetwork.org. Reproduced with permission of the Global Footprint Network, Oakland, CA, USA

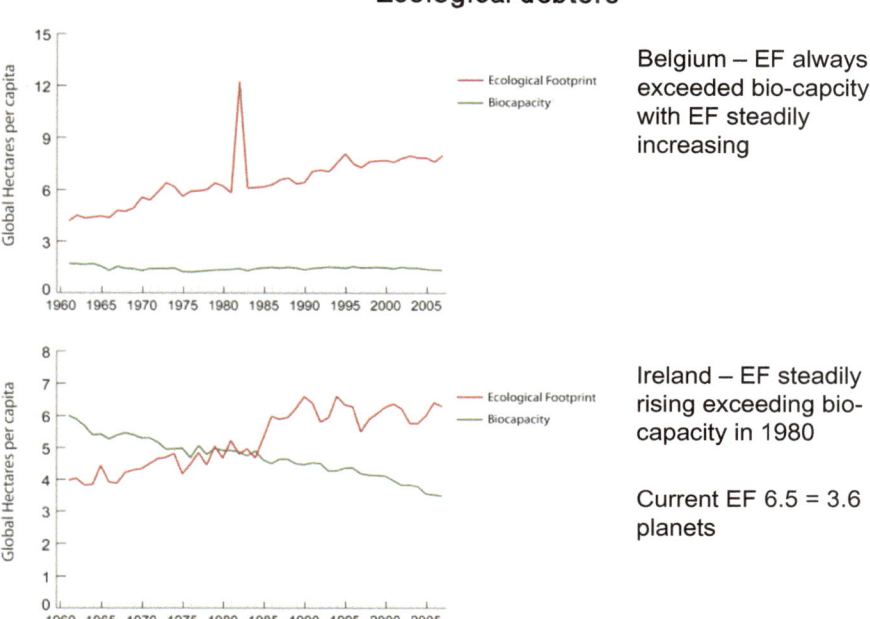

Fig. 7.4 Examples of countries that are ecological debtors. *Source*: http://www.footprintnetwork. org. Reproduced with permission of the Global Footprint Network, Oakland, CA, USA

Unlike carbon footprints which work well at the individual or community level, ecological footprints do have intrinsic problems when used at lower resolutions, and are best suited to national or large business evaluations. Figure 7.3 compares two countries that are ecological creditors. Both these countries have enormous reserve bio-capacity in relation to relatively low populations. In the case of Australia they have a very high but relatively constant ecological footprint which is rapidly eating into their bio-capacity reserves which are decreasing at an alarming rate. Factors such as global warming that has triggered massive droughts and wildfires have also played a part in reducing bio-capacity so there is an urgent need to reduce the ecological footprint as well as rebuild bio-capacity.

Most European countries fall into the ecological debtor category with Belgium having a slowly rising ecological footprint in relation to a small but static bio-capacity (Fig. 7.4). In contrast, Ireland has seen a steady rise in its ecological footprint since 1960, although it is beginning to stabilize, and a steadily declining bio-capacity. The country switched from being an ecological creditor to debtor around 1980. Unlike Belgium its bio-capacity is rapidly in decline so it needs to reduce demands coupled with a strategy to sustain its bio-capacity to halt its continuous decline. In 2012 Ireland had an ecological footprint of 6.22 gha per capita and a bio-capacity of 3.41 gha per capita giving an overall footprint of 3.5 Earths (Table 7.1).

The US, like Ireland is an ecological debtor country (Fig. 7.5). In 2007 it had an ecological footprint of 8.0 gha per capita and a bio-capacity of 3.9 gha per capita

Table 7.1 Calculation of an ecological footprint for Ireland

Factor	Ireland	
	Ecological footprint (gha per capita)	Bio-capacity (gha per capita)
Carbon	3.75	–
Grassland	0.45	0.79
Cropland	1.26	0.59
Fishing ground	0.04	1.64
Forests	0.53	0.24
Urban area	0.26	0.16
Total	**6.22**	**3.41**

In the new system employed in 2011 the carbon footprint was reduced by 27 % due to a revision in oceanic carbon sequestration. *Source*: http://www.footprintnetwork.org. Reproduced with permission of the Global Footprint Network, Oakland, CA, USA

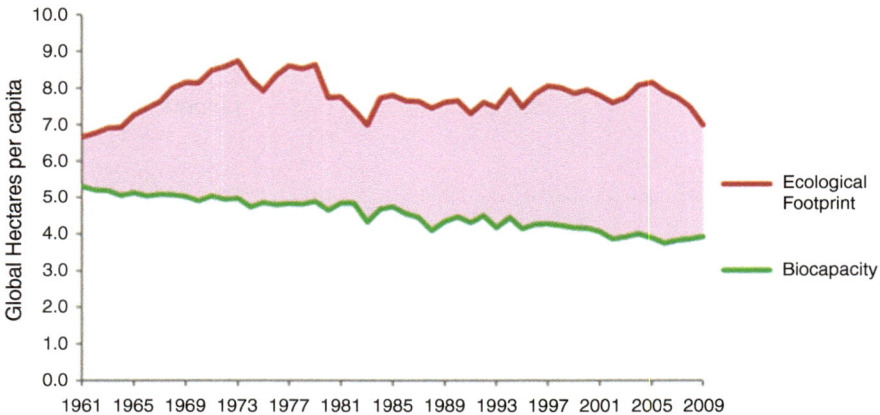

Fig. 7.5 The United States of America is an ecological debtor country. The shaded area shows the ecological deficit. *Source*: http://www.footprintnetwork.org. Reproduced with permission of the Global Footprint Network, Oakland, CA, USA

Table 7.2 Calculation of US ecological footprint in 2007

Factor	United States of America	
	Ecological footprint (gha per capita)	Bio-capacity (gha per capita)
Carbon	5.57	–
Grassland	0.14	0.26
Cropland	1.08	1.58
Fishing ground	0.10	0.41
Forests	1.03	1.55
Urban area	0.07	0.07
Total	**8.0**	**3.9**

Source: http://www.footprintnetwork.org. Reproduced with permission of the Global Footprint Network, Oakland, CA, USA

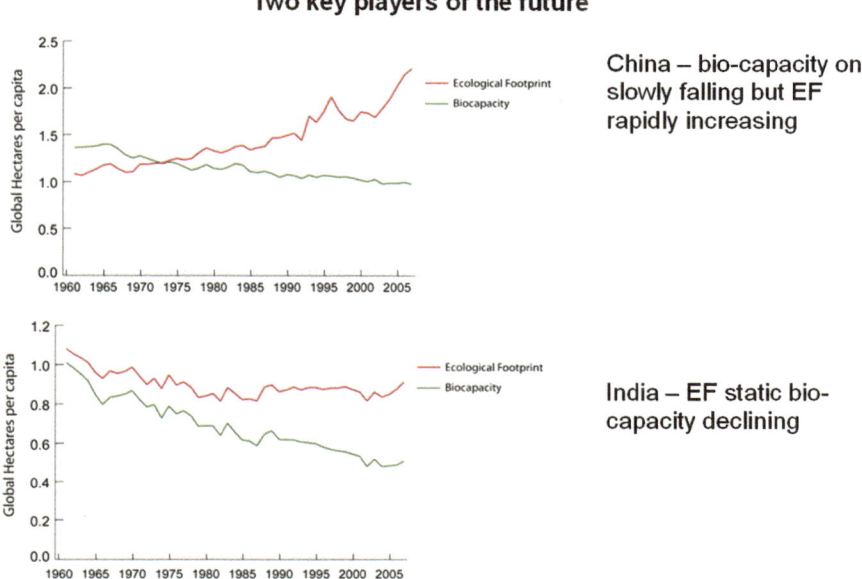

Fig. 7.6 The two most critical ecological footprints, China and India, both of which are net ecological debtors. *Source*: http://www.footprintnetwork.org. Reproduced with permission of the Global Footprint Network, Oakland, CA, USA

giving it a deficit of 4.1 gha per capita (Table 7.2). Details of the ecological footprint and biocapacity for each country can be found in GFN (2010).

China and India are both undergoing rapid economic development as well as large population increases. In China, wealth is increasing and this is reflected by a raid increase in ecological footprint, although still significantly below that of the USA or other developed countries. The country is vast and so we see a relatively low rate of decline in bio-capacity, but as the footprint reaches European levels or beyond, then the rate of decline in bio-capacity will inevitably increase (Fig. 7.6). In India wealth is static showing a much slower increase in ecological footprint, although bio-capacity is also steadily declining. The problem here is that the ecological footprint and bio-capacity per capita are both low, but the rapidly expanding population will rapidly reduce bio-capacity even if the ecological footprint is stabilized. As a rule of thumb, population dilutes bio-capacity while ecological footprint is linked to wealth and consumerism.

> **We need to sustain and improve our bio-capacity coupled with reducing our demands**

Today, more than 80 % of the world's population is resident in countries that use more resources than is renewably available within their own borders (Fig. 7.7).

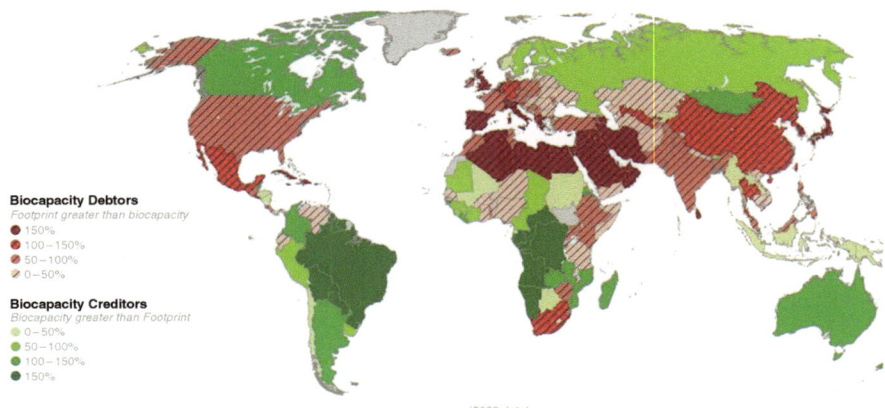

Fig. 7.7 Ecological Creditors and Debtors (2009-published 2012) showing where there is excess bio-capacity (*green*) and where it has been exceeded (*red*). Currently 151 % of global bio-capacity is being utilized. *Source*: http://www.footprintnetwork.org. Reproduced with permission of the Global Footprint Network, Oakland, CA, USA

These countries rely for their needs on resource surpluses concentrated in ecological creditor countries, which use less bio-capacity than they have. By comparison, in 1961, the vast majority of countries around the globe had ecological surpluses. Those numbers have slowly dwindled and the pressure on the remaining bio-capacity reserves continues to grow.

The ecological footprint varies within countries by region, cities, companies and between individuals. Even ecovillages and eco-communities still only achieve values of 2.5–3.0 gha per person, well above the 1.8 gha to achieve 1 Earth lifestyles. **The core to achieving sustainability is stabilizing and subsequently increasing bio-capacity, reducing the ecological footprint, and finally stabilizing and eventually reducing global population**.

More information: GFN (2010) *Ecological Footprint Atlas 2010*. Global Footprint Network, Oakland, CA, USA. http://www.footprintnetwork.org/images/uploads/Ecological_Footprint_Atlas_2010.pdf

> **The scope for personal action in achieving these goals is significant, and ecological footprinting allows us to interrogate our lifestyles, our use of resources and the management of our bioresources**.

The **per capita ecological footprint**, or ecological footprint analysis (EFA), is a means of comparing consumption and lifestyles, and checking this against nature's ability to provide for this consumption. EFA is a tool that can inform policy by examining to what extent a nation uses more (or less) than is available within its territory or to what extent the nation's lifestyle would be replicable worldwide.

Ecological footprints can be used to demonstrate that many current lifestyles are clearly non-sustainable. Such a global comparison also clearly shows the inequalities of resource use on this planet at the beginning of the twenty-first century. The footprint is a useful tool to educate people about carrying capacity and over-consumption, with the aim of altering personal behaviour.

The Greater London Authority commissioned a study to determine the ecological footprint for London in 2002 which was estimated at 48,868,000 gha which is equivalent to 6.63 gha per capita (GLA 2003) (Fig. 7.8). In contrast its biocapacity was only 1,210,000 which is 0.16 gha well below the global average per capita biocapacity of 2.18 gha at the time of the study. The report concluded that to be sustainable London will have to reduce their ecological footprint by 80 % by 2050 to meet the global average footprint which is expected to be 1.3 gha per capita by mid-century. In order to achieve an interim reduction of 35 % by 2020 they suggest that every Londoner should: (i) reduce their consumption of natural gas from 9.5 to 6.2 MWh per annum; (ii) Install 11 m^2 of solar panels; (iii) Reduce their travel by 3000 km per year or switch 3500 km of car travel to bicycle; (iv) Consume 70 % less meat and reduce food waste by 100 kg per year; (v) Eat a minimum of 40 % locally sourced unprocessed food; and (vi) produce 1 tonne less waste per year. A similar very detailed study has been done for Vancouver in Canada (Rees and Moore 2013).

The bio-capacity of planet Earth in 2005 was equivalent to 2.1 gha per capita and had fallen to below 1.8 gha per capita by 2014. This will continue to fall as the population increases and environmental degradation increases. However, there is a worrying relationship between human welfare and ecological footprint (Fig. 7.9) which shows that welfare generally increases the higher the ecological footprint. This is a fundamental issue that must be tackled. **It must be possible to increase the quality of life without exceeding one Earth … this is a challenge for all of us**.

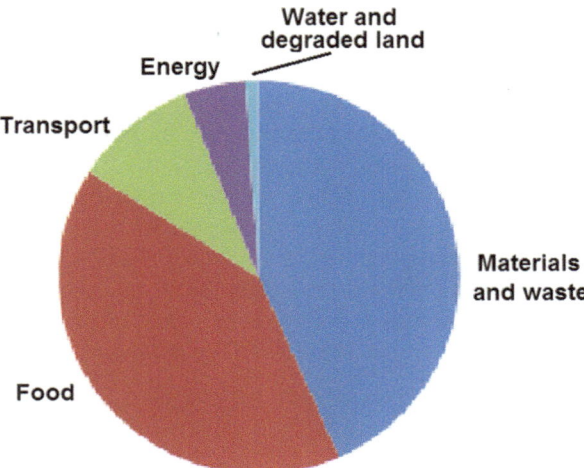

Fig. 7.8 Breakdown of Londoner's ecological footprint. *Source*: GLA (2003). Reproduced with permission of the Greater London Authority, London

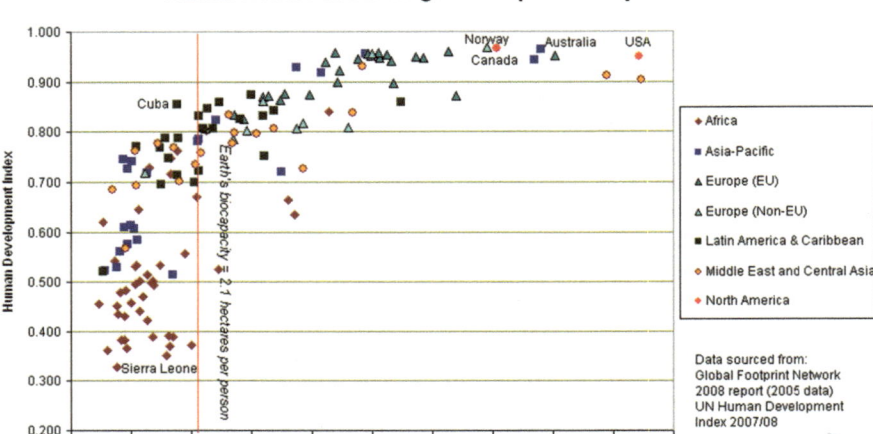

Fig. 7.9 The relationship between ecological footprint in global hectares per capita and human welfare as measured using the Human Development Index. *Source*: GFN (2010). Reproduced with permission of the Global Footprint Network, Oakland, CA, USA

7.3 Global Living Planet Index

One of the problems identified with ecological footprinting is the exclusion of bio-diversity and its preservation. There has been a 28 % decline in the diversity of vertebrate species alone over the period 1970–2007. The Living Planet Index (LPI), originally developed by the WWF, is often used in conjunction with ecological foot-printing. The index is an indicator of the state of global biological diversity based on trends in vertebrate populations.

The *Living Planet Database* (*LPD*) is maintained by the Zoological Society of London and contains over 10,000 population trends for more than 2500 species of fish, amphibians, reptiles, birds and mammals (Fig. 7.10). The global LPI is calcu-lated using nearly 8000 of these population time-series, examples of which are shown below. The conclusion of WWF's most recent *Living Planet Report* (2008) is that we are now living in severe ecological overshoot. Worldwide, people are con-suming about 30 % more natural resources than the planet can replace. Freshwater species are most vulnerable due to pollution and exploitation of water resources which is reflected in a rapid decline in the LPI (Fig. 7.11).

More information: http://wwf.panda.org/about_our_earth/all_publications/living_
    planet_report/

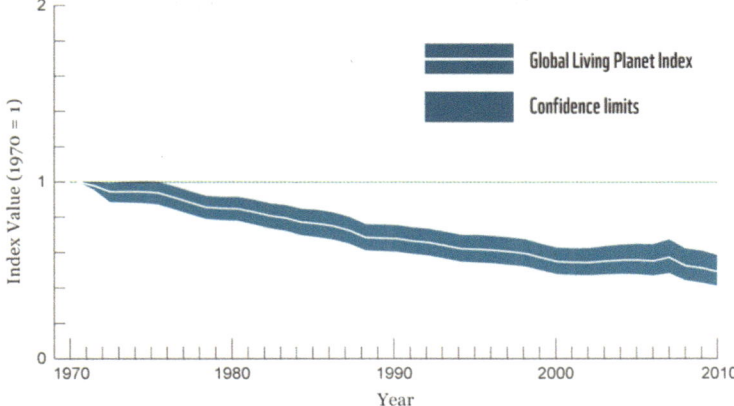

Fig. 7.10 There has been a decline in the global living planet index (LPI) of 52 % between 1970 and 2010. *Source*: McRae et al. (2014). Reproduced with permission the World Wildlife Fund, Gland, Switzerland

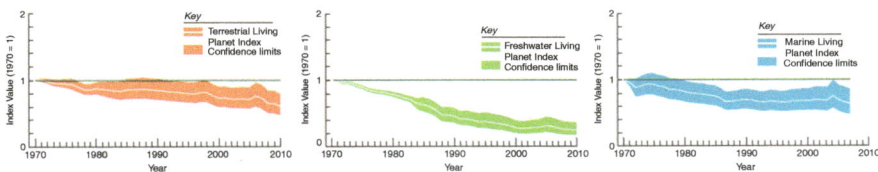

Fig. 7.11 Decline in the global living planet index (LPI) for terrestrial, freshwater and marine organisms since 1970. The increase seen during the period 1970–1980 are largely new specie recorded. *Source*: McRae et al. (2014). Reproduced with permission the World Wildlife Fund, Gland, Switzerland

7.4   One Planet Economy Network

The key to achieving a sustainable and stable culture is to understand the relationship between the use of natural resources and the resultant environmental impacts that arise from their use. The European Union has adopted a Sustainable Development Strategy (SDS), whose aim is 'to safeguard the Earth's capacity to support life in all its diversity, respect the limits of the planet's natural resources and ensure a high level of protection and improvement of the quality of the environment, prevent and reduce environmental pollution,' and '**promote sustainable consumption and production in order to break the link between economic growth and environmental degradation**.' The key element of the strategy is to reduce the negative environmental impacts associated with the use of resources in a growing economy. This is to be achieved through investment into research to find ways of decoupling the use of resources for economic growth and also reduce

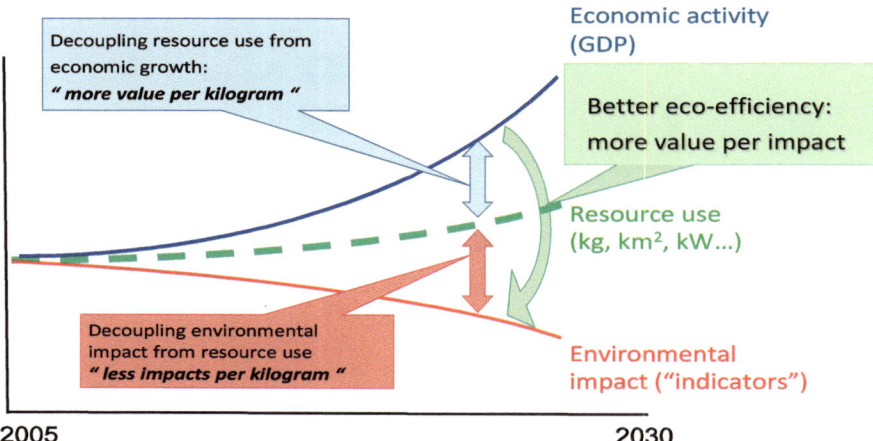

Fig. 7.12 The concept of double decoupling to break the link between economic growth and resource use, and environmental damage arising from the use of resources. *Source*: European Commission (2005). Reproduced with permission of the European Union

environmental damage arising from resource use, which has been described as double decoupling (Fig. 7.12) (Galli et al. 2011).

Indicators are being used in new ways to help develop sustainability programmes such as the EU SDS. One new methodology, the One Planet Economy Network, has developed a 'Footprint Family' of indicators: Ecological, Carbon and Water footprints. The objective is to create a community, regional or national plan to enable a transformation to a one Earth economy. The method is able to track the multiple and often hidden demands that human consumption makes on the planet's resources and also to accurately measure their impacts on the planet. With this information decision makers can develop an informed response to issues such as limits to natural resource or freshwater consumption, and sustainable use of natural capital across the globe. One of the primary outputs of the project was the development of an online scenario-modelling and policy assessment tool (i.e. EUREAPA). EUREAPA translates complex science into practical information which decision makers can use to tackle complex issues simultaneously and better deal with tradeoffs (Briggs 2013).

More information: http://www.oneplaneteconomynetwork.org/index.html
EUREAPA: http://www.oneplaneteconomynetwork.org/eureapa.html

The key theme of the One Planet Economy is the transformation to a low-carbon and energy society. We need a transformation strategy which increases resource efficiency and reduces environmental pressure. Such strategies need to be gradual, long term, reliable and equitable, to ensure a stable economic transformation, so the strategy needs to be based on sound business concepts, viable investment, and the right incentives for each of the stakeholders, similar to that described by the Stern Report (Sect. 2.2.2).

It envisages a sector by sector industrial transformation to the use of low impact technologies, integrated logistics, more equitable distribution and importantly sustainable consumption patterns. In order for this strategy to work it must be financially viable, socially responsible and above all practical, that is the economic, social and environmental costs/benefits must all stack up. Such a plan has been developed for Wales (i.e. One Planet Wales) and is being used for a long term transition to a sustainable region based on a one Earth per person lifestyle. This is the way to really implement the ideology of ecological footprinting.

More information on One Planet Wales: http://wales.wwf.org.uk/what_we_do/ changing_the_way_we_live/one_planet_wales/

One Planet Economy is the process of converting ecological footprints in action to achieve a sustainable world.

7.5 Setting Sustainability Targets

I always remember early in my career when a particularly aggressive Dean asked me to justify my job. My immediate response was 'tell me where the net is and I will try and kick for goal'. It is rather similar to global warming which really is a moving target, so it is very hard to predict with certainty where targets should be set. By 2050 the global population may have increased by 50 %, while there may be additions to bio-capacity due to land reclamation and reforestation, we simply don't know. But using the mid range IPCC estimate for growth, global bio-capacity will decrease from 1.8 to 1.3 gha per person by 2050 putting huge strains on existing resources and natural systems. To reduce the current footprint for Ireland to a fair share bio-capacity of 1.3 will require approximately a 75 % reduction. This reduction factor will continue to grow as we approach 2050 unless we act now to reduce our personal and national footprints.

Factor 4 is a theory based book written by Ernst von Weizsacker and colleagues which explores the idea of doing more with less reducing resource depletion without reducing the quality of life. The basic concept is if you **do twice as much with half the resources** you've achieved a factor 4 improvement.

The book gives 50 examples of best practice, argues that natural resources can be used more efficiently in all domains of daily life, either by generating more products, services and quality of life from the available resources, or by using fewer resources to maintain the same standard. This is similar to the idea of using resources more effectively to increase the value and at the same time reducing their environmental impact (Fig. 7.12). As the study commissioned by the Greater London Authority showed, even a 35 % reduction will require significant changes in personal attitudes and behaviour (Sect. 7.2) (GLA 2003). Personal targets are discussed in details in Chap. 15.

7.6   Conclusions

- Ecological footprinting gives us a far more holistic approach to tackling global warming than just using carbon or water footprint models.
- Governments, especially within Europe, now have the technical and management tools to start to tackle global warming seriously.
- Some Governments consider tackling climate change more important than others, some fear losing markets, others see developing nations becoming richer as a threat to their own economic stability. Whatever the reason, the longer they delay in tackling GHG emissions then the more severe climate change will be
- Key actions require us to (i) stabilize and increase bio-capacity, (ii) reduce our ecological footprint (through reduced consumerism and greater efficiency), and (iii) stabilize population.
- Sustainability targets will require a minimum factor four reduction and possibly higher for Ireland and most European countries.

The seventh step is accepting the concept of ecological footprinting as a holistic approach to achieving sustainability and dealing with global warming. This involves us all looking at the three key mechanisms and how we can make positive contributions.

Your aim should be:

- To do twice as much with half of the resources (Factor Four Reduction) and thereby reduce your ecological footprint;
- To stabilize and increase bio-capacity by using your space, garden and land better and by becoming involved with conserving and transforming your immediate environment through community projects.
- To create a sustainable family unit in terms of size and carbon footprint

Homework!

I want you to consider ways in which you can use resources more effectively. Some examples will be explored later such as combining trips in your car, car sharing, ensuring the washing machine is full when used, sharing trips to recycling centres to ensure the car is fully loaded. I want you to spend a little time thinking about this and devising six ways in which you can reduce your footprint by doing more with less.

References and Further Reading

Briggs, J. (2013). *The EUREAPA technical report (methodology)*. Godalming, England: One Planet Economy Network. Retrieved from http://www.oneplaneteconomynetwork.org/resources/programme-documents/EUREAPA_Technical_Report.pdf

European Commission. (2005). *Thematic strategy on the sustainable use of natural resources—Commission staff working document*. Annexes to the Communication from the Commission to the Council, the European Parliament, the European Economic and Social Committee and the Committee of the Regions. COM(2005) 670 final. Retrieved from http://ec.europa.eu/environment/natres/pdf/annex_com_en.pdf

Galli, A., Wiedmann, T., Ercin, E., Knoblauch, D., Ewing, B., & Giljum, S. (2011). *Integrating ecological, carbon and water footprint: Defining the footprint family and its application in tracking human pressure on the planet* (Technical Document 28). Godalming, England: One Planet Economy Network. Retrieved from http://www.oneplaneteconomynetwork.org/resources/programmedocuments/WP8_Integrating_Ecological_Carbon_Water_Footprint.pdf

GFN. (2010). *Ecological footprint Atlas 2010*. Oakland, CA: Global Footprint Network. Retrieved from http://www.footprintnetwork.org/images/uploads/Ecological_Footprint_Atlas_2010.pdf

GLA. (2003). *London's ecological footprint: A review*. London, England: Greater London Authority.

McRae, L., Freeman, R., & Deinet, S. (2014). The living planet index. In: R. McLellan, L. Iyengar, B. Jeffries, & N. Oerlemans (Eds.), *Living Planet Report 2014: Species and spaces, people and places*. Gland, Switzerland: WWF. Retrieved from http://wwf.panda.org/about_our_earth/all_publications/living_planet_report/living_planet_index2/

Rees, W., & Moore, J. (2013). Ecological footprints, fair Earth-shares and urbanization. In R. Vale & B. Vale (Eds.), *Living within a fair share ecological footprint* (pp. 3–32). Oxford, England: Earthscan.

Wakernagel, M., & Rees, W. (1996). *Our ecological footprint: Reducing human impact on earth*. Philadelphia, PA: New Society.

Home of Ecological Footprinting

http://www.footprintnetwork.org/en/index.php/GFN/

Key Players

http://www.oneplaneteconomynetwork.org/index.html
http://www.carbontrust.co.uk/Pages/Default.aspx
http://www.antaisce.org/naturalenvironment/Sustainability/EcologicalFootprint.aspx

Living Planet Report

http://awsassets.panda.org/downloads/lpr2010.pdf

Integrating Ecological, Carbon and Water Footprints

http://www.oneplaneteconomynetwork.org/resources/programme-documents/WP8_Integrating_
  Ecological_Carbon_Water_Footprint.pdf

Finance and Sustainable Development

http://www.ecologicalbudget.org.uk/

Calculation of Personal Ecological Footprint

http://myfootprint.org/en/
http://www.footprintnetwork.org/en/index.php/GFN/page/calculators/
http://www.ecologicalfootprint.com/

Problems with Ecological Footprinting

http://blog.nature.org/science/2013/11/05/science-ecological-footprint-plos-biology-critique-
  blomqvist/

Part III
Our Use of Resources

Chapter 8
Energy: Green or Otherwise

Energy demand is expected to rise by 30 % globally by 2040, so how will this extra energy be generated and what effect will this have on greenhouse gas emissions? In this chapter we discuss energy use and the future of specific energy sources, including how energy consumption can be minimized. Image by Frank O'Brien. *Source*: http://www.askaboutireland.ie/enfo/irelands-environment/county-focus/wexford/wind-farming/. Reproduced with permission of Frank O'Brien

The Wind Farm developed at Carnsore Point shown above was the first wind farm completed on the east coast of Ireland. It was constructed by Hibernian Wind Power Limited and opened in November 2002. Carnsore Wind Farm uses 14 wind turbines which have the total capacity to generate a maximum of 11,900 kW of renewable energy. These green energy supplies can power an average of 10,000 homes and

© Springer International Publishing Switzerland 2015   179
N.F. Gray, *Facing Up to Global Warming*, DOI 10.1007/978-3-319-20146-7_8

save over 30,000 tonnes of CO_2 annually or can offset 20–25 long haul flights to Australia from Europe each year, so long as the wind is blowing! However, this is now a very small installation in comparison to current and future projects. For example, in April 2013 the largest offshore wind farm in the world came on stream. Situated in the Thames estuary it comprises 175 turbines with a maximum generating capacity of 630 MW enough to power 470,000 homes as well as save 925,000 tonnes of CO_2 per annum. The UK's current generating capacity from offshore wind farms is around 3.6 GW but the aim is to raise this to 18 GW by 2020.

8.1   How Much Energy Do We Use?

The global total energy consumption is equivalent to 12,730,400,000 tonnes of oil equivalent (toe). A kgoe is a kilogram of oil equivalent and is a universal unit of energy usage. This has risen steadily at an annual rate of 2.5 % over the past decade, although the increase in the period 2012–2013 showed a slight decline at 2.3 % growth. While the use of energy has remained static within the OECD countries over the past decade, with the EU showing a small but steady decline, non-OECD countries have experienced a rapid growth in energy use over the same period. This is reflected by the static energy growth in the US compared to that in China (Table 8.1). In terms of the percentage share of global fuel consumption then the non-OECD countries currently use 56.5 % showing that developing countries now exceed the industrialized nations in terms of total energy consumption. The main consumers of energy, and growth during 2012–2013 shown in parentheses, are China 22.4 % of global energy consumed (4.7 % growth), USA 17.8 % (2.9 %), Russia 5.5 % (0.2 %) and Japan 3.7 % (−0.6 %). Consumption within the EU is 13.2 % overall with Germany 2.6 % (2.8 %), UK 1.6 % (−0.5 %) while Ireland is <0.1 % of global fuel consumption (1.7 % growth). Largest growth in the period 2012–2013 occurred in Asia and in particular China, the Philippines and Indonesia at around 5 %, with South America and in particular Argentina, Brazil, Columbia and Venezuela having increases of between 3 and 4 %. Other countries saw large decreases in consumption such as Sweden (−5.2 %), Norway (−5.9 %), Greece (−6.7 %), Hungary (−6.9 %) and Lithuania (−7.2 %).

Table 8.1 Total energy consumption by country or region over the period 2001–2013 in million tonnes of oil equivalent (Mtoe) in 2013

Country	2003	2004	2005	2006	2007	2008	2009	2010	2011	2012	2013
China	1245.3	1466.8	1601.1	1767.9	1880.1	1971.4	2104.3	2339.6	2544.8	2731.1	**2852.4**
US	2302.3	2349.1	2351.3	2333.1	2371.7	2320.1	2205.9	2284.9	2265.4	2208.0	**2265.8**
India	320.8	345.1	366.8	390.0	420.1	446.5	483.8	510.2	534.6	573.3	**595.0**
United Kingdom	225.4	227.3	228.2	225.4	218.2	214.5	203.9	209.2	196.3	201.6	**200.0**
Republic of Ireland	14.2	14.6	15.1	15.4	15.9	15.7	14.4	14.4	13.3	13.2	**13.3**
European Union	1789.6	1819.2	1818.7	1826.3	1801.1	1794.0	1691.2	1752.8	1691.2	1685.5	**1675.9**
OECD	5520.9	5633.6	5679.0	5687.8	5727.5	5671.5	5398.3	5598.2	5535.8	5484.4	**5533.1**
Non-OECD	4423.0	4794.6	5035.4	5333.0	5592.0	5794.7	5927.5	6357.3	6695.7	6998.9	**7197.3**
Total World	9943.8	10,428.2	10,714.4	11,020.8	11,319.5	11,466.2	11,325.9	11,955.6	12,231.5	12,483.2	**12,730.4**

Source: BP (2014). Reproduced with permission of *BP Statistical Review of World Energy 2014*

A Mtoe is a million tonnes of oil equivalent. A tonne of oil equivalent (toe) is a unit of energy roughly equivalent to the energy content of 1 tonne of crude oil. So 1 toe equals 10^7 kilocalories or 41.87 gigajoules (GJ).

It is interesting that while economic growth in developing countries is low, they collectively account for 80 % of the global increase in energy consumption seen in 2013. Rapid economic growth in the US during 2013 accounts for the net increase of 1.8 % in energy consumption in the OECD group of countries, with other members such as Japan, the EU and Spain showing significant reduction in energy usage at −0.3 %, −0.6 % and 5.0 % respectively.

Energy demand is expected to rise by 30 % globally by 2040, although it is anticipated that the demand in the OECD countries will remain at current levels with energy conservation offsetting increases in population, economic growth and consumerism which normally drive demand upwards. Current forecasts suggest that the overall energy demand of China will exceed that of the US by 70 % by the year 2035.

Globally oil and coal remain the major sources of primary energy at 32.9 % and 30.1 % respectively (Table 8.2), although natural gas is also a major source (30.1 %). So fossil fuels are still supplying 87 % of the global energy demand. Of the non-fossil fuel sources hydro-electricity is the main source of energy at 6.7 % with nuclear and renewable energy sources just 4.4 and 2.2 % overall. There is clear evidence that developed countries are moving away from the most polluting fossil fuels in terms of CO_2e emissions, with the developing nations having a high reliance on coal in particular. All countries rely heavily on oil primarily for transportation,

Table 8.2 Total energy consumption by fuel source in million tonnes of oil equivalent (Mtoe) in 2013 for selected countries and regions

Country	Oil	Natural gas	coal	Nuclear	Hydro-electricity	Renewables	Total
USA	831.0	671.0	455.7	187.9	61.5	58.6	2265.8
China	507.4	145.4	1925.3	25.0	206.3	42.9	2852.4
India	175.2	46.3	324.3	7.5	29.8	11.7	595.0
Germany	112.1	75.3	81.3	22.0	4.6	29.7	325.0
Ireland	6.7	4.0	1.3	0.0	0.1	1.1	13.3
UK	69.8	65.8	36.5	16.0	1.1	10.9	200.0
OECD countries	2059.9	1444.4	1066.9	447.0	319.3	195.6	5533.1
Non-OECD	2125.1	1576.0	2759.8	116.1	536.5	83.7	7197.3
European Union	605.2	394.3	285.4	198.5	81.9	110.6	1675.9
World	4185.1	3020.4	3826.7	563.2	855.8	279.3	12,730.4

Source: BP (2014). Reproduced with permission of *BP Statistical Review of World Energy 2014*

Table 8.3 Total energy consumption per capita (2005)

	kgoe
World average	1,778
Low income country average	492
High income average	5,524
Bangladesh	171
Ireland	3,656
UK	3,895
USA	7,886
Canada	8,473
Iceland	12,209

Source: World Resources Institute http://earthtrends.wri.org/text/energy-resources/variable-351.html. Reproduced with permission of the WRI, Washington, DC, USA

although its percentage share in terms of global fuel use has fallen steadily since 1965 when records first began (BP 2014). The demand for oil in China is about half of that of the US which used 18.835 million barrels per day in 2011.

Energy consumption data: http://www.enerdata.net/enerdatauk/; http://www.eia.gov/countries/

World energy Council: http://www.worldenergy.org/

Summary of world energy consumption and resources: http://www.bp.com/content/dam/bp/pdf/Energy-economics/statistical-review-2014/BP-statistical-review-of-world-energy-2014-full-report.pdf

The per capita energy usage varies significantly globally from 171 kgoe in Bangladesh to 12,209 kgoe in Iceland (Table 8.3). As one would expect there is a huge difference between the averages for low income countries (492 kgoe) compared to high income countries (5525 kgoe) showing a massive energy divide.

While the world trend in energy usage is rising like most European countries there has been a downward trend in energy usage in Ireland since 2007 (Table 8.1). While primary energy consumption has stabilized at around 13.2 Mtoe this is still above the Kyoto target as energy usage is still 32 % above that in 1990. Energy related CO_2 emissions during 2010 fell by 1.0 % to 42 Mt CO_2e in Ireland. Countries are able to use carbon trading in order to reach the first target under Kyoto. The first commitment period was 2008–2012 and the second commitment period, set in 2012, runs from 2013 to 2020. This latest target for non-traded emissions is 20 % below the 2005 level which is to be achieved by 2020. The latest prediction by the Economic and Social Research Institute (ESRI) based in Dublin is that Ireland will be 5 % above the 2005 ceiling by that time.

World energy demand rises by on average 2.5 % per year

Carbon dioxide equivalent emissions from China have increased by a factor of 3.6 since 1990 and they are now the largest single emitter of GHGs at 8502 Mt CO_2e per annum (2013). After China the big emitters are the US 5101 Mt CO_2e per annum, India 2011, Russia 1661, Japan 1186, Germany 746 South Korea 584 and Iran 561 Mt CO_2e per annum based on 2013 data. Most of the larger EU countries are in the 400–500 Mt CO_2e category for example, the UK at 449 Mt CO_2e per annum. At the other end of the emissions scale are New Zealand 32, Sweden 39, Norway 40, Portugal 45 and Finland 49 Mt CO_2e per annum. Of course emissions per country are controlled by many factors including population size, economic growth and the type of energy used.

8.1.1 Electricity

China has an electricity generating capacity of 12,470,000 MW which is similar to that of the United States at 1,054,000 MW. In comparison Ireland has a generating capacity of just 8000 MW with an overall production of 25 TWh per year which works out to be about **5.6 MWh per capita per year,** making it one of the lowest consumers of energy per capita in Europe. The UK per capita usage of electricity is almost identical to Ireland at 5.5 MWh while Norway, being the highest within the EU, is just under 25 MWh per capita per year compared to **the United States at 13.25 MWh per capita per year**. Like all countries the fuel which Ireland uses to produce its electricity varies and is based on a mixture of coal, peat, oil, natural gas, electricity and renewable energies. As can be seen in Table 8.2, the country relies heavily on natural gas and oil, with oil representing 50 % of overall energy supply, while peat consumption has remained steady since 2000 at around 7 %. Renewable energy consumption has been increasing since 2002, as has the consumption of imported electricity. Imported energy dependency fell from 90 % (2007) to 86 % in 2010. In 2008 Ireland imported 92.1 % of the natural gas which it consumes, 100 % of its coal and crude oil but produces 100 % of its peat and current biomass demand. Overall 82 % of imported energy at that time was in the form of either oil or natural gas.

Generating capacities of all countries: http://www.eia.gov/cfapps/ipdbproject/
    IEDIndex3.cfm?tid=2&pid=2&aid=7
https://www.cia.gov/library/publications/the-world-factbook/rankorder/2236rank.
    html
Electricity usage per country: http://data.worldbank.org/indicator/EG.USE.ELEC.
    KH.PC

The GHG emissions vary from country to country depending on the fuel mix used. So for the period January to December 2012 the electricity in Ireland was generated from the following fuels: 19.9 % coal, 47.7 % gas, 6.9 % peat, 23.7 % renewable and 1.8 % other. This gives a conversion factor of 0.481 kg of CO_2e per kWh of electricity supplied, calculated over the period January to December, 2013. Ireland has continued to reduce its GHG emissions from electricity through careful

Table 8.4 Fuel mix for the generation of electricity in Ireland from all companies

Fuel type	2005	2006	2007	2008	2009	2010	2012
Coal (%)	24	19	18	17	14.2	16.0	20
Gas (%)	46	50	55	60	61.9	64.1	48
Oil (%)	12	9	6	4	2.5	1.6	0
Peat (%)	8	7	6	7	6.7	5.8	7
Renewables (%)	9	11	11	11	14.2	12.1	24
Other (%)	1	4	4	1	0.5	0.4	2
Total (%)	100	100	100	100	100	100	100
CO_2 emissions (Kg per kWh)	**0.576**	**0.549**	**0.538**	**0.533**	**0.504**	**0.519**	**0.481**

Source: SEAI (2013). Reproduced with permission of the Sustainable Energy Authority of Ireland, Dublin, Ireland

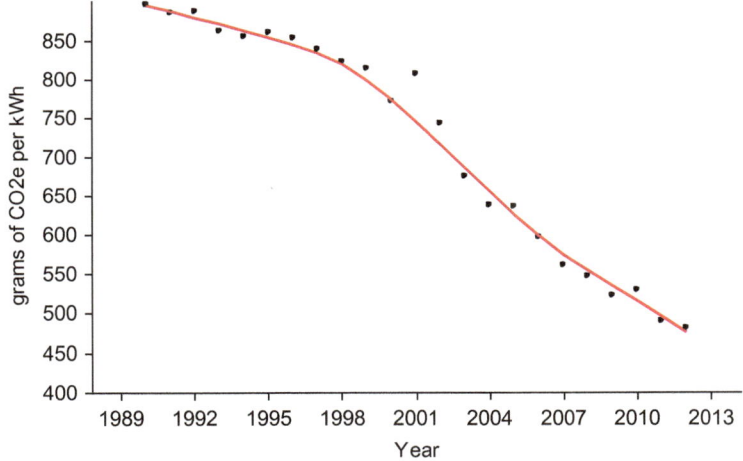

Fig. 8.1 Decline in greenhouse gas emissions (g CO_2e) per kWh of electricity generated in Ireland during the period 1990–2012. *Source*: SEAI (2013). Reproduced with permission of the Sustainable Energy Authority of Ireland, Dublin, Ireland

planning of fuel mixes and investment in more efficient power stations. By 2012 Electric Ireland had phased out its use of oil and also reduced its dependence on gas but increased its use both of coal and peat, two inefficient fossil fuels. However, key savings in emissions came from the continued development of renewable energy (Table 8.4).

Other emission factors: http://www.seai.ie/Publications/Statistics_Publications/ Energy_in_Ireland/Energy_in_Ireland_1990_-_2012_Report.pdf

Here is some more good news. The CO_2e emissions per kWh of electricity generated in Ireland has fallen by 49.6 % from 0.896 kg CO_2e per kWh in 1990 to just 0.452 kg in 2013. This reflects the phasing out of inefficient power stations, replacing dirty fuels with cleaner and investment in renewable energy sources (Fig. 8.1).

Most of this decrease has occurred since Kyoto and reflects how effective the process has been. This has a significant knock on effect to householders who are seeing a year on year decrease in their GHG emissions from their use of electricity without actually having to reduce their energy usage. It is unlikely that this can fall much further in Ireland due to the ban on the development of nuclear power, but tidal and estuary barriers could bring this conversion value down even further. This reduction has been seen throughout the OECD countries. So Governments are playing their part in reducing emissions in relation to power generation, so the next step must be for us to reduce our own use of energy.

> **Electricity generation is getting increasingly efficient in reducing the GHG emissions per kW generated year on year.**
> **It is now our turn to reduce emissions even further by adopting a positive conservation approach to electricity use.**

It is difficult to compare countries directly due to importation of fuel and often difficulty in obtaining accurate data. So the EU has set up the RE-DISS Project which stands for Reliable Disclosure System for Europe. The project aims to improve the accuracy and reliability of information on electricity generation and associated emissions by harmonizing information Europe-wide. They have produced a table showing the conversion factor for each country (Table 8.5). What is interesting is that countries with high renewable and nuclear sectors have significantly lower carbon emissions per unit of electricity supplied reflecting the lower use of fossil fuels. For example, Switzerland has the lowest emission for electricity at just 16.9 g CO_2e per kWh compared to the European average of 410 g CO_2e per kWh. The Swiss fuel mix is 48 % renewable, 48 % nuclear and just 4.6 % fossil fuels. Likewise Sweden has emissions of just 37.5 g CO_2e per kWh from their fuel mix of 38.3 % renewable, 57 % nuclear and 4.7 % fossil fuels. The downside is of course with nuclear is that you do generate radioactive waste that requires significant investment in treatment and storage. The table below gives more examples compared to the European average, showing that a low renewable component to the energy mix always results in high emission values. So for many countries **the only way to reduce emissions from electricity generation is to invest in renewable or nuclear**. Ireland like many other countries is nuclear-free putting increased pressure to invest in renewables, but unlike Scandinavian and Alpine countries that are large producers of hydro-electric power (HEP), Ireland will have to develop tidal or wave power in order to deliver reliable long term supply. However, by careful selection of cleaner fossil fuels it has managed to reduce emissions significantly while still having a high dependence on such fuel types. Germany and Japan have both recently begun to phase out their nuclear generating capacity, which has resulted in a sharp rise in their use of fossil fuels and a significant increase in their GHG emissions per unit of electricity supplied to consumers.

Table 8.5 Examples of fuel mixes and greenhouse gas (GHG) emissions per kWh of electricity supplied in (2012)

Country	Renewable energy sources (%)	Nuclear energy sources (%)	Fossil and other energy sources (%)	GHG emissions (g CO$_2$/kWh)	Radioactive waste figures (mg NW/kWh)
European average	10.88	38.35	50.77	**409.92**	1.15
France	14.1	76.8	9.1	**37.3**	2.30
UK	10.3	9.6	80.1	**563.4**	0.29
Poland	9.2	0.3	90.5	**1016.7**	0.01
Ireland	3.7	0.0	96.3	**481.0**	0.00

Also given is the amount of nuclear waste (NW) produced per kWh. *Source*: REDISS (2013). Reproduced with permission of REDISS (Reliable Disclosure Systems for Europe), European Union, Brussels, Belgium

More information: http://www.reliable-disclosure.org/

A full breakdown of emissions per kWh of electricity generated throughout Europe for 2013 is given at: http://www.reliable-disclosure.org/upload/65-RE-DISS_2013_Residual_Mix_Results_v1-0_2014-05-15.pdf

More information: http://www.reliable-disclosure.org

> **Current conversion factor for GHG emissions form electricity use in Ireland was 0.637 kg CO$_2$ per kWh in 2004**
> **It is now 0.452 kg CO$_2$ per kWh (2013) and will fall even further during the decade**

The conversion factor for GHG emissions from electricity in the United States for 2010 was 0.690 kg CO$_2$ per kWh. The fuel mix for the country is quite complex but is summarized in Table 8.6 with coal and natural gas making up 66 % of the generating capacity.

US Energy: http://www.epa.gov/cleanenergy/index.html
US Energy emission factors per fuel type: http://www.epa.gov/climateleadership/inventory/ghg-emissions.html
Energy calculator for US: http://www.epa.gov/cleanenergy/energy-resources/calculator.html

The problem with all fuels and electricity in particular is that losses occur during the transformation of fossil fuel into another form of energy including electricity (generation). Under the second law of thermodynamics conversion of heat into another energy or work will result in losses of up to 60 % of that heat! Also there are significant losses during distribution (transmission). For example, even the best gas turbines are only 50–60 % efficient. So electricity generation may not be the most efficient use of fossil fuels for a range of uses such as heating your home.

Table 8.6 Fuel mix for electricity generation in the US

Fuel	MWh (×1000) equivalent	%	Fuel	MWh (×1000) equivalent	%
Coal	1,585,998	39	Geothermal	16,517	<1
Natural gas	1,113,665	27	Petroleum oil	13,453	<1
Nuclear	789,017	19	Oil	13,410	<1
HEP	264,712	7	Other gas	12,271	<1
Wind	167,865	4	Solar	9,253	<1
Biomass	59,894	1	Other	12,355	<1
Total generating capacity 4,058,210,000 MWh					

Source: US EPA, http://www.epa.gov/cleanenergy/index.html. Reproduced with permission of the US Environmental protection Agency, Washington, DC, USA

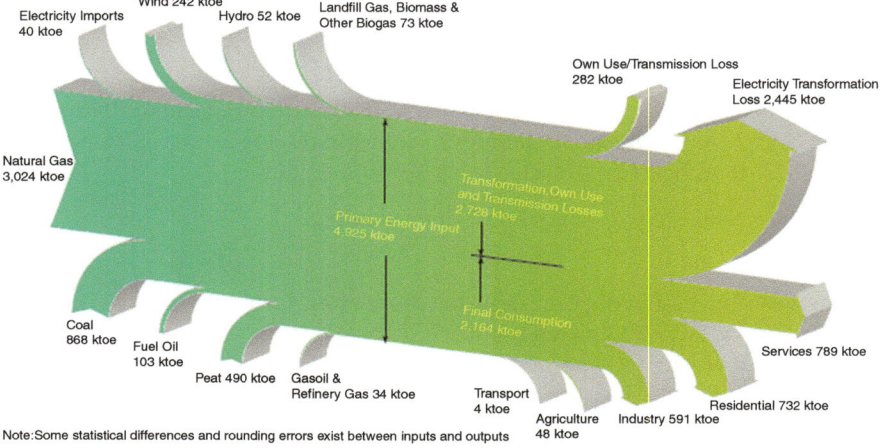

Fig. 8.2 Energy flows in electricity generation in Ireland 2010. *Source*: SEAI (2013). Reproduced with permission of the Sustainable Energy Authority of Ireland, Dublin, Ireland

The mass balance of energy flows in the generation and supply of electricity shows that in Ireland for every 4925 ktoe of primary energy used 2728 ktoe is lost and only 2164 ktoe (44 %) is actually consumed (Fig. 8.2).

> **Problem is that electricity generation using fossil fuels is a very inefficient use of these resources**

8.1.2   All Fuels

Globally we still have a massive dependency on fossil fuels (Fig. 8.3) with Ireland with a typically high dependence (95 %) with natural gas relatively less polluting in terms of GHG and other emissions, followed by oil, peat and coal.

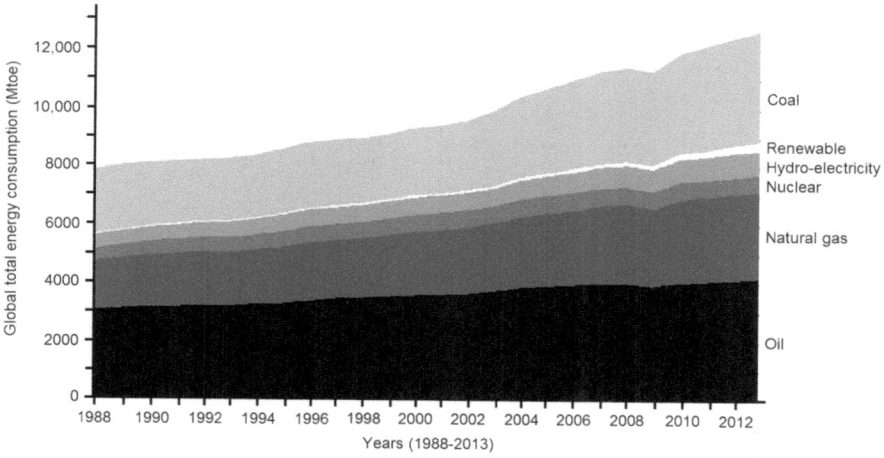

Fig. 8.3 Global energy consumption in Mtoe based on fuel type 1988–2013. *Data source*: BP (2014). Reproduced with permission of BP Statistical Review of World Energy 2014

Ireland has very little in the way of natural fossil fuel resources resulting in the country being very vulnerable to fuel dependency making the economy even more susceptible to energy price fluctuations. Compared to Norway, which has North Sea oil reserves, extensive hydro-electricity and natural gas, Ireland, as the rest of Europe, is very dependent on maintaining friendly relations with fuel supplying nations, although it is currently developing its natural gas reserves off the West coast and its inland shale gas deposits (Table 8.7; Fig. 8.4). This makes the development of renewables and energy conservation vital in terms of sovereignty and creating a long term sustainable and affordable energy supply. With North Sea gas production declining, the UK is also becoming increasingly reliant on gas imports. So shale gas could increase energy security by cutting those imports. This scenario is repeated throughout the world, with shale gas and oil sands decreasing energy dependence as well as providing a secure long term supply of energy.

Residential demand for energy in Ireland has remained more or less stable with energy efficiency offsetting an increase in population, better living standards and increased housing. The transport sector on the other hand has grown steadily reaching a peak at the height of the boom and then declining, although this trend is now in reverse. These trends can be seen in Table 8.8. In terms of share there has been a shift away from industry so that transport and residential energy usage together make up 60.1 % of all energy use. Transport energy increased by 132 % between 1990 and 2010 which is equivalent to a 4.3 % increase per annum, although during the period 1995–2000 the rate of increase peaked at 11.5 % per annum. Residential energy use increased by 6.7 % in 2010 due to cold winter but when this is seasonally adjusted we see a decrease of 2.9 %. This reflects improved thermal performance of buildings. The Kyoto agreement was originally based on capping emissions at the

Table 8.7 Where is European energy coming from?

	2001	2002	2003	2004	2005	2006	2007	2008	2009
Hard coal									
Russia	11.5	13.1	13.5	18.7	24.1	25.4	25.1	26.3	30.2
Colombia	12.5	12.6	12.5	12.1	12.1	12.0	13.0	12.5	17.6
South Africa	27.0	31.4	31.5	26.6	25.7	24.3	20.8	17.1	16.0
United States	11.2	8.2	7.0	7.5	7.8	8.0	9.3	14.3	13.7
Australia	16.3	16.9	17.0	15.3	13.5	12.4	13.5	12.0	7.6
Indonesia	5.7	6.7	7.1	7.0	7.4	9.7	7.9	7.4	7.1
Ukraine	1.6	2.0	1.3	2.0	2.1	1.6	1.7	2.2	1.6
Canada	3.8	3.2	2.9	2.5	3.3	2.8	3.1	2.7	1.4
Norway	0.9	1.0	1.2	0.6	0.6	0.3	0.6	0.6	0.8
Others	9.7	5.0	6.0	7.8	3.5	3.7	5.0	4.8	3.9
Crude oil									
Russia	25.5	29.2	31.1	32.2	32.5	33.4	33.2	31.4	33.1
Norway	20.1	19.4	19.2	18.8	16.9	15.5	15.1	15.1	15.2
Libya	8.2	7.5	8.4	8.8	8.8	9.2	9.8	9.9	9.0
Saudi Arabia	10.8	10.1	11.3	11.3	10.6	9.1	7.2	6.9	5.7
Kazakhstan	1.6	2.4	2.7	3.4	4.5	4.6	4.6	4.8	5.4
Iran	5.9	4.9	6.4	6.3	6.1	6.2	6.2	5.4	4.7
Nigeria	4.8	3.5	4.3	2.6	3.2	3.6	2.7	4.0	4.5
Azerbaijan	0.9	1.0	1.0	0.9	1.3	2.2	3.0	3.2	4.0
Iraq	3.8	3.0	1.6	2.2	2.1	2.9	3.4	3.3	3.8
Others	18.3	18.8	14.2	13.4	14.0	13.2	14.7	16.1	14.6
Natural gas									
Russia	47.7	45.0	45.1	43.8	40.6	39.3	38.4	37.6	34.2
Norway	22.8	26.2	25.5	25.0	24.4	25.5	28.2	28.9	30.7
Algeria	21.2	21.2	20.0	18.2	18.0	16.4	15.4	14.7	14.1
Qatar	0.3	0.9	0.7	1.4	1.6	1.8	2.2	2.2	4.6
Libya	0.4	0.3	0.3	0.4	1.7	2.5	3.0	2.9	2.9
Nigeria	2.3	2.2	3.1	3.7	3.5	4.3	4.7	4.0	2.4
Trinidad and Tobago	0.3	0.2	0.0	0.0	0.2	1.3	0.8	1.6	2.2
Egypt	0.0	0.0	0.0	0.0	1.6	2.5	1.8	1.7	2.1
Oman	0.4	0.4	0.2	0.5	0.6	0.3	0.1	0.1	0.4
Others	4.6	3.7	5.1	7.0	7.8	6.1	5.4	6.3	6.4

The EU and Ireland are very dependent on Russia for all its main fossil fuels who were supplying 30 % of our coal, 33 % of oil and 34 % of natural gas in 2009 with this dependence increasing.
Source: European Environment Agency, http://www.eea.europa.eu/data-and-maps. Reproduced with permission of the European Environment Agency, Copenhagen, Denmark
Source: Eurostat (online data codes: nrg_122a, nrg_123a and nrg_124a)

Net (extra-EU) imports as a % of total GIEC

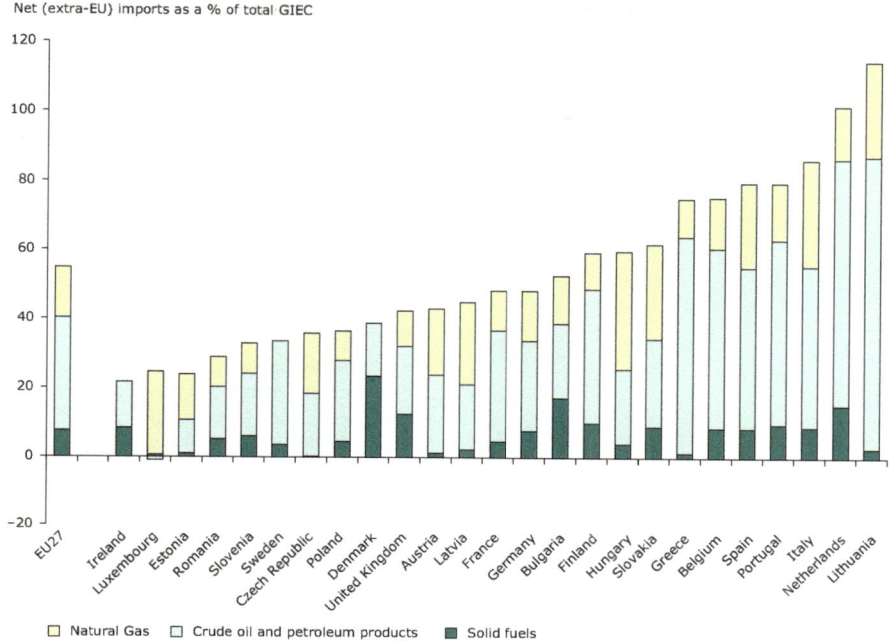

Fig. 8.4 Dependency on imported energy within Europe as a percentage of Total Gross Inland Energy Consumption (TGEC) during 2008. *Source*: European Environment Agency, http://www. eea.europa.eu/data-and-maps. Reproduced with permission of the European Environment Agency, Copenhagen, Denmark

Table 8.8 Breakdown of the total primary energy requirement (TPER) for Ireland by sector 1990–2010

	Growth %	Average annual growth rates %						Shares %	
	1990–2010	1990–2010	1990–1995	1995–2000	2000–2005	2005–2010	2010	1990	2010
Industry	13.7	0.6	3.5	4.7	−0.6	−4.7	−5.6	26.8	19.2
Transport	132.0	4.3	3.2	11.5	4.0	−1.2	−7.9	21.8	31.7
Residential	42.0	1.8	0.6	2.7	2.3	1.5	6.7	31.8	28.4
Commercial/public	83.2	3.1	2.7	5.5	4.1	0.1	13.2	16.1	18.5
Agriculture	2.3	0.1	5.4	−1.0	0.5	−4.3	5.3	3.5	2.3
Total	**55.4**	**2.2**	**2.2**	**5.4**	**2.8**	**−1.3**	**−0.3**		

SEAI (2013) Energy in Ireland 1990–2012: 2013 Report. *Source*: SEAI (2013). Reproduced with permission of the Sustainable Energy Authority of Ireland, Dublin, Ireland

Table 8.9 The energy
trilemma index for selected
countries for the year 2014

Trilemma index 2014		
Country	Score	Overall rank
Switzerland	AAA	1
Sweden	AAA	2
UK	AAA	4
USA	AAC	12
Australia	AAD	13
Ireland	ABC	22
Russia	ABD	50
China	ACD	74
India	CDD	122

The three scores are indicative of the success of
three indicators energy security, energy equity
and environmental sustainability respectively.
Source: World Energy Council http://www.
worldenergy.org/. Reproduced with permission
the World Energy Council, London, UK

1990 level, but it is clear that this is an impossible target for Ireland with only
agriculture and industry sectors showing small reductions in growth.

> **With residential and transportation representing 60 % of all energy use
> this can only be tackled by direct personal action.**

The World Energy Council has developed the concept of an energy sustainability
index, which they have called *the energy trilemma index*. The index is based on
three key energy indicators: energy security, energy equity and environmental sustainability of energy supplies. Using these indicators countries are ranked A for
excellent through to D for poor performance in each area (Table 8.9).

More information: http://www.worldenergy.org/data/trilemma-index/

8.2 Renewable Energy

Current renewable energy production in the US is just under 7 % of total energy
demand with 2.5 % of this from older hydro-electric power (HEP) systems, while
globally the use of renewable energy is less than 3 % (Table 8.1). In Ireland 4.6 %
of all energy (not just electricity) came from renewable sources in 2010 with targets
set under the EU Renewable Energy Directive of 16 % by 2020, and for 40 % renewable electricity generation by 2020. In 2007 Ireland was producing 9.4 % of electricity by renewable energy primarily by wind and HEP. By 2010 electricity

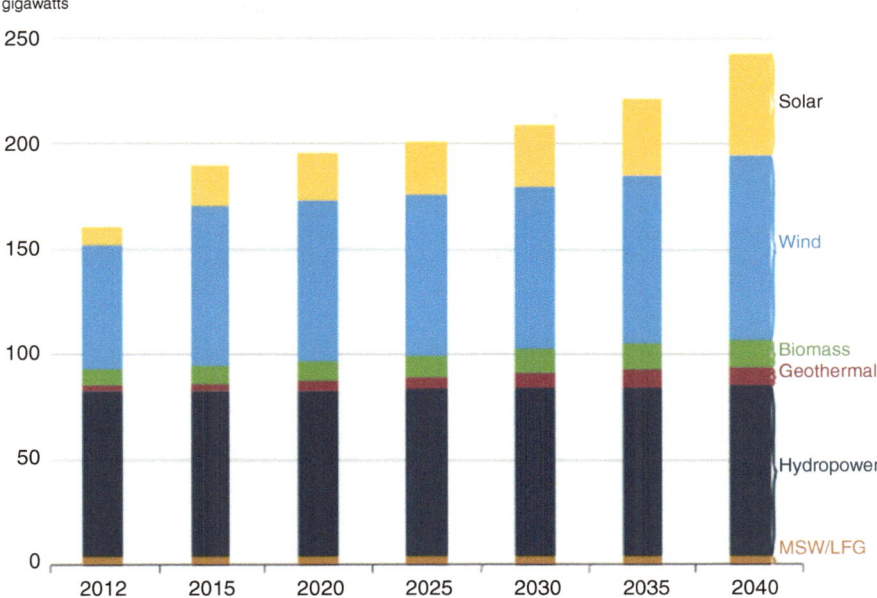

gigawatts

Fig. 8.5 Renewable energy generating capacity in the United States in 2012 and predicted growth up to 2040. Hydroelectricity generation is expected to remain static and to be overtaken as the main renewable resource in 2014. MSW/LFG is energy derived from is municipal solid waste and land-fill gas. *Source*: EIA (2014). Reproduced with permission US Energy Information Administration, Washington, DC, USA

production by renewables had risen to 14.8 % split equally between wind, biomass and solar panels, with HEP producing just 6.5 % of renewable electricity. The EU average for the use of renewables in the generation of electricity is 28 % while globally this figure is just 6.7 % with HEP making up two thirds of this (Table 8.10).

Predictions by the US Energy Information Administration (EIA) show that generating capacity from renewables is expected to rise by 52 % by 2040 with most of that due to expansion in wind and solar power which in 2012 were generating 60 and 8 GW per annum increasing to 87 and 48 GW by 2040 respectively (EIA, 2014) (Fig. 8.5). Globally, renewables are generating 502 billion kWh per year (2012).

More information on US energy predictions: *EIA (2014) Annual energy outlook 2014 with projections to 2040. Report: DOE/EIA-0383(2014), US Energy Information Administration, Washington, DC.* http://www.eia.gov/forecasts/aeo/pdf/0383(2014).pdf

Renewable energy has a number of fundamental problems that need to be addressed. First just how do we capture energy from wind, sun, waves, or crops economically and at high enough amounts to sustain all of our energy needs? Where we do have large renewable installations these are often in remote or inhospitable

Table 8.10 Percentage of renewable energy in total electricity production (2013)

Country	%	Country	%
Norway	97.9	Canada	62.7
Brazil	77.1	Portugal	62.5
New Zealand	74.0	Sweden	53.2
Columbia	73.7	Spain	40.8
Venezuela	68.8	Italy	38.8

The table shows the top 10 countries and includes all renewables including HEP. *Source*: World Energy Council http://www.worldenergy.org/. Reproduced with permission the World Energy Council, London, UK

areas, so another problem is delivering this energy from areas of production (i.e. from solar farm located in deserts, tidal power from exposed coasts, and offshore wind farms) to the place of demand. Another problem is the form of the energy generated. Wind farms are ideal as they drive turbines that generate electricity, but often it is difficult or costly to convert renewable energy into a useable form. For example, even electricity has to be loaded into cars via batteries or as hydrogen. Of course renewable energy is dependant largely on weather conditions so we need a base supply when it is not windy, too dry or dark. This was reflected in a fall in generation from wind turbines in Ireland due to a lack of wind in 2010. When competing directly with fossil fuels, renewables are only competitive at the present time when given subsidies and the tendency by governments is not to currently subsidize new projects due to a proliferation of fossil fuel generating capacity.

What are the current renewable options? In terms of generating we have **solar thermal** which is heating water via a radiator system often located on the roof, which reduces the energy used in heating water in the home and can also sometimes help in heating the house. **Wind power** generates electricity directly but is only effective when the wind is blowing. **Solar photovoltaic** is relatively expensive and has a low emission rate. It trickle feeds the electricity distribution network and on a sunny day can operate basic things such as lights and the fridge. Storage is via batteries which can be expensive in terms of life compared to recovery of investment. **Ocean wave** power is still in its infancy but shows significant promise. **Geothermal** collected heat from the soil and is the third most popular household generating system but problems have been encountered with permafrost development around the extraction system in some areas. Like wind turbines, **hydro-electric** schemes are increasingly common and offer a reliable source of continuous electricity. They vary in size to huge national schemes to small schemes operated by farmers and other landowners. These are efficient systems if used one way, the older pump storage systems where the water is released from an upper storage reservoir to a lower one to generate power at periods of peak demand, and then the water is pumped back up to the upper reservoir during periods of low demand. These are only efficient when renewable energy can be used to operate the return pumps. The use of estuarine barriers to use tidal power to replenish storage areas is perhaps one of the most exciting prospects for reliable, on-demand electricity generation, but may come at high environmental costs in terms of biodiversity losses, especially migratory species and wildfowl. **Biomass** is currently the preferred option for power generation and many larger power stations are being converted to biomass. The actual efficacy of this in terms of GHG balances is still under debate, but biomass and wood pellet boilers are commonly used to heat individual homes, hotels, schools and small businesses. Liquid transportation fuels (**ethanol/hydrogen**) do offer a small saving in terms of emissions over oil, but the balance of taking prime land out of food production for growing biofuels is a contentious one.

Apart from generating energy, like electricity, storing and delivering energy from renewables is difficult with most renewables except for biomass, ethanol and hydrogen unable to be stored efficiently. Just like electricity they have to be generated on demand. Currently electricity can only be stored in vehicle batteries, fuel cells or stationary batteries, alternatively the energy could be stored as compressed air or ice storage with the embedded energy being converted back to electricity when needed. All storage systems are inefficient with transmission always a problem with minimum losses of 5–10 % likely.

It is estimated that by 2020 the EU will reach its interim target of producing 20 % of its energy by renewables. However, when the current renewable sectors are examined (Fig. 8.6) then it is clear that there is a huge dependence on biomass, mainly for heat and to a lesser extent for the generation of electricity. While willow

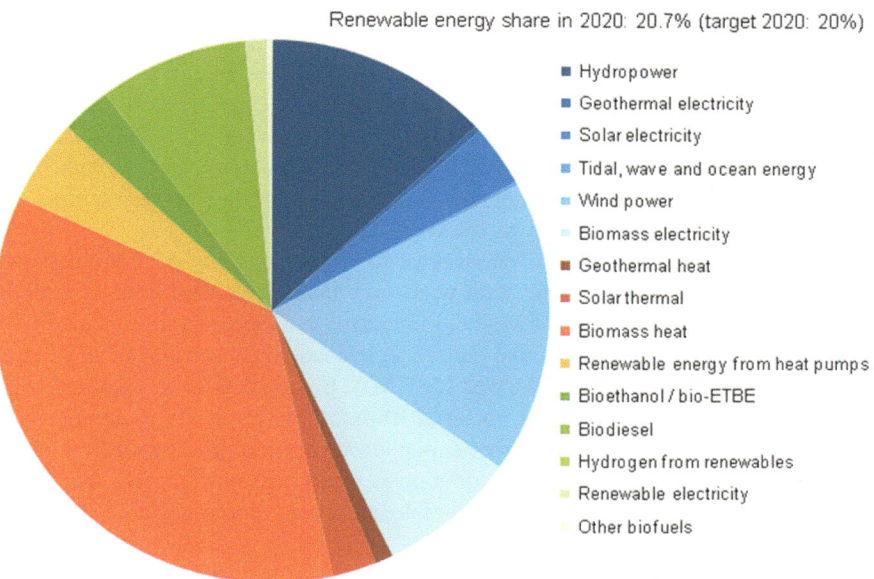

Renewable energy share in 2020: 20.7% (target 2020: 20%)

- Hydropower
- Geothermal electricity
- Solar electricity
- Tidal, wave and ocean energy
- Wind power
- Biomass electricity
- Geothermal heat
- Solar thermal
- Biomass heat
- Renewable energy from heat pumps
- Bioethanol / bio-ETBE
- Biodiesel
- Hydrogen from renewables
- Renewable electricity
- Other biofuels

Fig. 8.6 Expected renewable energy share in EU by 2020 The 20 % target is expected to be easily exceeded. *Source*: European Environment Agency, http://www.eea.europa.eu/data-and-maps. Reproduced with permission of the European Environment Agency, Copenhagen, Denmark

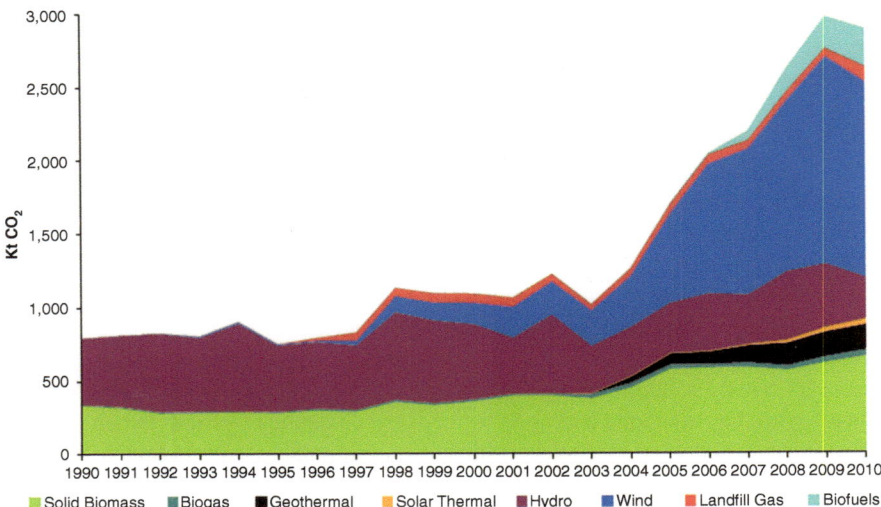

Fig. 8.7 Greenhouse gas emissions saved by the use of renewable energy systems in Ireland 1990–2010 in thousands of tonnes of CO_2e (Kt CO_2e). *Source*: SEAI (2013). Reproduced with permission of the Sustainable Energy Authority of Ireland, Dublin, Ireland

and the grass miscanthus are both widely grown for biomass, the majority of biomass comes from forestry.

However, don't underestimate the usefulness of renewable energy in our fight to reduce GHG emissions. In Ireland during 2010 some 2.89 million tonnes of CO_2e (Mt CO_2e) were saved by the use of renewables: 1.33 Mt CO_2e by wind generation, 0.63 Mt CO_2e from solid biomass and 0.28 Mt CO_2e from HEP (Fig. 8.7). The biomass sector is growing rapidly creating an enormous demand for timber, something reflected globally. In the UK, biomass electricity generating power stations are sourcing timber from as far away as the USA.

In Ireland the contribution of renewables to gross electricity generation shows that HEP output is largely static with wind now taken over as the major source. However, the unpredictability of wind output means that there is a significant difference between theoretical and actual output. The installed capacity of wind generation in Ireland by September 2011 was 1585 MW. Since 2006 the capacity of wind generated electricity has doubled but in 2010 output was only 25 % compared to relative output in 2006, so this makes dependence on wind power problematic for suppliers (Fig. 8.8). In contrast sea, tidal and estuary barrages all give a constant reliable output which is why such schemes are so attractive as renewable electricity sources.

Renewables can only really affect electricity usage at present, but we are still very dependent on other fuels apart from electricity at the household level for space heating and transportation. Although that dependence has been reduced in most countries, countries that have extensive HEP still retain a high dependence on electricity. Ireland is typical in the high level of electricity used (Fig. 8.9).

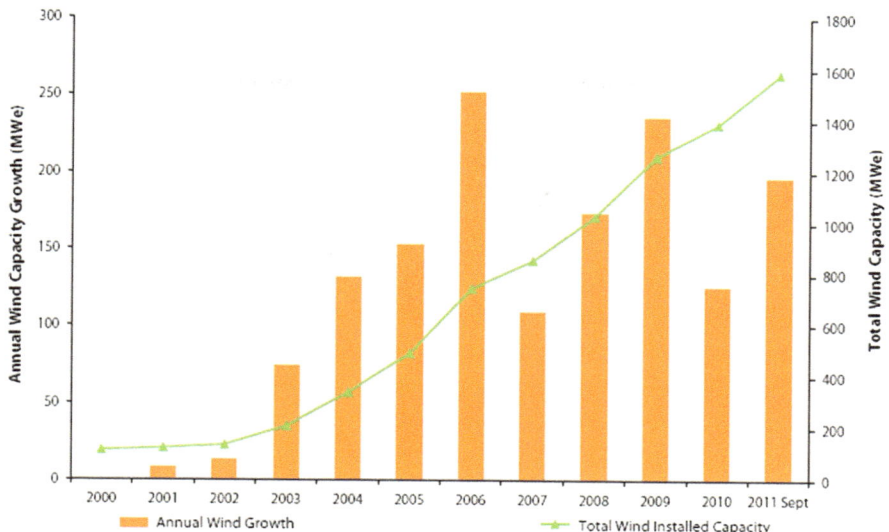

Fig. 8.8 Comparison of annual growth in wind generation and growth in total capacity. *Source*: SEAI (2013). Reproduced with permission of the Sustainable Energy Authority of Ireland, Dublin, Ireland

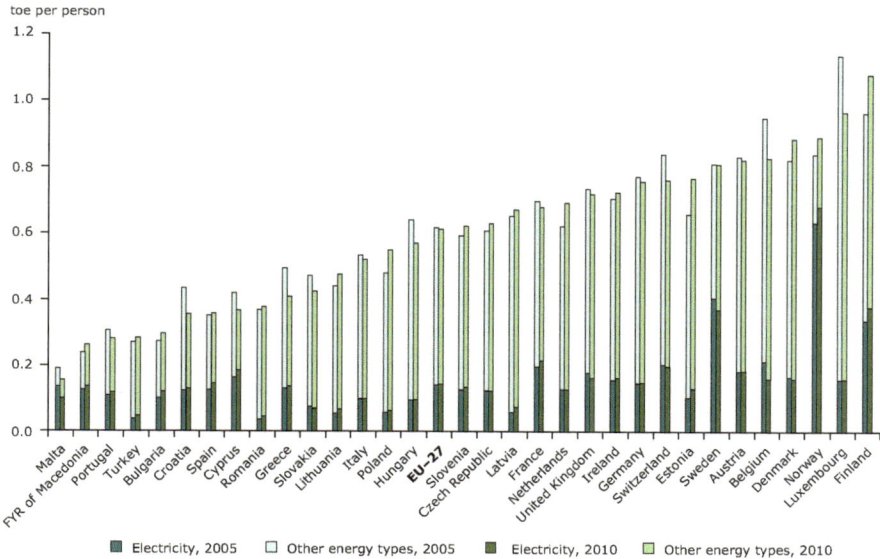

Fig. 8.9 The average per capita final energy consumption (in tonnes of oil equivalent) of households in 2005 and 2010, divided into electricity consumption and other energy types. *Source*: European Environment Agency, http://www.eea.europa.eu/data-and-maps. Reproduced with permission of the European Environment Agency, Copenhagen, Denmark

A UK Government survey published by the UK Energy Research Centre in September 2013 has shown that support by the public for renewable energy seems to be declining, although it remains the most favoured method of generating electricity. Over the period 2005–2013 support for wind generation had fallen from 82 to 64 % and for solar from 87 to 77 %. In Japan 23 % of people want to see nuclear energy phased out immediately while a further 53 % want to see it gradually phased out, with only 17 % in favour of generating electricity using nuclear compared to 33 % in 2007. This is not surprising since the major accident in 2011 at Fukushima. However the incident has had no effect on how the UK public perceive nuclear energy with those supporting it rising from 26 % in 2005 to 32 % in 2013. Even those who totally oppose it have fallen over the same period from 37 to 29 % splitting the national into two halves in terms of their support for nuclear power.

8.3   The Nuclear Debate

The electricity industry throughout Europe is undergoing modernization and decarbonising (i.e. reduction and capture of CO_2 emissions of coal and oil-fired power stations). Many coal fire power stations have already closed in the UK. The idea is that the generating capacity lost will be replaced by four nuclear power stations due to be decommissioned by 2019. However, in the interim the UK, like many other countries, is facing a generating shortfall that may cause power shortages at peak periods. Currently the UK generating power is 77.9 gigawatts (GW) which is well above the maximum peak demand of 60 GW, although some of this capacity is wind dependent. However a raft of new European legislation limiting emissions from power stations may well undercut this current safety margin in capacity. The Large Combustion Plant Directive (LCPD) which covers all major generating power plants within the European Union is designed to limit the emissions of SO_2, NO_X and dust during operation. The investment to retrofit these plants to reduce emissions is very high and so the Directive allows for operators to 'opt out' which many countries have done; but this involves closing these plants by 31 December 2015 or after operating for 20,000 h from the 1 January 2008. So in the UK a number of oil and coal plants are scheduled to close under the opt out clause under the LCPD, with 11.5 GW of capacity being lost. Further legislation is on its way including the Industrial Emissions Directive (IED) which consolidates all existing directives and sets even more stringent emissions standards which comes into force in 2016 requiring all plants to fully comply by 2020. Again, plants can opt out and will be allowed to operate for a further maximum of 17,500 h or until 2023 followed by plant closure.

Currently the UK produces 19 % of its electricity from nuclear energy exactly the same as the United States. However a significant portion of these plants are also ending their operational life. Many have already been given extended operational deadlines but four plants will have to close by 2019 taking 3950 MW out of generating capacity alone. To offset this loss and also the loss in generating power due to

Table 8.11 Current UK nuclear power stations and scheduled closure dates

Plant	Capacity (MW)	Published closure date
Wylfa	490	2014
Dungeness B	1110	2018
Heysham 1	1160	2019
Hartlepool	1190	2019
Torness	1250	2023
Heysham 2	1250	2023
Hinkley point B	1220	2023
Hunterston B	1190	2023
Sizewell B	1188	2035

Source: Department of Energy and Climate Change https://www.gov.uk/government/policies/maintaining-uk-energy-security--2. Reproduced with permission H.M. Government, UK

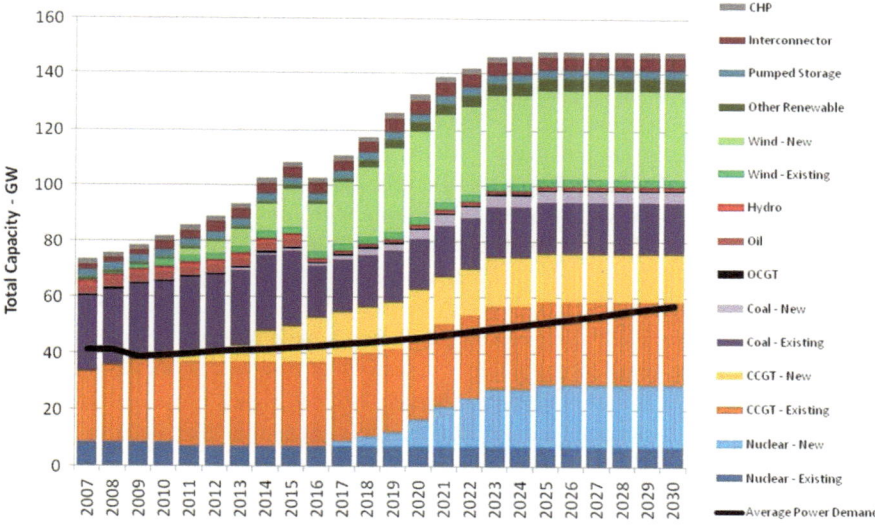

Fig. 8.10 Expected sources of power generation in the UK up to 2030. *Source*: National Grid (2011). Reproduced with permission of the National Grid plc, Warwick, UK

oil and coal plants having to close under the LCPD, the UK plans to invest £930 bn over the next twelve years until 2025 in order to generate 16 GW of new electricity by nuclear power (Table 8.11; Fig. 8.10). Although once a leader in the field with its first Magnox reactor opened at Calder Hall in 1956 no new plant has been built since the new reactor at Sizewell B started generating electricity in 1995. So the UK is now reliant on countries such as France, Japan and China for the technical expertise to develop its largely neglected nuclear industry. The French state-backed company EDF, along with Chinese partners, is to build two new reactors at Hinkley Point costing £16 billion which will be able to supply approximately 5 % of the UK electricity demand. It will use the proven and reliable European Pressurised Reactor design which is also used at the Flamanville Plant in France and also at Olkiluoto in Finland and should come on stream by 2023, although the cost of the energy will be 50 % more than the current wholesale price for electricity. However, the rest of the world seems to be going in different directions in regards to nuclear energy. Germany have begun withdrawing from nuclear power as has Japan, since the Fukushima disaster, while China, which built its first reactor just 20 years ago, now has 17 reactors in operation with a further 28 under construction. Ireland in contrast has no nuclear power plants and has declared itself nuclear free as far as electricity generation is concerned. It is also concerned about potential safety issues arising from any new plants that are planned in the UK as well as existing installations and has an ongoing dispute with the UK Government over the radioactive emissions from the nuclear fuel reprocessing plant at Sellafield into the Irish Sea

More information on the future of power supplies in UK: http://www.raeng.org. uk/news/publications/list/reports/RAEng_GB_Electricity_capacity_margin_ report.pdf

In my opinion it is not possible to meet carbon reduction targets and keep the lights on without nuclear energy. My students for the past 30 years will attest that as a committed environmentalist I have always been pro-nuclear energy for that very reason. We need a reliable backup energy source for renewables that can be geared up and down rapidly, which is not really possible with fossil fuel power stations. It all comes down to risk and at the moment the risk from global warming far outweighs the risks associated with nuclear energy. We have been overwhelmed by negativity about nuclear power rather nicely summed up by the ongoing Simpsons cartoon. It is possible that The Simpsons alone may have wrongly informed a whole generation about the safety of the nuclear industry? Of course it is not without risks as we saw in 1986 at Chernobyl and in March 2011 at Fuskushima following the Great Tohoku earthquake and tsunami. But the facts tell us that the risk from nuclear energy is far less compared to fossil fuel energy generation. Every day 3500 people, mainly young children, die due to atmospheric contamination released from fossil fuelled power stations, and a report by NASA suggests that 1.8 million people have been saved by replacing such plants with nuclear generation. Regardless of what we would like to believe the United Nations, who has extensively studied the effects of the Chernobyl incident, has concluded that 50 people died from the incident in total, mainly those working to cap the reactor, as well as a number of children who subsequently tragically developed thyroid cancer. There were no abnormal births or developmental cancers. There was however, widespread post traumatic stress and other psychological problems for those living close to the plant, similar to that seen in a war zone. So there were problems but these are the same sort of problems that are encountered after any tragic event of this scale. No-one died from radiation after the earthquake and tsunami hit Fuskushima. In fact over the past 50 years three times more people have died from incidents associated from the generation of energy from wind than from nuclear. We are constantly exposed to scare stories about radiation and end up with a completely misguided and incorrect idea about the real risks and benefits associated with nuclear energy generation. Even a coal fired power station will emit more radiation than a nuclear plant due to its production of fly ash, and some experts believe that it could be as much as a 100 times more radiation that is eventually emitted. I am not trying to trivialize people's concerns, but we have to realize that if we want to continue with something similar to our current economic model and lifestyle, then we need to either accept an enormous reduction in the availability of energy or we invest in a large range of low-carbon technologies, including nuclear.

> **We have to decarbonize our energy supplies globally if we are to stop the worst effects of global warming.**
> **It is impossible to replace fossil fuels with renewables without using nuclear energy in the timescale we have left to us.**

Our problem is that we have poured billions of dollars, pounds, yen and euro into physics research, several billion into CERN alone, but have hardly invested anything in comparison in developing more efficient and safer reactors, nor in the reuse of spent fuels. Why, because it's not as exciting as other areas of physics and, remember The Simpsons cartoon, nuclear research has a high negativity associated with it. Today the much of modern nuclear waste doesn't come from power stations it comes from medical and industrial uses, so we are always going to be generating nuclear waste of some kind, we just have to start investing more money in finding appropriate solutions. Nuclear isn't the threat to the survival humankind here, it CO_2 emissions from fossil fuel power stations, but it could be the only way to produce near zero carbon electricity. So we need to get positive about the possibilities.

For more information watch the film *Pandora's Promise* a documentary film produced by Robert Stone and released in 2013 which looks at the nuclear debate.

Film link: http://pandoraspromise.com/

8.4   Household Energy Use and CO₂e Emissions

The majority of fossil fuel consumption in the domestic sector is for **thermal use**, which is predominantly for space and water heating with the remainder used for cooking (Fig. 8.11). Electricity accounts for all **non-thermal uses** including lighting, domestic and kitchen appliances, TV and other household goods.

Oil, gas and solid fuels are mainly used for space and water heating being generally more efficient in terms of GHG emissions for thermal uses. However there are exceptions where electricity is used for this purpose. For example, in apartments and maisonettes predominantly located in larger urban areas. Electricity is also used

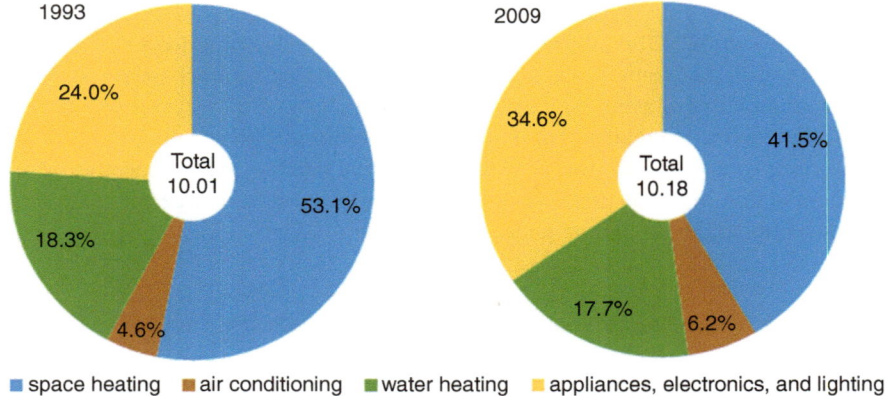

Fig. 8.11 Space heating remains the major energy use as can be seen from household energy usage figures for the United States. The comparison between 1993 and 2009 shows that while space heating takes up less of the energy household budget the use of appliances has increased. However, overall total energy use per household is on the decline. Total energy is in quadrillion BTU. *Source*: EIA http://www.eia.gov/todayinenergy/. Reproduced with permission US Energy Information Administration, Washington, DC, USA

for space and water heating in older buildings constructed during periods when electricity was relatively cheap and in regions where other fuels, particularly natural gas, was not a viable option. HVAC (Heating Ventilation, Air Conditioning) can be considered to be part thermal, part non-thermal, although it is not widely used in temperate countries such as the UK and Ireland.

A household emission survey of 103 Irish homes with occupancy rates of between 1 and 6 people (average 2.9) recorded that 42.2 % of GHG emissions were derived from home energy use (Kenny and Gray, 2009). The study showed that household energy consumption per capita became more efficient as the occupancy rate increased which was also true for transportation.

More information: Kenny, T. and Gray, N.F. (2009) A preliminary survey of household and personal carbon dioxide emissions in Ireland. *Environment International*, **35**, 259–272.

According to the Department of the Environment, Food and Rural Affairs (Defra), the average annual electricity consumption in the UK is 4800 kWh per household. This varies from 3000 kWh for a small household to 7000 kWh for a large one. In the UK the CO_2 emission conversion factor used for electricity is 0.527 kg CO_2e per kWh which includes an allowance for the 7.5 % of losses on the national grid and other inefficiencies that occur before electricity reaches the end user. So excluding space and water (thermal) heating the average domestic electricity use can be broken down as:

19 % Consumer electronics
19 % Lighting

18 % Cold appliances
15 % Cooking appliances
15 % Wet appliances
9 % Domestic ICT (Information and Communications Technology)
5 % Other

The average UK annual gas consumption is 16,000 kWh per household with a GHG conversion factor for natural gas of 0.203 kg CO$_2$e per kWh. The conversion factor for heating oil (kerosene or paraffin) is 2.96 kg CO$_2$e per litre

---

UK GHG conversion factors
Electricity **0.527 kg CO$_2$e per kWh**
Natural gas **0.203 kg CO$_2$e per kWh**
Heating oil **2.96 kg CO$_2$e per litre**

8.4.1   Home Energy Measurements

Energy usage in the home can be calculated by reading the energy rating label on the back or underside of the appliance. The label should give the amount of energy an appliance uses in Watts. You may remember from school that wattage = current × voltage, but we don't have to worry about that here, it's just the wattage that is important.

There is a thousand Watts in a kilowatt (kW) and consumption is measured in kilowatts used per hour (kWh), so the consumption (C) in kWh per year is:

$$C = \left(\text{Watts} \times \text{hours used per day} \times 365\right)/1000 \, \text{kWh per year}$$

We have divided the total amount of watts used per year by 1000 in order to transform Watts into kilowatts.

So if you use a plasma TV with an energy rating of 450 W and you watch it for on average 2 h a night the your consumption is:

$$\left(\text{Watts} \times \text{hours used per day} \times 365\right)/1000$$

$$\left(450 \times 2 \times 365\right)/1000 = \textbf{328.5} \text{kWh per year}$$

The conversion factor for turning kWh into GHG emissions (kg CO$_2$e) in the UK is 0.527 kg CO$_2$e per kWh. Thus we multiply kWh by the conversion factor, so the energy used watching your plasma TV for 2 h every night for a year generates:

$$328.5 \times 0.527 = \textbf{173.1 kg CO}_2\textbf{e}$$

The energy used by electrical appliances can also be measured directly using a plug in Watt Electricity Usage Monitor or similar meter. Alternatively calculators that give the energy usage of different appliances are available online or as apps.

Most energy suppliers offer advice on saving energy and so saving money. In Ireland the main electricity supplier, Electric Ireland, is very proactive in encouraging customers to save energy through a wide range of energy saving advice online. One of their most useful tools is the Electric Ireland energy calculator. This lists all appliances and heating options (non-thermal and thermal) and provides a detailed breakdown on how much each of your home electrical appliances and lights cost to run. It is completely interactive and allows you to compare the cost of using appliances in different ways, for example, washing clothes at 40 °C rather than 60 °C. It also allows you to explore the positive impacts of using your appliances wisely (e.g. using a small ring on your electric cooker rather than a larger one).

Electric Ireland energy calculator: http://www.esb.ie/esbcustomersupply/
   residential-energy-services/reduce-your-costs/web-calculator.jsp
Also available via Smartphone: http://www.esb.ie/esbcustomersupply/residential-
   energy-services/reduce-your-costs/smartphone-calculator.jsp

Electric Ireland also provides other interactive tools and energy saving advice on line including the **Energy Efficient House** which gives you an interactive tour of a typical two story house and indicates where energy can be saved. The **ESB Energy Saving Wizard** is an interactive energy conservation package with a range of online tool that guides you through a series of questions about the house and the answers generate energy saving recommendations personalised to your home. The recommendations also indicate costs, grants available and potential savings.

Energy Efficient House: http://www.esb.ie/esbcustomersupply/residential-energy-
   services/reduce-your-costs/energy-efficient-house.jsp
ESB Energy Saving Wizard: http://www.esb.ie/esbcustomersupply/residential-
   energy-services/energy-saving-wizard/index.jsp

8.4.2 Is Standby Really a Problem?

The straight answer is globally yes. When appliances are left plugged in then they usually are using a very small amount of energy, even when they appear completely shut down. This can account for between 5 and 10 % of residential electricity usage in most developed countries. Standby power in commercial buildings is smaller but still significant, so approximately 1 % of global CO_2e emissions are used for standby functions.

So why don't we turn all appliances off? Is standby power use really necessary?
Certain appliance functions do require to be left on permanently in standby mode and so use small amounts of electricity all the time. These include maintaining signal reception capability (for remote control, telephone or network signals), monitoring temperature or other conditions (such as in a refrigerator), powering an internal clock (although in personal computers this is done by a battery), battery charging and continuous display. In reality very few appliances except for your freezer and refrigerator need to be left on. Indeed there is a real risk of fire from leaving certain electrical appliances plugged in when not in use. A personal computer is not worn out quicker nor is the life expectancy of the monitor reduced if it is turned off after use. My PCs always seem to last years longer than my colleagues and I turn mine off at least once a day if not more frequently. One thing that confuses me is my TV satellite box. In the earlier version of the box the introduction channel which it defaulted into when I switched it on used to tell me to not leave the box on standby … 'if we all turned off our box it would save enough energy to power all the lights in Birmingham'… it used to say repeatedly or something to that effect. My new HD box doesn't work … why? I am told by my installer that it is because if I don't leave in on all night in standby then it is unable to download regular software updates! This means that the box slowly stops working properly if not permanently on. I notice that they no longer suggest on the introduction channel that we could save lots of energy by not using standby! Making products that require us to leave them indefinitely plugged in continuously using electricity is undesirable in terms of safety as well as adding significantly to greenhouse gas emissions.

More information: http://standby.lbl.gov/standby.html

Energy Star products are a global approach to minimizing standby energy use, but does not eliminate them. It is important to ensure that any products that do need to be left on carries this logo as it ensures that minimum electricity will be used. http://www.energystar.gov/

8.4.3   Turning Off Desktop PCs

When it comes to computers then substantial power savings can only be made by powering down desktop PCs when they are not needed. Ideally this is done simply through cultural change, that is ensuring desktop users turn off their PCs at the end of the day to support an organizational goal of reducing unnecessary power consumption.

A desktop computer uses approximately 80 W and an LCD screen 25 W that's just 105 W or 0.105 kW so on an individual basis it is not a huge amount. But just like the issues with leaving equipment on standby or leaving lights on in a room when no-one is using it, a balance needs to be struck between leaving the machine on and the actual time it is used. So if 1 kWh results in 0.537 kg CO_2e (i.e. the conversion factor) and if we assume an office computer is used for 8 h per day for 226 days a year (assuming 25 days holidays, weekends and bank holidays) then we can make a simple estimate of the amount of emissions generated per PC.

So the gross saving when a single PC is shut down after the full daily 8 h shift is 392 kg CO_2e per year (CO_2e year^{-1}), which is over a third of a tonne of CO_2e. So does this make a difference?

Well there are 48 PCs in the main library office here at Trinity College which have to be left on 24 h a day, 7 days a week, although the staff work normal 8 h shift patterns. If these were closed down at the end of each shift then the potential savings in that one office area would 18.8 tonnes CO_2e year^{-1}. There are some 4000 desktops in Trinity College which offers a theoretical potential maximum saving of 1568 tonnes CO_2e year^{-1}. Currently Electricity Ireland charges €0.1699 per kWh (November, 2013) so the potential saving to College is €496,108, so everyone wins. The number of desktop PC's in UK universities is estimated at 1,470,000 with a potential annual GHG emissions saving if they were all turned off when not in use approximately 570,000 tonnes CO_2e year^{-1} and a potential saving at 0.12p per kWh of £120 million per year.

Desktop computers normally consume between 1 and 5 W when switched off, in standby (ACPI S3) or full hibernate modes. The actual savings that an organization can achieve will be dependent on local conditions such as the power that desktop computers run at; the cost of a unit (kWh) of electricity; the amount of time computers are actually used over a year; whether power management is currently implemented (e.g. policies may already be in place to switch computers into sleep/standby (ACPI-S3), hibernate and off); and the need for continuous running (hospital and scientific equipment etc.). Power management monitoring (PMM) (e.g. FiDo software) and wake on LAN (WOL) facilities that automatically turn on computers and turn them off significantly reduce emissions.

PC screen savers do not save any electricity.

To calculate the power used by a PC:

$$\text{Power used} = \text{power consumed} \times \text{run time}$$

Where the power used by the computer is 0.105 W and the conversion factor from kWh to Kg CO_2e is 0.537 kg CO_2e kWh^{-1}

If **left on all the time** the actual power used per computer is:

$$\text{Power used} = 0.105 \text{ Watts} \times 24\text{hours} \times 365 \text{ days}$$

$$= \textbf{920kWh per annum per PC}$$

If **turned off when not in use** the actual power used per computer is:

$$\text{Power used} = 0.105 \text{ Watts} \times 8\text{hours} \times 226 \text{ days}$$

$$= \textbf{190 kWh per annum per PC}$$

To convert these values to GHG emissions then:
Emissions = power consumed × conversion factor

PC permanently left on emissions = 920kWh × 0.537 kg CO_2e kWh^{-1}

$$= \textbf{494 kg CO}_2\textbf{e per annum per PC}$$

PC turned off when not in use emissions = 190kWh × 0.537 kg CO_2e kWh^{-1}

$$= \textbf{102 kg CO}_2\textbf{e per annum per PC}$$

8.5   Energy Targets

Energy reduction targets have been set by the IPCC and agreed under the Kyoto and other protocols (Sect. 4.5). Every country has adopted a policy framework to help achieve reduction targets, whether a signatory or not. In Ireland, the Government's energy policy framework (2007–2020) '*Delivering a Sustainable Energy Future for Ireland*' (March 2007) placed sustainability of energy supply and its use at the heart of Irish energy policy objectives. One of the key strategic goals to achieve sustainability is maximising energy efficiency. In order to meet Ireland's sustainable energy goals a number of targets have been set:

➢ An overall national energy-saving target of 9 % for 2016, as part of the *EU Energy End Use Efficiency and Energy Services Directive* (ESD).
➢ A 20 % savings in energy across the electricity, transport, and heating sectors by 2020 as outlined in Ireland's *National Energy Efficiency Action Plan* (NEEAP).
➢ An indicative 30 % energy-saving target for Ireland by 2020 to surpass the EU goals.

While these are very challenging targets requiring significant investment in energy-efficient equipment, it also **requires a massive change in customer attitudes to the use and conservation of energy**. The Government has also proposed

Table 8.12 Proposed energy saving actions to reduce household energy demand

Market	Fuel	Key cost-effective measures
Existing homes	Electric	Compact Fluorescent Lamp (CFLs), energy efficient floor lamps, proper sizing of Central Air Conditioning, high efficiency appliances, towel warmer timers, low flow measures, tank wraps, ceiling insulation, ceiling and wall insulation, duct[a] diagnostics and repair
Existing homes	Oil and gas	Condensing boiler, ceiling insulation, duct insulation, duct repair, Heating Ventilation and Air Conditioning (HVAC) diagnostics and repair, programmable thermostats, water heater blankets, high-efficiency water heaters, low-flow showerheads
New construction	Electric	CFLs, energy efficient floor lamps, proper HVAC sizing, high-efficiency water heater, high efficiency clothes washer, high efficiency dishwasher, pipe wrap
New and existing homes	Oil and gas	High-efficiency appliances, water heater blankets, programmable thermostats, low-flow measures, tank and pipe wraps, ceiling insulation, high-efficiency and condensing boilers, HVAC diagnostics and repair, duct

Source: SEAI (2013). Reproduced with permission of the Sustainable Energy Authority of Ireland, Dublin, Ireland
[a]Ducts are used in HVAC systems to deliver and remove air

energy saving actions to reduce household energy demand, however, the problem with these measures is that they have largely been implemented either in new builds or as retrofits (Table 8.12). **Without accepting the fact that actual use of energy must significantly be reduced then meaningful emissions reduction targets are impossible to achieve in the medium to long term**. Little has actually been achieved in reducing demand for energy. Once these relatively easy actions have been implemented reductions can only be achieved by increasingly expensive and challenging steps.

> **Ireland's sustainable energy policy is based on maximizing energy efficiency rather than reducing actual use and dependency.**

The Friends of the Earth in Ireland have suggested that massive cuts in emissions as well as financial savings could be made by adopting more rigorous energy saving in schools, hospitals and other public offices. It especially highlights buildings that are heated when empty during weekends and leaving computers switched on when not in use. It estimates that it could save the state 1.75 billion euro over 10 years and calls for retrofitting such buildings with better insulation, reducing dependence on air conditioning and using low-carbon heating systems. However, small changes can make a difference. A school in Clara in County Clare was able to reduce its electricity usage by 70 % by simple actions such as lowering the heating setting on the central heating, using time switches for heating rooms that were not used all the time, turning off the boiler at night and weekends, and encouraging staff and pupils

Table 8.13 Possible personal emission reduction targets in tonnes CO_2e for Irish individuals based on current footprint

Emission source	Current usage		20 % reduction	50 % reduction	80 % reduction
	%	Tonnes	Tonnes		
Home energy	**42.2**	**2.4**	**1.92**	**1.20**	**0.48**
Transportation	35.1	2.0	1.60	1.00	0.40
Aviation	20.6	1.2	0.96	0.60	0.24
Waste	2.1	0.1	0.08	0.05	0.02
Total		**5.70**	**4.56**	**2.85**	**1.14**

to close doors and turn off lights. Plans are already in place by the Government to reduce electricity use in over 10,000 buildings run by central and local government under the National Energy Efficiency Action Plan. It is claimed that the plan has already reduce energy usage by 33 % across the public sector.

More information: *Friends of the Earth (2013) 'Cuts that don't hurt'. Friends of the Earth, Dublin.* http://www.foe.ie/download/pdf/dublin_foe_cuts_that_dont_hurt.pdf

More information: *The second National Energy Plan launched in February 2013.* http://www.dcenr.gov.ie/NR/rdonlyres/B18E125F-66B1-4715-9B72-70F0284AEE42/0/2013_0206_NEEAP_PublishedversionforWeb.pdf

8.5.1   Personal Targets

How do we reduce our emissions? Governments will eventually have to pass on the requirement to meet emission targets to us either directly or indirectly. So over time it will become increasingly difficult to meet higher reduction targets. Also these reductions must be real and not simply offsets. So how do we set initial limits? For some limits will be too low and for other too high, so how can it be done?

The problem is inequality as the degree of difficulty for individuals to meet interim targets really depends on your original energy usage. Therefore this approach favours rich countries and disadvantages poorly developed countries. So, should we go for percentage reduction of carbon limit per capita as in Table 8.13? Whatever method we adopt it is clear that a major area for reduction is household energy usage. This is something everyone must take part in. Selfishness will only lead to central Government taking action and that can only mean rationing of energy which could mean switching the power off at periods of high demand. This is something we look at in detail in Chap. 15.

> **Effective energy conservation starts by setting personal reduction targets for electricity use in the home**

8.6 Conclusions

❖ Great strides in reducing emissions has already been made in the fuel mixes and efficiency of power stations reducing generating emissions in Ireland for example by 46 % from 0.896 kg CO_2e per kWh in 1990 to just 0.452 kg CO_2e per kWh in 2013. This isn't just an Irish phenomenon; it is being done throughout most of the developed world.

❖ Demand however just goes on increasing with more than half of the energy that has been consumed since the industrial revolution having been used in the last two decades, despite advances in efficiency and sustainability.

❖ We have to reconsider our concerns over nuclear energy and invest more in its development into an even more efficient and cleaner fuel.

❖ To achieve reduction targets for GHG emissions we need to bridge the gap between current usage and required reductions, this requires everyone to be involved.

❖ We have to move away from supply-side management to demand-side management (Sect. 11.2) which means living within a finite amount of energy shared equally amongst all of us.

❖ Conservation is achieved using a mixture of structural (smart lighting/heating, CFC lights) and behaviour actions.

  ○ Using cleaner energy is vital, and we must avoid dependence on high emission fuels
  ○ Using less energy overall by careful planning and investment by householders, starting with identifying wastage.
  ○ Using energy more effectively means conducting detailed surveys of how we individually use energy and identifying the most cost effective ways of reducing usage.

➢ The major energy sectors are transportation, space heating and water heating. In the future we will have to survive using far less energy and that means addressing the issues that will arise from living within a slowly reducing energy budget. We have two major challenges:

  ○ Moving towards a low-energy (carbon) lifestyle
  ○ Moving towards a low-energy (carbon) economy

The eighth step is reducing the use of energy by adopting a proactive low-energy lifestyle. Start by conducting an energy usage survey of your home and lifestyle, then set personal reduction targets for electricity used in the home, taking time to select and investment in the best energy efficient appliances, turning lights and appliances off when not required, and always selecting the most efficient energy sources. Use this ethos to influence others in your family and work.

Table 8.14 National conversion factors for Ireland at 2011

	g CO_2 per kWh equivalent
Liquid fuels	
Motor spirit (gasoline)	252
Jet kerosene	257
Other kerosene	258
Gas/diesel oil	264
Residual oil	274
LPG	229
Naphtha	264
Petroleum coke	335
Solid fuels and derivatives	
Coal	341
Milled peat	420
Sod peat	374
Peat briquettes	356
Gas	
Natural gas	205
Electricity (2012)	481

Source: SEAI (2013). Reproduced with permission of the Sustainable Energy Authority of Ireland, Dublin, Ireland

Homework!

This is an important exercise that is going to save you money and make you a little more adaptable when it comes to energy dependence. First you need to find out just how much energy you have been using, so track down all those electricity, gas and oil bills. Don't forget the other fuels such as coal, bottled gas, peat etc. Then try and reconstruct how much energy you have been using over the past few years. You will see seasonal variations. Use actual energy usage not the amount you spent as this will generally have been going up over the years. Now for each year calculate the total and then take an average for the last 2 or 3 years. This is going to be your annual maximum. **The first decision you have to make is that you will not exceed this total.**

Next is convert your household energy usage to a CO_2e emissions value and this can be done by multiplying by the conversion factors in Table 8.14. Remember to convert grams (g) to kilograms (kg) by dividing your answer by 1000.

These values are from 2011 except for electricity and I want to use these so that we can compare your energy usage to the national average of 2.4 tonnes per person per year. Are you above or below the average? **If you are above the national average then reducing your emissions to the national average should be your immediate target.**

Next is to concentrate just on your electricity usage. So look at the various online resources at Energy Ireland and calculate the energy usage of your household and try to recreate your electricity bill.

(http://www.esb.ie/esbcustomersupply/residential-energy-services/reduce-your-costs/web-calculator.jsp)

The final part of the exercise is to create a list of what actions use most energy and where energy usage can be reduced in the future.

References and Further Reading

BP. (2014). *BP statistical review of world energy June 2014*. London, England: BP plc. Retrieved from http://www.bp.com/content/dam/bp/pdf/Energy-economics/statistical-review-2014/BP-statistical-review-of-world-energy-2014-full-report.pdf

EIA. (2014). *Annual energy outlook 2014 with projections to 2040*. Report: DOE/EIA-0383(2014). Washington, DC: US Energy Information Administration. Retrieved from http://www.eia.gov/forecasts/aeo/pdf/0383(2014).pdf

National Grid. (2011). *UK future energy scenarios*. Warwick, England: National Grid plc. Retrieved from http://www.nationalgrid.com/NR/rdonlyres/86C815F5-0EAD-46B5-A580-A0A516562B3E/50819/10312_1_NG_Futureenergyscenarios_WEB1.pdf

REDISS. (2013). *European residual mixes 2012. Results of the calculation of residual mixes for purposes of electricity disclosure in Europe for the calendar year 2012*. Brussels, Belgium: Reliable Disclosure Systems for Europe, European Union. Retrieved from http://www.reliable-disclosure.org/upload/34-RE-DISS_2012_Residual_Mix_Results_v1_1.pdf

SEAI. (2013). *Energy in Ireland 1990-2012: 2013 Report*. Dublin, Ireland: Sustainable Energy Authority of Ireland. Retrieved from http://www.seai.ie/Publications/Statistics_Publications/Energy_in_Ireland/Energy_in_Ireland_1990_-_2012_Report.pdf

Global Energy Consumption

http://www.enerdata.net/enerdatauk/; http://www.eia.gov/countries/
http://www.worldenergy.org/

Energy Use Ireland

http://www.seai.ie/Publications/Statistics_Publications/EPSSU_Publications/Commissioned_Research/Energy%20End-Use%20in%20Ireland.pdf

http://www.seai.ie/Publications/Statistics_Publications/EPSSU_Publications/Energy%20In%20Ireland%201990%20-2010%20-%202011%20report.PDF

http://www.reliable-disclosure.org/

Saving Energy

http://www.energysavers.gov/

Energy Efficient Homes

http://www.epa.gov/greenhomes/ReduceEnergy.htm

Home Energy Measurements

https://www.esb.ie/esbcustomersupply/residential-energy-services/home/index.jsp

Energy Calculator

http://www.esb.ie/esbcustomersupply/residential-energy-services/reduce-your-costs/smartphone-calculator.jsp

Chapter 9
Travel: Here, There, Everywhere

Travel and transportation is a major component of our personal and business footprints as well as a key source of global greenhouse gases. In this chapter we explore travel emissions in detail and how to minimize them. Image: White exhaust contrails from a DC-8 testing a mix of standard and a plant-derived biofuel. *Source*: https://www.flickr.com/photos/nasacommons/. Reproduced by permission of NASA

9.1 Introduction

When I was a teenager someone told me that the village butcher had never travelled further than Gloucester which was about 10 km away. I couldn't believe this and asked him if this was true. He explained that apart from being in France between 1915 and 1918 he had indeed never been further than the local town. 'Why do I need to travel any further?' He asked confused, 'I have everything I need here and I didn't have a great time in France.' I suppose he was not untypical of that era and indeed for

© Springer International Publishing Switzerland 2015
N.F. Gray, *Facing Up to Global Warming*, DOI 10.1007/978-3-319-20146-7_9

the majority up until the 1950s war service would have been the only time they did go abroad. Foreign travel was mainly the preserve of the wealthy or the very intrepid; although most people were mobile using the extensive network of buses and trains and so did get to cities and the coast. But this travel was a treat and often something that was only enjoyed on bank holidays or on annual holidays, and when I look at my own family I see a similar trend. My grandparents rarely travelled outside their local village except for my Grandfather's similar visit to France during the First World War. They did take several day trips by coach to holiday resorts or to London each year, usually just for the day, and apart from half a dozen week-long holidays during their lifetime spent at some nearby seaside resort, that was the extent of their experience of travel. So by modern standards they didn't actually travel very much. My parents were far more mobile and my father brought his first car when he was in his 30s and enjoyed day trips a couple of times a month. They were part of that expanding group of visitors in the mid 1950s to late 1960s that flocked to stately homes and the growing number of attractions that were sprouting up for the independent traveller. They also usually had an annual holiday by the seaside generally on the south coast, and apart from my father serving in North Africa, Italy and then Germany during the Second World War, they never went abroad, except for couple of trips to see me after I had moved to Ireland towards the end of their lives in the 1980s. Again this was typical, but during their lifetime foreign travel was increasing at a rapid rate. In contrast my generation are inveterate travellers often investing in a holiday home abroad, but during the 1970s up to the early 1980s air travel in particular was still expensive, but with the advent of cheaper flights in particular there was a steady increase in overseas travel. So holidays in Spain or Greece became the norm and we saw the demise of our own rather colder holiday coastal destinations throughout that period. Our younger generations born since the 1980s expect to travel and to visit every corner of the globe, something often achieved before they are in their mid-twenties. So our mobility is something that is relatively new. We now travel farther to work, we travel more as part of our occupation and we travel more for recreation and for holidays than ever before, and it is a trend that is continuing to rise. So how is that affecting our carbon footprint and global warming? How does it affect us?

9.2   Travel as Part of Our Carbon Footprint

Travel and transportation is a necessary component of economic activity as well as being critical for vital activities such as welfare, humanitarian aid, conservation, and security. It is rapidly growing in all countries, especially the developing nations. Transportation of commercial goods (i.e. freight transport) is growing at a faster rate than personal transport, with trucks the main mode of transport in Europe although in the US trains are still widely used to move freight. Access to personal transportation is also rapidly increasing, especially in Asia. The problem is that 95 % of global energy consumption for transportation is based on oil, predominately diesel and petrol (gasoline), which is equivalent to 23 % of global energy related GHG emissions in 2004 at 6.3 Gt CO_2e. Road transport is responsible for 74 % of these emissions which continues to grow faster than any other energy consuming sector.

Transportation is a major factor in many areas of consumerism which is embedded in the carbon footprint of all products that we use including consumables such as food, clothing and electrical good. While transportation represents 13 % of our global footprint, it is only the direct transportation emissions that we can account for within a personal carbon footprint.

We looked at household CO_2e emissions in detail in Chap. 5. The household emission survey of 103 Irish homes carried out by Kenny and Gray in 2009 showed that on average 42.2 % of emissions were associated with household energy use, 35.1 % with transport, 20.6 % with air travel and 2.1 % with waste disposal. So **on average over half of our household or personal GHG emissions arise from travel**. The average household carbon footprint was 16.55 t CO_2e year^{-1} which is equivalent to an average personal carbon footprint of 5.70 t CO_2e year^{-1}. So 3.2 t (56 %) of CO_2e year^{-1} of our personal footprint arises from travel. The survey showed that household energy consumption was more efficient as occupancy rate increases and this is also true for transportation as car trips are more likely to be combined and holidays shared by all the family members.

On average 3.2 t of CO_2e per year of our personal footprint of 5.7 t (56 %) arises from travel

More information: Kenny, T. and Gray, N.F. (2009) A preliminary survey of household and personal carbon dioxide emissions in Ireland. *Environment International*, **35**, 259–272.

9.3 Aviation

It is hard to believe that 20.6 % of the average Irish household carbon footprint is associated with air travel equivalent to 1.15 t CO_2e per person per year (ca^{-1} year^{-1}). This figure rises for single person households to 1.69 t CO_2e ca^{-1} year^{-1} and 2.23 t CO_2e ca^{-1} year^{-1} in two person households respectively and then starts to decline with household size. The greatest aviation users in terms of emissions per year are individuals and couples without children. Aviation impact is only measured in terms of fuel used and is expressed as grams of CO_2e per journey per passenger based on the flight distance. This figure does not take into account hidden infrastructural emissions which are potentially extremely large. Heathrow airport had a carbon footprint of 2.32 million t CO_2e in 2011 including some offsetting. According to them 15 % is for the direct operation of the airport and the rest is associated with aircraft movement, take off and landing, passenger and staff travel. This illustrates a serious problem in the allocation of emissions. In reality the travel to and from the airport is the business of the passenger or employee as is the fuel used during the flight, including take off and landing. Why this has been included in the airport footprint is unclear. Airports such as Heathrow have made major strides to reduce their emissions by energy conservation such as the use of renewable energy, but each passenger who flies also is responsible for his or her share of the airport's operation. Heathrow have indicated that this could be 5.4 kg CO_2e per passenger, although it is unclear as to what is included in this figure.

> **Aviation is a major and often avoidable component of an individual's footprint**.

What is certain is that the rate of air travel continues to increase, particularly driven by high mobility in both the 16–30 age group and the elderly, supported by cheaper flights. In essence the more people who fly the cheaper it becomes, and the cheaper it becomes the more people fly. In 2010 the total distance flown by passengers was just under 5000 billion km, equivalent to every man woman and child on the planet travelling 679 km on a plane (Fig. 9.1)!

There has been a threefold increase in passengers arriving at Dublin Airport since 1995 reaching a peak of 23.5 million in 2008 (Fig. 9.2). This began to decline in 2009 due to the economic crisis and was reflected at all European airports, reaching a trough of 18.4 million passengers in 2010. Since that time passenger numbers have slowly increased once again to 18.7 and 19.1 million in 2011 and 2012 respectively (Table 9.1). Dublin receives 80 % of all international passengers to Ireland, 1.64 million of which were transatlantic. In 2013 passenger numbers rose by 6 % to 20.17 million and by a further 8 % in 2014 to 21.5 million showing a steady increase.

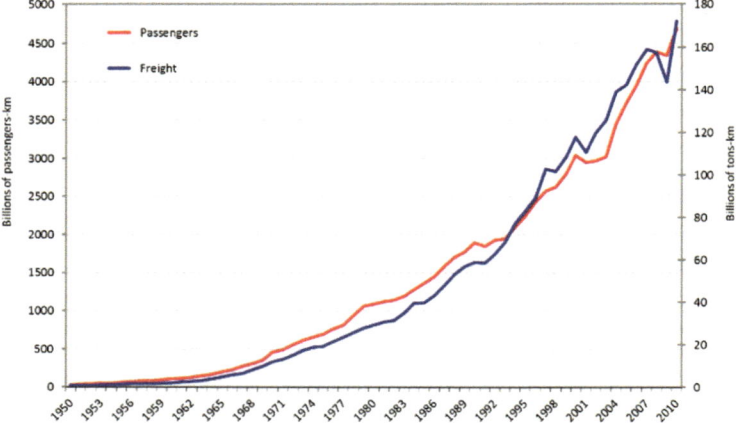

Fig. 9.1 Increase in passenger and freight carried by air since 1950 expressed as billions of passengers per km and billions of tons per km. *Source*: International Civil Aviation Authority, http://www.icao.int. Reproduced with permission of the ICAO, Neuilly-sur-Seine, France

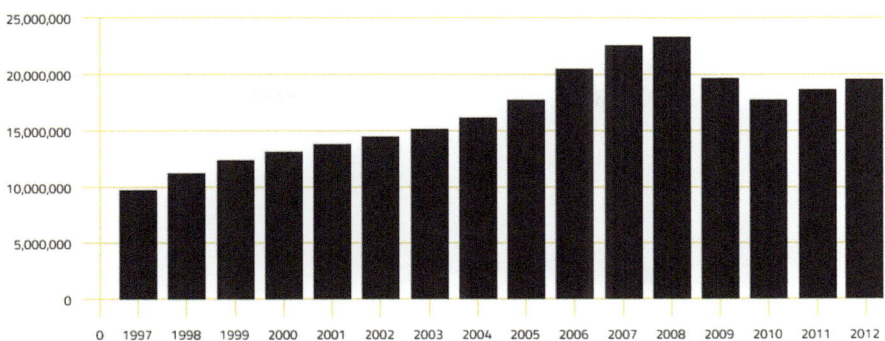

Fig. 9.2 Passenger numbers at Dublin airport 1997–2012. *Source*: DAA, http://dubplus.ie/. Reproduced with permission of the Dublin Airport Authority, Dublin, Ireland

Table 9.1 Passenger numbers at Dublin airport showing that the global recession had a relatively small impact on numbers travelling by air

Flight type	Passengers in thousands (000's)					
	2013	2012	2011	2010	2009	2008
Transatlantic	1864	1644	1567	1489	1614	1748
UK	7181	6894	6970	6727	7575	8560
Continental Europe	10,513	10,016	9816	9570	10,416	11,997
Other International	541	478	260	266	243	252
Domestic	68	61	120	369	635	845
Transit	<1	7	8	10	19	65
Total	*20,167*	*19,100*	*18,741*	*18,431*	*20,502*	*23,467*

Source: DAA, http://www.flytodublin.com/market-profile/passenger-traffic/. Reproduced with permission of the Dublin Airport Authority, Dublin, Ireland

9.3.1 Emissions from Flying

The most convenient and accurate way of measuring emissions from flying is by expressing them in terms of grams of CO_2e per passenger per kilometre travelled (g CO_2e per passenger per km). There is a significant difference in the emissions per passenger kilometre when flying a short distance compared to a medium or long haul flight due to the energy used during take-off and landing, the type of plane used and the weight of fuel and cargo carried. A domestic or short haul flight (<500 km) produces the highest emissions at 259 g CO_2e per passenger per km compared to medium haul (500–1500 km) or long haul (>1500 km) flights that generate lower emissions at 178 g and 114 g CO_2e per passenger per km respectively. Interestingly emissions from flying can be equated to passengers driving the same distance in a car. So for short, medium and long-haul flights emissions are almost identical to each passenger driving the same distance in a top of the range petrol fuelled car, a mid-range petrol car or a small diesel car respectively.

Fig. 9.3 The planes that fly on long haul flights from Europe to America or Asia and beyond are very large. The two main planes are the Boeing 747-8 (above on maiden flight) which can seat 467 passengers and the 747-400 which can seat between 416 and 524 passengers. Image by Jim Anderson. *Source*: boeingdreamscape, http://flickr.com/photos/49902951@N02/4584756327. Reproduced under licence under the terms of http://creativecommons.org/licenses/by/2.0/deed.en

The 747-8 is the most efficient jumbo on the market and can carry 230,000 L of fuel weighing 185 t (Fig. 9.3). Planes use a lot of fossil fuels and have until recently paid no tax on their fuel at all. Private aviation in the EU had a tax imposed in 1st November 2008 and commercial aviation was brought into the Emissions Trading Scheme (ETS) in 2012 and so have had to buy and trade in carbon credits to cover emissions since 2013, with the price passed onto customers. The Aviation Emissions Directive (2008/101/EC) amended the ETS Directive 2003/87/EC in order to include a system for emissions trading within the aviation sector. The legislation covers all operators both EU and non-EU who fly to or from a member state of the European Union as long as the aircraft weighs over 5700 kg although smaller operators (i.e. those with annual emissions of less than 10,000 t CO_2e or who have fewer than 729 flights per year equally distributed over 4 month assessment periods) are excluded.

Aviation Emissions Directive: http://eur-lex.europa.eu/LexUriServ/LexUriServ. do?uri=OJ:L:2009:008:0003:0021:EN:PDF

The EU has also allocated free allowances to over 900 operators based on historical data for 2004–2006 which showed that emissions were on average 221,420,279 t per annum. So the cap has been set at 95 % of this value. Operators receive 82 % of the allowance free with 15 % being auctioned. Interestingly the remaining 3 % is kept in reserve and will be allocated to new operators or those that are rapidly expanding.

Aviation allowances: http://ec.europa.eu/clima/policies/transport/aviation/allowances/ index_en.htm

Table 9.2 Examples of European travel distances by plane from Dublin

Dublin to specified destinations	Single km	Return km
Amsterdam	757	1514
Athens	2860	5720
Belfast	141	282
Berlin	1319	2638
Copenhagen	1241	2482
Glasgow	309	618
Hamburg	1076	2152
London	463	926
Paris	782	1564
Rome	1889	3778
Vienna	1685	3370
Warsaw	1829	3658

To fly from Dublin to New York is a distance of 5100 km which is a carbon footprint of 581.4 kg CO_2e per passenger or 1.163 t of CO_2e per passenger for a return flight. So a full 747-8 return flight is equivalent to **543 t of CO_2e**. Dublin to Sydney (Australia) is 17,200 km which is equivalent to 1.961 t CO_2e per passenger or 3.92 t CO_2e per passenger return. So a full 747-8 return flight is equivalent to **1831 t of CO_2e**.

So to calculate your emissions from flying simply determine the distance flown. You can use the simple table above (Table 9.2) or go to one of the many websites that give distances between airports along standard flight paths. Examples of calculators are: http://www.webflyer.com/travel/mileage_calculator/ http://www. airmilescalculator.com/. Airport codes and locations can be found using: http:// www.world-airport-codes.com/.

Multiply the single or return flight distance in kilometres by the relative conversion factor to give your emissions in grams of CO_2e (for distances of less than 500 km multiply distance in km by 259; 500–1500 km use 178; and for trips greater than 1500 km use 114).

So if two people fly to Amsterdam for the weekend from Dublin which is a distance of 757 km (Table 9.2), then each of them would generate:

$$\text{Distance (km)} \times \text{conversion factor} = \text{grams of } CO_2\text{e emitted}$$

$$757 \text{ km} \times 178 \text{ g } CO_2\text{e per passenger} = 134{,}746 \text{ g } CO_2$$

If we divide grams by 1000 we get kilograms, which in this case is 135 kg CO_2e.

So the return trip is 270 kg CO_2e per passenger and double this for two people then this is equivalent to 540 kg CO_2e (which is over half a tonne of GHG). Then you have travel to and from the airport as well…so flying, and travel in general, is a big emitter of GHGs.

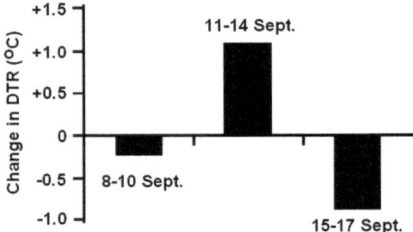

Fig. 9.4 The effect on surface temperature of contrails shown as the change in the average diurnal temperature (DTR) 3 days before and 3 days after the no-fly period (11–14 September, 2001) imposed immediately after the 9/11 attack when there were no contrails. *Source*: Travis et al. (2002). Reproduced with permission by Nature Publishing Group

9.3.2   Contrials

Aircraft have some interesting environmental effects as well as creating GHG emissions from burning fuel. They also produce contrails which are made up of condensed water vapour discharged from the aircraft's exhaust (Fig. 9.4). These form expanding lines of water vapour that appear as long thin clouds. These contrails do two things. First they reduce solar energy reaching the surface of our planet by reflecting the energy back, but secondly they also prevent infrared radiation being reflected back into space thereby trapping the heat. Although at first sight it appears that these two outcomes cancel each other out, in reality contrails have an overall cooling effect. This was observed when all flights in the USA were grounded due to the 9/11 terrorist attack (11–14 September 2001). During the no-fly period the surface diurnal temperature increased by over 1 °C directly due to there being no contrails. Figure 9.4 shows the 3 days before and the 3 days immediately after the no fly period when aircraft were grounded and there were therefore no contrails. The subsequent 3 days when air traffic was much heavier than normal a greater cooling effect was recorded. So the greater the air traffic the greater the cooling effect.

The Carbon Trust is a key provider of information to businesses to help them assess and reduce their carbon footprints. Businesses that carry out such assessments often find interesting and often unexpected areas where emissions are high. For example, at Trinity College Dublin where I work there are 1364 academic and research staff (2009). In a detailed survey it was revealed that 85 % of the staff annually travelled by air to conferences and/or research meetings. The average distance flown by these staff members was 15,367 km per annum, although this figure was skewed slightly by a few staff members with very high travel footprints. The median distance travelled, which is a better indicator of the distance that most people flew was in fact half this at 6470 km. The range of those who actually did fly in that year was from 450 to 189,524 km. That works out as an average footprint of 1783 kg CO_2 per person per year, or 3.4 t CO_2e per person per year excluding radiative forcing. This is equivalent to a total of 2432 t CO_2 year^{-1} for the all the academic and research staff or driving a Toyota Yaris 38.5 million km each year. So each member of staff drove an equivalent distance of 33,218 km to attend conferences and meetings on average. This did not include personal air travel, holiday travel, nor travel by rail or car for business, commuting or pleasure, and excluded administrative staff some of whom had very large footprints indeed. Without exception all the staff were aware of the need to reduce travel in order to reduce emissions and many lamented the fact that they did have to travel as part of their work. But the reality is of course that many of these trips were probably unnecessary and that meetings can be done using other means such as Skype or video conferencing.

> **Aviation has a cooling affect as well as adding to GHG and atmospheric pollution**

More information: http://facstaff.uww.edu/travisd/pdf/jetcontrailsrecentresearch.pdf

It seems unfair to expose my colleagues in this way and I hope they will forgive me, but the truth is that all businesses including universities have enormous carbon footprints for travel. Many larger companies have attempted to mask this part of their carbon footprint by either ignoring it altogether or calling it a zero emission, a common practice, by paying an offsetting fee per tonne of carbon emitted (Sect. 6.6). In many cases, offsetting travel has actually increased the amount of travel undertaken by some businesses. Another interesting strategy is to bring in ancillary carbon emissions into company footprints and this raises the interesting question of **whose carbon is it**? So for example, should companies include the travel by its employees to and from its place of work? To me this is quite a serious question. We should all have an equal share in global energy and also an equal responsibility to reduce personal emissions. So if an employee wants to use his or her share in travelling to work then it is their choice, but having said that then they need to adjust their personal carbon footprint to take this into account and compensate for these emissions by making savings elsewhere. It should be up to us how we spend our carbon allowance so long as we do not exceed it. Ideally those who use less should be rewarded and interest in the concept of personal emissions trading is increasing. So in the case of Trinity College for example, those who exceed the average mileage flown each year could perhaps offset this by transferring money from research accounts to other staff research accounts who did not fly or who were below the average as a reward. However, this form of offsetting does not necessarily result in reduce emissions overall.

> **We need to identify exactly who is responsible for what and ring fence emissions so they are only counted once.**

9.4   Travel by Other Means

Aviation is of course just one of the transport categories and we tend to travel inside our own country primarily by road and rail. Over a third of average Irish household emissions (35.1 %) is due to transportation other than aviation. So transportation (non-aviation) is the second largest emissions sector of our primary carbon footprint after household energy usage. This equates to an average personal footprint of 2.0 t CO_2e year^{-1} or 5.81 t CO_2e year^{-1} for the average household.

What is interesting about this type of transportation is that it is often an unavoidable component of our footprint as it is generated through important activities such as commuting to work, college or school, or in other cases to go food shopping, visiting the doctor and all those other important activities. With rising rents and house prices in our cities then more and more people are forced to commute ever increasing distances. Of course holiday and recreational travel is also a component, and for some people, especially those in cities this can be the largest part of their travel emissions. Like aviation, impact is generally only measured in terms of fuel used. There are also hidden infrastructural costs such as road construction and maintenance, provision of servicing and fuel, which are potentially large. However, the road network is absolutely vital for all of us including industry and commerce, as well as for the transportation of food and other goods, and of course support services such as police fire and ambulance (Fig. 9.5).

Fig. 9.5 This is Beijing, China. In August 2010 National Highway 110 developed a 100 km long traffic jam which lasted 10 days! But wherever you live cars and other vehicles dominate our lives and change our landscape. Original copyright owner unknown

9.4.1 The Car

Car manufacturing is a surprisingly powerful group within the global economy with European car manufacturers supplying 25 % all of vehicles and 35 % of all passenger vehicles worldwide. With an annual turnover of 551 billion euro (2007) equivalent to 6.5 % of the European Union's GDP and employing two million people directly and a further ten million people indirectly, it has proven to be a difficult industry to regulate in terms of addressing global warming. Having said that, car manufacturers are currently under a lot of pressure due to the recent economic crisis which has resulted in a serious reduction in car sales and even threatening survival of car manufacturing in Europe. Other pressures include being forced to reduce GHG and other emissions due to global warming and air pollution, the problems of high oil prices and oil security, all of which is creating a demand for more fuel-efficient vehicles.

Road transport results in the release of significant volumes of pollutants. Lead, which is highly toxic, has been removed from petrol for several decades and sulphur which caused acid rain has also been significantly reduced from all fuels. However, 66 % of all nitrous oxides are still produced by road transport causing acid rain and equally importantly reacting with volatile organic compounds (VOCs) in the air to make ozone. This latter reaction is triggered by sunlight with low level ozone in particular causing damage to crops and other plants as well as humans. Diesel also produces PMs (particulate matter) which are tiny particles that cause a wide range of health problems. Together PMs and ozone cause an estimated 370,000 premature deaths globally every year. To combat this, the EU has introduced air quality legislation which requires fine particulate matter ($PM_{2.5}$—that is particles of less than 2.5 μm in diameter) to be reduced in urban areas by 20 % over the period 2010 and 2020. Also, average low-level ozone concentrations must not exceed 120 μg per cubic metre on more than 25 days per year, with low level ozone a particular problem in summer.

EU Air Quality Directive: http://ec.europa.eu/environment/air/quality/legislation/directive.htm

The transport sector is responsible for around one-fifth of the EU's greenhouse gas emissions with 70 % of emissions from road transport. To combat this, **the IPCC requires developed countries to reduce GHG emissions by 25–40 % by 2020 and by 80–95 % by 2050 to avoid the worst impacts of climate change**. So while car manufacturers have been forced to respond, it has taken a long time. A lot of this change has been due to the EU setting strict limits designed to dramatically cut greenhouse gas emissions from cars. In 2007 the average new car emitted approximately 160 g CO_2 per km and so manufacturers were set a new target to bring this figure down to 120 g CO_2 per km for at least 65 % of their new cars by 2012. After 2012 the proportion of new cars that must comply with this limit will increase gradually until it reaches 100 % by 2015. This emission limit is likely to fall further to 95 g CO_2 per km by 2020.

In 2010 the average CO_2 emissions of a new passenger car in the EU27 Member States were 140.3 gCO_2 per km. An average new passenger car emitted 5.4 g CO_2

per km less (3.7 %) than in 2009 when average emissions were 145.7 gCO_2 per km. The difference between average CO_2 emissions of new diesel and new petrol vehicles is just 3.3 g CO_2 per km. This gap is considerably lower than a decade ago, when the difference was 17 g CO_2 per km, which combined with better fuel consumption figures is why those who drove long distances tended to favour diesel cars, which is still the perception today.

How far do we drive? Well this depends on so many factors. People in rural areas often drive more than those in cities and towns due to a lack of public transport. Ireland has quite a high rural population which can be quite isolated. In 2005 an organization *Sustainable Energy Ireland* used National Car Testing data (NCT—which is equivalent to the UKs MoT test) to assess **average mileage of Irish car owners which was found to be 16,894 km per year** (10,498 miles per year). Petrol cars were driven on average only 15,969 km (9923 miles) compared to diesel cars with an average annual mileage of 23,817 km (14,799 miles). In contrast the **average EU passenger travels only 13,000 km per annum**, some 70 % in cars.

Car ownership in EU Member States has risen from 345 per 1000 in 1990 to 464 per 1000 in 2007 which is similar to the USA (Fig. 9.6). Ireland has seen a dramatic increase in car ownership but still below the EU average at 425 per 1000 with Luxembourg topping the list at just below 700 cars per 1000 of population.

EU vehicle registrations decreased by 2.3 million in 2010 compared to 2007, when European sales reached a peak of 16 million new vehicles. Since 2007 car sales have slumped due to the global recession. In 2013, 11,825,400 new cars were purchased and registered in the European Union (EU28). Of these 53 % were diesel, 1.4 % hybrids, and 1.8 % LPG, 0.42 % electric, and the remainder petrol driven. In terms of total European sales for new cars Germany was top of the list at 25 % of all new cars, followed by the UK (19 %), France (15 %) and Italy (7 %). Ireland represents less than 1 % with just 74,367 new cars registered in 2013 of which 0.06 % were electric. In terms of the numbers of electric cars purchased and registered in 2013, Ireland had the seconded lowest percentage after Greece (0.01 %), while at the top of the list is Norway at 5.79 % and the Netherlands 5.43 % of all new cars (ICCT 2014).

More information: http://www.theicct.org/

The World Bank keeps an updated table of the number of registered vehicles per 1000 of the population which includes all vehicles both commercial and private for all countries. The list makes fascinating reading and can be accessed using the link below with a small selection is given in Table 9.3.

More information: http://data.worldbank.org/indicator/IS.VEH.NVEH.P3
Current European vehicle data base: http://www.theicct.org/european-vehicle-market-statistics-2014

The number of vehicles is continuing to increase globally, but remain low per thousand of population throughout Africa and Asia, and in China and India in particular, where growth in the number of vehicles per 1000 people is in excess of 300 % over the period 1999–2009. Therefore, with the rapidly developing economies throughout Asia, then vehicle ownership is set to rise rapidly (Table 9.4).

Passenger cars per 1 000 inhabitants

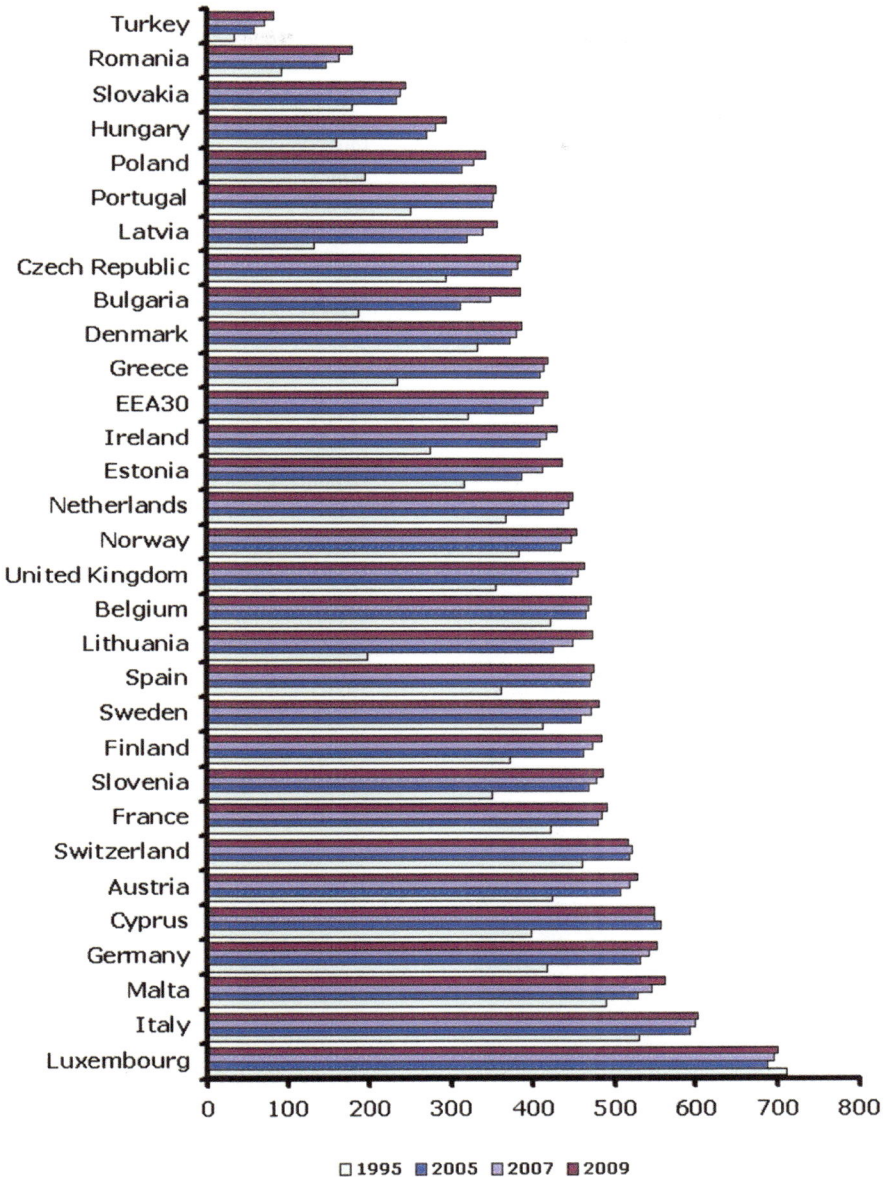

Fig. 9.6 The growth in car ownership in 30 EEA member countries (that is EU-27 plus Norway, Switzerland, Turkey) since Kyoto. *Source*: EEA, http://www.eea.europa.eu/data-and-maps. Reproduced with permission of the European Environmental Agency, Copenhagen, Denmark

Table 9.3 Examples of the number of registered vehicles per 1000 of the population in 2010

Country	Motor vehicles per 1000 people in 2010
San Marino	1263
USA	863
Iceland	745
Luxembourg	739
New Zealand	712
Australia	695
Italy	679
Canada	607
Spain	593
Germany	572
Netherlands	527
UK	519
Ireland	513
Romania	235
Ukraine	173
Botswana	133
China	85
Kenya	24
India	18
Uganda	8
Bangladesh	3
Togo	2

Data source: The World Bank, http://data.worldbank.org/indicator/IS.VEH.NVEH.P3. Reproduced with permission The World Bank, Washington, DC, USA

Table 9.4 Rate of increase of vehicle ownership over the decade 1999–2009

Country/region	1999	2009	Increase (%)
Africa	20.9	24.9	19
Asia, Far East	*39.1*	*157.7*	*303*
Asia, Middle East	66.2	101.2	54
Canada	560.0	620.9	11
Central and South America	133.6	169.7	27
Europe, East	370.0	363.9	−1.7
Europe, West	528.8	583.3	10.4
Pacific	513.9	560.9	9.2
United States	790.07	828.04	4.8

Data source: The World Bank, http://data.worldbank.org/indicator. Reproduced with permission The World Bank, Washington, DC, USA

Table 9.5 Trend in global vehicle numbers in thousands (000 s) 1960–2009

Type of vehicle	1960	1970	1980	1990	2000	2005	2009
Car registrations	98,305	193,479	320,390	444,900	548,558	617,914	681,154
Truck and bus registrations	28,583	52,899	90,592	138,082	203,272	245,798	284,101
World total	126,888	246,378	410,982	582,982	751,830	863,712	**965,255**

SUVs and other similar vehicles such as 4-cab pickups used for private travel are not included in car registrations but under commercial vehicles (i.e. truck and bus registrations). *Data source*: The World Bank, http://data.worldbank.org/indicator. Reproduced with permission The World Bank, Washington, DC, USA

As a child I used to wonder where all the car fumes and smoke went to…I still do.

There are currently over a one billion vehicles using almost exclusively fossil fuels polluting our atmosphere. These are spread over 15 billion km of roads. Of that billion vehicles three quarters (740,000,000) are cars. Total vehicle numbers in 2010 were USA 239.8 million, China 78.0 million, Japan 73.9 million, Brazil 64.8 million (up from 29.5 million in 2000) and India 20.8 million. The global trend continues to rise even though more and more of us live in cities (Table 9.5).

9.4.2 Commuting

Commuting varies from country to country. In Ireland 69 % of all commuters used a car in 2011 to get to work, spending on average 26.6 min travelling an average distance of 18 km (CSO 2014). Sixty per cent of children are driven to primary school compared to 40 % to secondary school, with fewer children than ever before walking, cycling or using a bus to get to school. This trend has carried onto third level education as well with 29 % driving, 28 % walking, and just 5 % cycling with the remainder using public transport. In terms of all travel (excluding aviation) 73 % of journeys were made by car, 16 % on foot, 4 % by bus, and 6 % by rail systems. On average people in employment made 19 journeys per week travelling 278 km. However, when these values are broken down then those in rural areas travel on average 286 km per week compared to urban dwellers at 180 km with those in rural areas making fewer individual trips and spending less time travelling than those in urban areas (CSO 2014).

Similar trends are seen in the UK with the average distance travelled to work being 15 km (9.32 miles), an increase of 10.7 % since 2001 with the numbers travelling less than 5 km to work dropping by 43 % over the same period. The shortest journeys were made by those living in London with an average of 11 km (6.83 miles)

(ONS 2014). Independent surveys have shown that workers spend 41 min each day on average commuting with around two million commuters spending 3 h or more a day commuting to and from work.

There are estimated to be 128.3 million commuters in the US with 29 % travelling less than 5 miles to work. Twenty-two per cent travel 6–10 miles, 17 % 11–15 miles and 10 % 16–20 miles. Eight per cent travel in excess of 35 miles to work each day. In terms of the transport mode of commuters in the US 75.7 % use a car, 12.2 % carpooling, 2.52 % use the bus, 1.45 % the subway and just 0.35 % cycle to work (BTS 2014).

1 mile is equivalent to 1.609 km

Of course there are many different ways of travelling, and many opt to make their daily commute by public transport, but just like aviation and the car there is a carbon footprint attached. Emissions are normally expressed as grams of CO_2e per passenger kilometre travelled (g CO_2e per passenger km) and of course trains and buses also cause pollution just like cars and lorries. It is very difficult to calculate public transport emissions accurately as they depend on the actual number of passengers and the fuel used, rather than theoretical carrying capacity. The fact that a train or bus may be filled to capacity or even overcrowded during rush hour it is most likely to be only partially full for the rest of the day, so efficiency in terms of g CO_2e per passenger km varies by trip. So an average value is required. Kenny and Gray (2009) determined Irish public transport emissions per passenger km and these are listed below. Commuter and intercity trains in Ireland are diesel, while the DART is an electrified suburban train service. The LUAS is an electrified light rail system similar to trams found in other European cities. Ireland being an Island is very dependent on both aviation and vehicle ferries to the UK and Europe (Table 9.6).

Table 9.6 Irish travel emissions per passenger km for public transport and motorcycles (Kenny and Gray 2009)

Mode	Emissions
Train (intercity)	54 g CO_2e/passenger km
Train (commuter)	49 g CO_2e/passenger km
Bus	74 g CO_2e/passenger km
LUAS	55 g CO_2e/passenger km
DART	43 g CO_2e/passenger km
Ferry[a]	20.35 kg CO_2e/passenger
Taxis	Same as passenger vehicles
Motorcycles	
Small <125 cc	73 g CO_2e/passenger km
Medium 125–500 cc	94 g CO_2e/passenger km
Large >500 cc	129 g CO_2e/passenger km
Average for motorcycles	106.7 g CO_2e/passenger km

[a]One way Dublin Holyhead High Speed Ferry

Table 9.7 Comparison of commuting options by emissions per passenger kilometre travelled (Kenny and Gray 2009)

Kilometre travelled per kg CO$_2$ emitted	Transport mode
24.5	Small car (driver + two passengers)
23.3	DART (suburban electric rail)
20.4	Train (intercity diesel rail)
18.1	LUAS (electric light rail)
16.5	Small car (driver + single passenger)
13.7	Scooter—petrol
13.5	Bus—diesel
12.3	Taxi—medium car (two passengers)
8.3	Small car (driver only)
6.25	Taxi—medium car (160 g km^{-1}) (single passenger)
3.69	2005—E210 Mercedes-Benz petrol
3.46	Typical 3.0 L diesel SUV

So what is the most effective way of commuting? Well walking and cycling are by far the most efficient methods although as we can see from Table 9.7 car sharing does reduce per passenger emissions significantly. Commuting is often complex. For example, it may require the person to drive to a train station, then travel by train maybe by a non-direct route and then take a bus to their final destination. In suburban and city areas then the bus network usually feeds into the rail network as well as providing direct routes themselves. So it is always going to be preferable to use public transport where it is available and reasonably direct, and by supporting public transport you reduce the number of cars on the road, reduce congestion, reduce local pollution, reduce emissions, ensure that public transport services are maintained and are at their most efficient in terms of emissions. However, cost and time factors are also important issues for individuals, so if we are to reduce the number of cars on our roads public transport has to be frequent, reliable, affordable and safe.

9.4.3 Are Modern Cars Really That Efficient?

I am always surprised when I see large fast modern cars with very small emissions. There is no doubt that cars are becoming increasingly efficient but some cars seem to be miraculous in terms of the emissions declared in order to pay a lower road tax. How can a car that weighs twice as heavy as a small engine car such as the Toyota Yaris be in the same emissions category or even have the same emissions? There seems to be a fundamental problem in how we calculate the GHG emissions from cars in the first place and also how the theoretical emissions compare to those in practice.

Every car make and type that is sold within the EU is required to be tested officially for both fuel consumption and environmentally dangerous emissions such as carbon dioxide (CO_2) or nitrogen oxides (NOx). The test conditions to calculate emissions are quite artificial. For example, there is just the driver plus 25 kg (no luggage or passengers) in the car when it is tested, which is carried out at a temperature between 20 and 30 °C on a flat road in the absence of wind with all ancillary equipment turned off (air conditioning compressor and fan, lights, heated rear window, etc.) and with the windows closed. To achieve these ideal conditions they are generally now performed on a roller test bench indoors. The current test is in two parts and calculates both the combined fuel consumption and CO_2 emissions. The urban cycle is a series of 12 starts and stops at an average speed of 20 km h^{-1} (12 mph) and never exceeding 50 km h^{-1} (31 mph), while the extra urban cycle is a sequence of acceleration, deceleration and steady-speed driving, never exceeding 120 km h^{-1} (75 mph).

A recent report by Transport and Environment, an NGO that studies sustainable transport polices and is based in Brussels, suggested that these tests are increasingly being manipulated even further in order for cars to secure better emission ratings. The report highlights examples of bending the rules by the use of slick tyres that have been pumped rock hard in order to minimize resistance, using special fuel mixes, disconnecting the alternator so that the battery is not charged during the test, adjusting or even disconnecting brakes to reduce friction, removing wing mirrors and taping up or sealing panel joints on the body of the car, including the windows, to reduce air resistance. These small adjustments, along with those already allowed, all help to significantly reduce emissions during the test. A study in Germany has shown that the fall in average emissions from 180 g km^{-1} to less than 150 g km^{-1} between 2001 and 2011 reported by the car industry, in fact only fell from 190 g km^{-1} to about 180 g km^{-1} when real cars were tested under normal driving conditions (Fig. 9.7). The average gap between real-world use and laboratory-generated data across all manufacturers widened to 31 % in 2013 from just 8 % in 2001. The US EPA fined the Hyundai Motor Company $100 million in November 2014 for overstating the fuel economy of their cars and has asked some other manufacturers to revise their fuel economy figures. However, car emissions have been falling in all European countries as fuel efficiency of new vehicles continues to increase. For example, CO_2e emissions from vehicles in the Netherlands have fallen by 34.7 % over the period 2006–2013 compared to a European average of 21.2 % (Table 9.8).

Transport and Environment Report: http://www.transportenvironment.org/publications/mind-gap-why-official-car-fuel-economy-figures-don%E2%80%99 t-match-reality

More information on Vehicle testing: http://www.transportenvironment.org/sites/te/files/publications/2014%20Mind%20the%20Gap_T%26E%20Briefing_FINAL.pdf

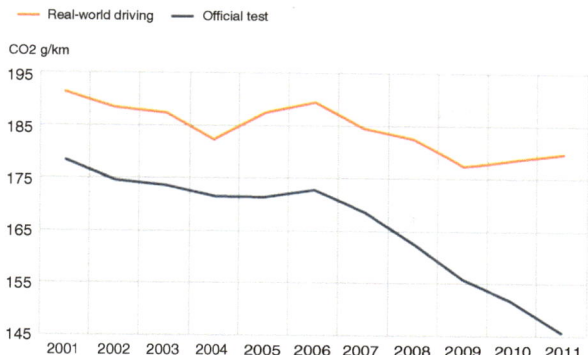

Fig. 9.7 The difference between emissions calculated under real conditions in Germany and those supplied by manufacturers has being increasing annually from a 7 % difference in 2001 to 23 % in 2011. *Source*: Transport & Environment. Reproduced with permission of the Transport & Environment Organization, http://www.transportenvironment.org

Table 9.8 Current average emissions (g CO_2e per km) for new cars purchased annually between 2006 and 2013 in selected European countries based on manufacturer's NEDC

Country	Average new car emissions per annum (g CO_2e km^{-1})								Per cent reduction over period (%)
	2013	2012	2011	2010	2009	2008	2007	2006	
EU	127	133	138	143	147	154	159	161	21.1
Austria	132	137	141	147	152	159	162	162	18.5
Belgium	124	132	137	136	143	148	153	154	19.4
Denmark	113	119	129	136	143	147	162	165	31.5
Finland	132	141	146	151	158	164	178	180	26.7
France	118	125	129	132	136	141	149	149	20.8
Germany	135	143	147	153	156	166	170	172	21.5
Greece	112	122	135	145	158	161	165	166	32.5
Italy	121	126	131	136	141	147	148	149	18.8
Ireland	122	126	129	133	144	158	164	165	26.1
Luxemburg	133	138	144	147	153	160	166	168	20.8
Portugal	114	120	126	130	135	138	143	145	21.4
Netherlands	109	120	128	138	149	158	165	167	34.7
Spain	124	130	137	141	145	151	156	156	20.5
Sweden	135	139	146	156	166	175	183	190	29.0
UK	129	134	140	146	151	158	164	167	22.8

Improvements in average car emissions began to increase in 2007 and have been steadily falling in all countries. This is not only due to more efficient cars but also due to customers choosing more efficient models. Source data abstracted from the data sets produced by the International Council on Clean Transportation Europe (ICCT 2014). Reproduced with permission of the International Council on Clean Transportation under a Creative Commons Attribution-ShareAlike 3.0 Unported License

Table 9.9 Irish car annual road taxation bands based on CO_2 emissions revised in 2013 and VRT rates

Band	Emissions (g CO_2 km^{-1})	Annual tax (€)	VRT (%)
A0	0	120	14
A1	1–80	170	14
A2	81–100	180	15
A3	101–110	190	16
A4	111–120	200	17
B1	121–130	270	18
B2	131–140	280	19
C	141–155	390	20
D	156–170	570	24
E	171–190	750	28
F	191–225	1200	32
G	226+	2300	36

Source: http://www.environ.ie/en/LocalGovernment/MotorTax/MotorTaxRates/MotorTaxRatesbased onCO2Emissions/. Reproduced with permission of the Department of Environment, Community and Local Government, Dublin, Ireland

To encourage the purchase of low emission cars, the Irish Government has changed their taxation bands from engine size to CO_2 emissions (Table 9.9). Likewise the initial vehicle registration tax (VRT) was also altered to encourage drivers to buy and imported fuel efficient vehicles (Table 9.9).

So how does real driving conditions affect those theoretical values? **Use of accessories** such as the lights, heater and air conditioning can added up to 5–10 % on the emissions per km. The **payload** is a critical as well which can include anything from heavy items in the boot to extra passengers, again increasing emissions by a further 5–10 %, although if vans are overloaded this can cause even higher increases. **Poor maintenance o**f the car's engine, underinflated tyres, or poor tracking can add another 5 %. Remember, the test on your prototype vehicle was most likely done indoors, so factors such as gradients, weather conditions, driving into the wind, and of course speed all add significantly to your emissions. Erratic and aggressive driving uses much more fuel and increases emissions by up to 10–20 %. Open windows, roof racks, bike racks can all increase emissions by 5–15 %. My emissions nightmare is seeing a SUV overtaking me at an excessive speed with a couple of bikes on the roof.

To try and compensate for this discrepancy Defra (UK) has assumed an uplift factor of +15 % to take account of these real world effects. This is an average, but equates to a decrease in miles per gallon or km per litre of 13 % for petrol vehicles and 9 % for diesel vehicles (Table 9.10).

So it would seem that the figures supplied by manufactures are sometimes misleading, and that we are mistaken if we think that if we buy a heavy, large engine

Table 9.10 Defra generalized CO_2 emission factors for cars based on fuel type and engine size

Vehicle type	Engine size	Size	EU conversion factors ($g\,CO_2\,km^{-1}$)	mpg	Revised DEFRA real world conversion factors ($g\,CO_2\,km^{-1}$)	mpg
Petrol cars	>1.4 L	Small	159.2	40.8	*183.1*	35.5
	1.4–2.0 L	Medium	188.0	34.6	*216.2*	30.2
	>2.0 L	Large	257.7	25.2	*296.4*	21.9
Diesel cars	<1.7 L	Small	131.0	56.7	*150.7*	49.3
	1.7–2.0 L	Medium	163.6	45.4	*188.1*	39.5
	>2.0 L	Large	229.1	32.4	*263.5*	28.2
Hybrid petrol-electric cars	Toyota Prius	Medium	109.7	59.3	*126.2*	51.5
	Honda Civic	Large	194.7	33.4	*224.0*	29.0

Data from Defra (2008). Reproduced with permission the Department for Environment, Food and Rural Affairs, London

1 mile per gallon (mpg) is equivalent to 0.425 kilometre per litre (km L^{-1})

vehicle it is going to have low emissions…it is unlikely to be true. However, the reality may even be more complex. A study conducted by Emissions Analytics of 500 vehicles in the UK, half petrol and half diesel, found that on average fuel efficiency was 18 % less in terms of mpg than suggested by the manufacturer. They found discrepancies for all vehicle types but in particular with small cars. Cars with engine sizes up to 1 L performed 36 % worse than expected based on the figures supplied by the manufacturers. The study observed that smaller engines have to work much harder when accelerating than cars with larger engines so that this group only averaged 16.4 km L^{-1} (38.6 mpg). Vehicle engines between 1 and 2 L showed a smaller average difference between expected and observed efficiency at 21 % (average 19.9 km L^{-1} or 46.7 mpg) and for 2–3 L engine cars the discrepancy was just 15 % less than the manufactures figures (average 19.1 km L^{-1} or 45 mpg). However, this depends on how these vehicles were driven and the study does not explore this. In the US the average fuel efficiency of cars is 23 mpg (9.8 km L^{-1}) (Sect. 5.4.1).

More information: http://emissionsanalytics.com/

Check measured emissions of your car: http://www.whatcar.com/truempg/my-true-mpg#

Table 9.11 Embedded CO_2e emissions form the manufacture of vehicles and how they might add to travel per kilometre emissions or annual carbon footprints based on vehicle lifetime in a personal carbon budget

Vehicle		GHG emissions as grams per kilometre (g km^{-1}) or tonnes per year (t year^{-1})				
	Embedded GHG	20,000 km (g km^{-1})	80,000 km (g km^{-1})	160,000 km (g km^{-1})	5 years (t year^{-1})	10 years (t year^{-1})
Yaris	6 t	300	75	37.5	1.2	0.6
Discovery	24 t	1200	300	150	4.8	2.4

9.4.4 Embedded Footprint of Car Manufacture

One thing we do not take into account in calculating travel footprints are the embedded GHG emissions from the manufacture of the car and its components. This varies from vehicle to vehicle as we can see in Table 9.11. But how do we allocate these emissions in personal footprints? We can see the comparison between the typical small car which has embedded CO_2e emissions equivalent to 6 t per vehicle to that of a typical SUV which has embedded emissions of 24 t of CO_2e. One way of allocating these could be to set them against actual mileage over the expected lifetime of the vehicle, so for 80,000 km this would be equivalent to an extra 37.5 and 150 g of CO_2e per km on top of the emissions for the fuel for the Yaris and Discovery respectively. Alternatively this can be allocated on a yearly basis, so for the Yaris this would be equivalent to 0.6 t CO_2e per year over a lifetime of 10 years (Table 9.11).

So should these emissions be allocated to personal (primary) footprints? Should the burden of embedded GHGs in manufactured objects be accountable to either the manufacturer or the purchaser/owner? This is a very difficult and emotive question, especially when the vehicle changes ownership. If the original purchaser was to be made accountable for emissions then this would probably seriously affect new car sales. Of course some of these original manufacturing emissions will be recoverable through recovery and reuse of vehicle parts and recycling, but this does illustrate the complexity of life cycle analysis and emissions allocation (Fig. 12.6).

9.4.5 Alternatives

Of course the obvious alternative to the car is to use public transport, cycle or walk. However, there are alternatives.

Car pooling is where trips are shared between normal commuters. It may be done on a rota basis each driver using his or her car on alternate days or weeks, or one person using their car, usually because its bigger, and the others contributing nominally towards the cost. It is this latter aspect which has caused problems in the past where insurance companies have been unhappy with the idea of actually making money from car pooling rather like a taxi. But the essence of car sharing is reducing the number of cars on the road with a single passenger (i.e. the driver).

Until recently insurance problems stopped private car owners receiving benefit from sharing their car, but most insurance companies now realize the importance of car pooling and have adjusted policies accordingly. Car pooling is now very organized with numerous online sites that allow drivers and passengers to find each other.

More information: http://www.carpooling.com/us/ http://www.carpool.ie/ http://www.carpooling.co.uk/

Car pooling does more than reduce our emissions, it significantly reducing all associated exhaust pollutants including hydrocarbons, volatile organic compounds, PM_{10} and $PM_{2.5}$ and NOx.

More information: http://www.epa.gov/otaq/consumer/420f08028.pdf

Car sharing schemes are becoming increasingly popular with people living in central urban areas who are served by an excellent public transport system and so do not need the use of a car very often. It can also often be difficult and expensive parking securely in the City Centre with many residential streets choked with parked cars, with front gardens often being turned in parking bays. So in many major cities car sharing schemes have been developed that provide a service to its members whereby they can hire a car for as little as 15 min. This means that you only pay for the use of a vehicle when you need it, so there is no maintenance, taxes, no insurance or the problem of having the vehicle tested. Also, unlike normal car hire, the cars are available close by throughout the city and can be used for very short periods and without the hassle of filling in forms and paying deposits each time a car is needed. The cars can be booked or simply just accessed on demand using a smart-card with members of the scheme billed each month.

In Dublin the car sharing scheme is called GoCars and has 50 cars which are available at 30 different pick up locations around the city centre. They can also be parked in the City free of charge which almost pays for the hire itself. If you add up how much time you actually need your car then such a scheme is very sensible and can save you a lot of money.

More information: http://www.gocar.ie/

Bike hire is widely available in cities throughout the world with a variety of private and public schemes available. Apart from private cycle hire companies, Dublin, like many European cities, also has a centralized bike rental scheme where bikes can be taken from over 44 public bike stations around the city with in excess of 550 bikes available for use. Bikes can be taken from one station and left at another with access to bikes by the use of smart cards which can be purchased by occasional use at €5 for 3 days or €20 for a year. The first 30 min hire is free and then after that a small charge is levied. The scheme was started in 2009 and since that time in excess of six million trips have been made 95 % less than 30 min in duration and so completely free. With 35,000 active long-term subscribers, the scheme has been a huge success and with the average trip just 13 min in duration. The next phase of the scheme will see the number of stations increased to 58 and the number of bikes to 1500 by 2015. Currently thanks to investment of dedicated cycle lanes 6 % of commuters in the City now cycle.

Logo reproduced with permission Dublin Bike Scheme.

More information: http://www.dublinbikes.ie/
http://www.dublincitycycling.ie/

> The ninth step is accepting that travel is a major factor in both personal
> and business GHG emissions and that you should try to minimize
> emissions from transportation and build them into your own personal
> carbon budgets.
> Always question when you plan a trip... Is this trip necessary? Is there a
> more efficient mode of transport I could take? Can I make long trips
> more effective (e.g. by staying longer)? Can I combine trips? Can I
> find alternatives to the need to travel (i.e. why are you going there)?

9.5 Conclusions

- Transportation is made up of private and public transport and also aviation. It represents, on average, 56 % of Irish personal and household primary carbon footprints
- Aviation alone represents 1.15 t per capita per year with cheap flights in particular having exacerbated aviation use through weekend breaks, multiple holidays, second homes, etc. So perhaps air travel should be considered more as a luxury and subject to higher carbon taxation
- Carbon emissions supplied by car manufacturers may be on average 23 % lower than we can expect under optimum road conditions. So it is better to look elsewhere for information the emissions before purchasing a new one.
- Transportation should always be included as part of the personal footprint and offers an important area where personal emission reductions can be made. So transportation should be minimized or offset against other emissions
- Public transport should always be used preferentially by those in urban areas, however where this is not possible then car sharing or carpooling should be considered.

Homework!

Travel related emissions make a huge dent in our personal carbon allowance and yet is one of the easiest areas to reduce emissions. So what I would like you to do is to keep a detailed record of your family travel footprint. Detail every trip from the home and the mode of transport used…even if its taking the dog for a walk or getting the newspaper. For each trip I want you to estimate the distance you travelled. At the end of the week I want you to do two things. First calculate the emissions from all your trips using Table 9.6 and then to explore how this could have been reduced. Was the trip necessary? Could you combine trips reducing the number of times you go into town each week? Could you car share that trip in your car with a friend or neighbour? Could you have left the car at home and used a different type of transport? Could those short trips be done on foot or by bicycle? How much could you have saved in emissions by being more selective in your frequency and choice of travel?

Finally if you don't walk or cycle why not have a go? Remember the health and environmental benefits from this mode of transport. The Dublin Bike scheme is excellent for example, and schemes like that can be found in many cities throughout Europe and beyond. But we don't all need to get out of our cars and walk or cycle, what we all need to do is to use transport carefully and make each mile or kilometre valuable.

References and Further Reading

BTS. (2014). *National household travel survey*. Washington, DC: Bureau of Transportation Statistics, US Department of Transportation. Retrieved from http://www.rita.dot.gov/bts/sites/rita.dot.gov.bts/files/subject_areas/national_household_travel_survey/index.html

CSO. (2014). *Profile 10: Door to door commuting in Ireland*. Dublin, Ireland: Central Statistical Office. Retrieved from http://www.cso.ie/en/media/csoie/census/documents/census2011profile10/Profile,10,Full,Document.pdf

Defra. (2008). *2008 guidelines to Defra's GHG conversion. Factors: Methodology paper for transport emission factors*. London, England: Department for Environment, Food and Rural Affairs. Retrieved from http://archive.defra.gov.uk/environment/business/reporting/pdf/passenger-transport.pdf

ICCT. (2014). *The European vehicle market statistics pocketbook 2014*. Berlin, Germany: The International Council on Clean Transportation. Retrieved from http://www.theicct.org/sites/default/files/publications/EU_pocketbook_2014.pdf

IPPC. (2007). *Climate change 2007: Synthesis report. IPCC fourth assessment report*. Geneva, Switzerland: Intergovernmental Panel on Climate Change. http://www.ipcc.ch/publications_and_data/ar4/syr/en/spm.html

Kenny, T., & Gray, N. F. (2009). A preliminary survey of household and personal carbon dioxide emissions in Ireland. *Environment International, 35*, 259–272.

ONS. (2014). *2011 Census analysis, distance travelled to work*. London, England: Office for National Statistics. Retrieved from http://www.ons.gov.uk/ons/dcp171776_357812.pdf

Travis, D. J., Carleton, A. M., & Lauritsen, R. G. (2002). Contrails reduce daily temperature range. *Nature, 418*, 601. doi:10.1038/418601a.

Air Travel

http://www.emiratesgroupcareers.com/english/about/environment/A380_environmental_facts.
  aspx

Contrails

http://science-edu.larc.nasa.gov/contrail-edu/science.html
http://facstaff.uww.edu/travisd/pdf/jetcontrailsrecentresearch.pdf

Car Emissions

http://www.direct.gov.uk/en/Dio11/DoItOnline/DG_10015994
http://www.fleetnews.co.uk/news/2011/5/12/ec-proposes-real-world-testing-for-co2-emissions/
  39571/
http://www.seai.ie/Power_of_One/Getting_Around/HCIYC/

Sustainable Travel

http://www.transportenvironment.org/

Chapter 10
Having Enough to Eat

The food that we eat and more importantly discard creates a very large personal carbon footprint that is hugely understated or ignored by most general carbon footprint calculators. In this chapter we explore where greenhouse gases arise in the food industry and how we can significantly reduce them at both production and consumption levels. The problem of food security and scarcity are also explored. Image by Yannick Beaudoin, UNEP/GRID-Arendal. *Source*: http://www.grida.no/photolib. Reproduced with permission Yannick Beaudoin, UNEP/GRID-Arendal

10.1 Introduction

Like all middle aged western men I am slightly obsessed with food. I am always worried that I get the balance right between getting enough protein and calories in my diet, reducing carbohydrates that may push up my glucose levels and result in

© Springer International Publishing Switzerland 2015
N.F. Gray, *Facing Up to Global Warming*, DOI 10.1007/978-3-319-20146-7_10

diabetes, as well as avoiding those saturated fats that may cause my cholesterol to soar. Of course I am constantly trying to resist all those tempting foods that may drive my weight above the critical point where I cross that line of normal to over-weight. The good thing about all this is that it has made me think more about food in general, where it comes from, how it is prepared and how to use it efficiently. In fact it has made me a person who is genuinely interested in and respectful of food from the person who grows it, to those who care for it during retail and of course three cheers for the cook. It has made me want to grow my own and learn more about its transformation from raw ingredients to that wonderful plate of food we eat at meal times. This is one of the greatest luxuries that we have, having access to plentiful supplies of good quality, diverse and healthy food. Most of us have the privilege to eat what we want, when we want to and as much as we want…which is probably why we are now suffering a global obesity crisis and a population comprised largely of people just like me worrying about their diet.

We are all familiar with the concept of food aid and famine and the fact that one person dies directly every 5 min from hunger, that means ten people will die from a lack of food by the time you have read this chapter or every time I give a lecture! In 2010, one in seven people were officially classified as undernourished with a third of the global population normally hungry or having problems in finding enough food each day (Fig. 10.1). The problem is that hunger and malnutrition results in other diseases, especially amongst the most vulnerable in the community, that is the very young, so every day 16,000 children are dying from hunger related causes.

It's not only in developing countries that we find hungry or malnourished people. Here in Ireland a quarter of the population are susceptible to food poverty. In a

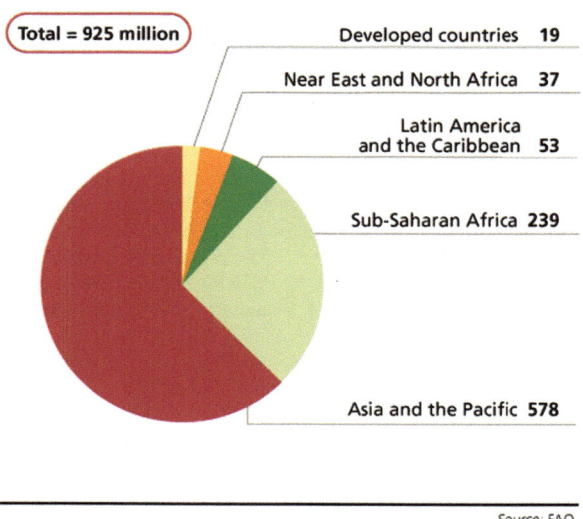

Source: FAO.

Fig. 10.1 Numbers of people classified as undernourished per continent in millions. *Source*: FAO, http://www.fao.org/docrep/007/y5650e/y5650e03.htm. Reproduced with permission of the Food and Agriculture Organization of the United Nations, Rome, Italy

research document prepared by the Labour party in 2011 '*Making food poverty history*: *Labour's Blueprint for Eliminating Food Poverty*' they showed that 19 % of male children and 14 % of female children in Ireland "*always or often go to bed hungry*". It suggests that a staggering 270,000 people in Ireland either experience food poverty or are extremely susceptible to food poverty, and that a further 800,000 people whose level of income means they are also susceptible to food poverty. Similarly food poverty is seen in all developed countries including the US where 14.9 % of households equivalent to 33 million adults and 16 million children are struggling to adequately feed themselves. This is not due to food scarcity but poverty. That is approximately one in six adults and one in five children. So keeping basic food prices as low as possible is very important. In 2012 the average spend on food and drink in the UK was 11.6 % of income rising to 16.6 % for low income families. Food prices in the UK rose in real terms by 11 % between 2007 and 2013 after a prolonged period of over 20 years with food prices steadily falling. Since 1980 food prices fell by over 25 % in real terms reaching a low in 2007 (Defra 2014).

More information on food poverty in the US: http://feedingamerica.org/

10.2   Climate Change and Agriculture

A complex relationship exists between climate change and agriculture. On one hand climate controls agricultural production globally, while on the other agriculture is a major source of GHG emissions. The IPCC estimates that agriculture contributes 13.5 % of total global GHG emissions (Fig. 10.2). Emissions are primarily the result of methane emissions from livestock and flooded rice cultivation with nitrous oxide

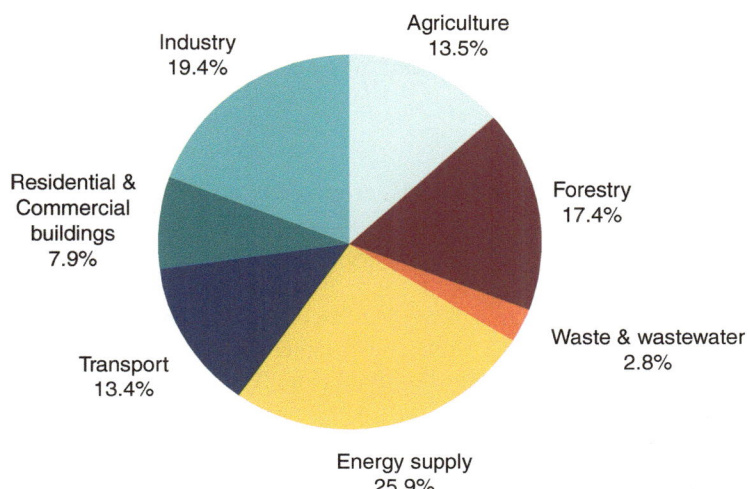

Fig. 10.2 GHG emissions by sector. *Source*: IPCC WG1-AR4 Report (IPCC 2007). Reproduced with permission of the Intergovernmental Panel on Climate Change, Geneva

emissions from fertilizer application also a significant contributor. Agriculture is also a leading driver in deforestation (another 17.4 % of the human-induced emissions).

The impact from global warming depends on how the local climate is altered and is dependent on the magnitude of warming. Two key factors primarily control crop yields, higher temperatures and changes in precipitation, both in terms of volume and pattern. Together these control soil moisture and the frequency, duration and intensity of drought and floods, both of which can significantly affect plant growth and yield. In some areas dangerous temperature thresholds will be reached that could destroy crops overnight and give rise to famine. Tropical countries (e.g. India) will soon face short periods of super-high temperatures well into the high 40 °C. These temperatures could completely destroy crops if they coincide with the flowering period. To some extent developed nations are able to cope with changes in rainfall patterns due to the ability to invest in irrigation or better drainage, and the acquisition of larger and quicker harvesting machinery that can take advantage of short dry periods. However, all this requires a high capital investment making food increasingly expensive.

Some scientists have suggested that the negative impacts of global warming will be mitigated by having increased atmospheric concentrations of carbon dioxide which will increase photosynthesis and hence plant growth, the so-called **CO_2 fertilization theory**. The extent to which such benefits will cancel out the expected negative impacts is uncertain and accounts for the wide range of predictions surrounding the future of agriculture productivity. The expectation is that CO_2 fertilisation would more than counteract crop losses from rising temperatures. However, field experiments have shown that for the world's main food crops (maize, rice, soybean and wheat) the fertilization effect will only be half as great as predicted. Also a 20 % increase in ozone concentrations, as a result of global warming, will cut yields by at least 20 % overall. Increases in ozone levels of this level are predicted for Europe, the USA, China, India and much of the Middle East by 2050. So in practice the fertilization effect will be more than offset by the impact of higher temperatures and differing precipitation, and will not prevent the world's crop yields from declining by 10–15 %.

It is most probable that the world's crop yields will decline by 10–15 % overall by 2050.

So globally climate change will result in an overall decline in agricultural production although there will be a slight rise in richer countries by the application of increasing investment to counteract the effects (Fig. 10.3). So those poorer countries that are already suffering from food scarcity will be hardest hit leading to migration and starvation.

Fig. 10.3 Change in overall agricultural output production predicted regionally from 2000 to 2080. *Source*: World Resources Institute, http://www.wri.org/resources. Reproduced with permission of the WRI, Washington, DC, USA

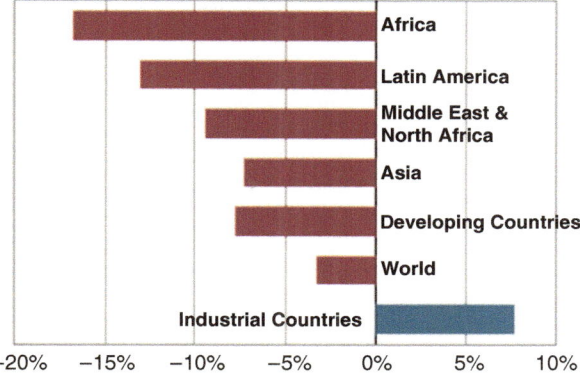

Change in output potential (2080s as % of 2000 potential)

10.3   Who Will Be Affected Most by Food Scarcity?

The worst impacts of global warming on food production will not be experienced equally around the world. At high latitudes and elevations, including areas like Scandinavia and Canada, higher temperatures will likely generate net benefits for agriculture, and could even create new agricultural land in areas previously too cold for farming. In the lower latitudes, **where most of the world's poor reside**, **impacts will be overwhelmingly negative**, including increased frequency of heat waves, heavy precipitation events and an expansion of the area affected by drought. The reason why the threat is mainly to developing countries is that the damage is greater closer to the equator where temperatures are already high. Many countries in the region are also vulnerable to flooding and sea level rise, so coupled with a poor adaptive capacity in developing countries, and particularly in Africa, the impacts on agriculture could have devastating consequences for food security, poverty, and social welfare (Fig. 10.3; Table 10.1).

In developed countries problems will also occur but may in some circumstances be mitigated by greater investment. There will be significant impacts to agriculture in the Southern United States, Southern Europe, and especially Australia, where temperatures are already extreme and rainfall low or unpredictable. The economy's ability to adapt to global warming may not always be sufficient. For example, Australia is a country where climate and location cannot fully be offset by having more resources for adaptation, a similar situation occurs in the Great Plains of the USA.

Even without climate change, population growth is already putting pressure on food supplies increasing food scarcity and raising food prices. The expected rise will be much higher where climate change is greatest. For example, predictions estimate that **without climate change** wheat prices will rise from $113 per tonne (2000) to $158 per tonne by 2050 (**39** %), with higher rises for both rice 62 %, maize

Table 10.1   Possible impacts on agriculture, forestry and ecosystems by region in the developing world

Region	Potential impacts
Africa	By 2020, in some countries, yields from rain-fed agriculture could be reduced by up to 50 %. Agricultural production, including access to food, in many African countries is projected to be severely compromised. This would further adversely affect food security and exacerbate malnutrition
Asia	By the 2050s, freshwater availability in Central, South, East, and Southeast Asia, particularly in large river basins, is projected to decrease
Latin America	By mid century, increases in temperature and associated decreases in soil water are projected to lead to gradual replacement of tropical forest by savannah in eastern Amazonia. Semi-arid vegetation will tend to be replaced by arid-land vegetation. Productivity of some important crops is projected to decrease and livestock productivity to decline, with adverse consequences for food security. In temperate zones, soybean yields are projected to increase. Overall, the number of people at risk of hunger is projected to increase

Source: FAO, http://www.fao.org/climatechange/en/. Reproduced with permission of the Food and Agriculture Organization of the United Nations, Rome, Italy

Fig. 10.4   Drought and crop failure in Australia and North West China during 2007–2008 led to a global rise in wheat prices of 83 %

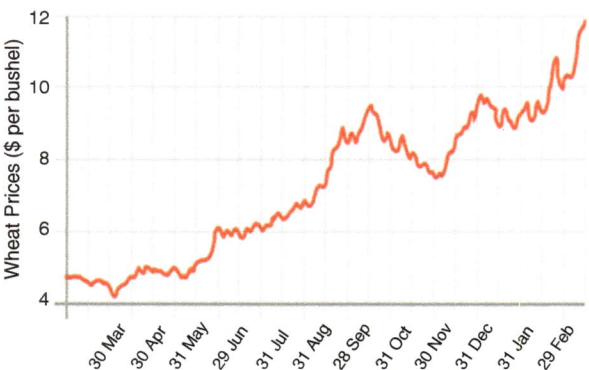

63 %. **With climate change** then wheat prices will rise by 170 %, rice 113 %, maize 148 %. The relationship between food supply and pricing is complex, but demand continues to soar as do prices as the big economic powers such as China (especially after the 2011 rice harvest failure) look towards the global markets for food. In recent years predicted rises have been overshadowed by such massive crop failures forcing prices for these staples to new heights as countries struggle to purchase cereals in particular (Fig. 10.4). Also some countries like Russia have held back supplies due to growing internal demand and poor yields. Current demand is still outstripping availability in most commodities including wheat, rice, and sugar. Yields of other important commodities such as maize, soya bean and cotton will drop significantly over the coming decades due to additional days when the temperature is above 30 °C, forcing production to other regions.

Prices are driven by growing demand, drought and reduced outputs, and more recently by demand for land to supply biofuels.

This area has been studied by the International Food Policy Research Institute (IFPRI) based in Washington, DC who are concerned that food scarcity will eventually effect the whole economic stability of countries. The aim of the IFPRI is to end hunger and poverty through appropriate local, National and International agricultural policies. Biological impacts on crop yields work through the economic system resulting in reduced production, higher crop and meat prices, and a reduction in cereal consumption. This reduction means reduced calorie intake and increased childhood malnutrition which forms an ever worsening loop (Fig. 10.5). It estimates that 25 million more children will be malnourished by 2050 due to the impact of climate change on global agriculture.

More information: http://www.fao.org/news/story/en/item/44570/icode/
More information: IFPRI (2009) *Climate change*: *Impact on agriculture and costs of adaptation*, Food Policy Research Institute, Washington, DC.

To avoid this scenario we need to transform our agricultural practices not only to be able to feed a growing population but to feed that population under the constraints imposed by a changing climate. We need to change the crops we grow, where we grow them, the methods of harvesting, but at the same time without causing environmental degradation and further loss of habitat and biodiversity. We need new agricultural practices that not only achieve food security goals but also help mitigate the negative effects of climate change. So we need the development of

Fig. 10.5 Climate change is driving food scarcity in areas where population increase is often greatest creating an impending major social disaster

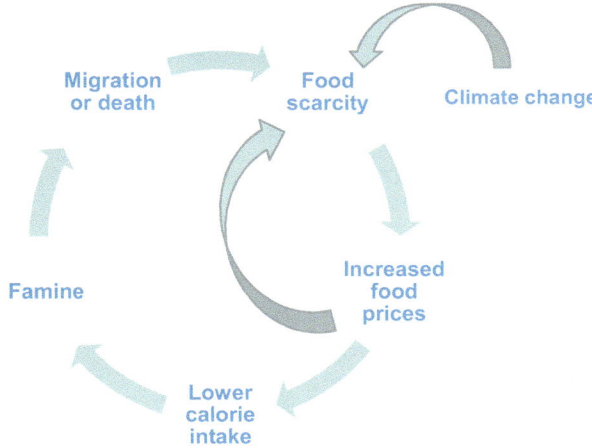

more productive and resilient agriculture, better management of natural resources, such as land, water, soil and genetic resources through practices such as conservation agriculture, integrated pest management, agroforestry and sustainable diets. This requires huge investment in R&D to improve crop varieties (especially drought-resistance), to enhanced crop and livestock productivity. Biotechnology and other new technologies will also play a major role as will genetic modification (GM) of crops. When we think about malnutrition we often forget that hunger is not the only problem, but a poor or incomplete diet that lacks vital minerals or supplements can also lead to health problems. For example, over two billion people globally lack sufficient iron in their diets leading to problems of fatigue. It is in these areas that GM foods can be particularly important. What is certain is that we must not take any more high diversity land into production, or sacrifice high productive land used for food crops to grow biofuels or biomass.

FAO logo. *Source*: FAO. Reproduced with permission of the Food and Agriculture Organization of the United Nations, Rome, Italy

This transformation of agriculture is being promoted by the FAO along with other partners under the term **Climate-Smart Agriculture**, which is agriculture that: '*sustainably increases productivity, resilience (adaptation), reduces/removes greenhouse gases (mitigation) while enhancing the achievement of national food security and development goals*'

More information: http://www.fao.org/climatechange/climatesmart/en/

So what are the key problems that need to be faced? Glacial melt water needed for irrigation is falling. These glaciers, particularly in the Himalayas, may disappear and cause some of the major rivers in the region to become much more unreliable in terms of flow, which will have negative effects on food yields in south Asia. Likewise the monsoons are also being affected leading to annual rains failing to occur or at different times of the year and in reduced amounts. Traditional seed varieties and livestock breeds are being lost at an increasing rate due to global warming and population expansion. These traditional food sources may provide a genetic resource that could more readily adapt to climate change. Crop diseases and insect pests will also thrive in a hotter or more humid climate, expanding their areas of infection to new regions where less natural resilience and natural predators are to be found. Portions of the most productive agricultural land in many regions will also be lost to sea level rise including 30 % of the land area of the Netherlands which is primarily used for agriculture (Sect. 13.5).

New areas will benefit from global warming bringing new land into production (e.g. Scandinavia, Canada). However it is unlikely that enough land will become available to replace that which is lost, also much of this new land will be unsuitable for production having thin soils and being very rocky resulting in low productivity. Much of these areas are currently covered by forest which means that bringing this land into production will release vast quantities of stored GHGs. So in many cases expanding agriculture to feed more people may simply exacerbate climate change. Of course the main problem is that we are assuming those in developing countries can afford food grown in northern Europe/Canada, which at current prices is unlikely.

Irish farming will also be affected by climate change but not as severely as more southern countries. The problems will primarily be the reduction in summer rainfall by up to 25 % and an increase in winter rainfall by 17 % within the next half century, with Southern and Eastern areas most affected. This will result is extended dry periods leading to drought, flooding, heavy rainfall events and extreme temperatures, in fact far more unpredictable weather patterns than we currently enjoy making farming activities increasingly difficult to plan and execute. While this will primarily impact on tillage farming from the sowing to harvesting of crops, livestock farmers will also feel the pressure from increased stress to animals and providing water in the summer to ensuring sufficient winter fodder. The wet summer of 2012 led to a significant fodder crisis in the winter and spring of 2012/13, requiring massive imports of hay and silage from the UK and Europe to meet the shortfall. The increased prevalence of pests and diseases, especially new pests as their ranges increase, due higher temperatures, will affect all farming sectors. It is not all negative, because the higher temperatures will increase cereal and beet yields, which is stark comparison to countries such as Romania and Hungary which will see a large decline in crop yields due to water scarcity. The solution for the Irish farming sector is to increase crop diversity, alter planting and harvesting regimes, develop more climate resilient crops, and introduce water management strategies.

More   information: http://www.stopclimatechaos.ie/download/pdf/projected_
economic_impacts_of_climate_change_on_irish_agriculture_oct_2013.pdf

The question that is constantly being asked is: Will the world be able to feed
itself when the population soars to nine billion during 2040–2050? The answer at
the moment is yes due to the improvements in farming outlined earlier. However,
there is another factor that may change the answer to no. The increase in personal
wealth in many developing countries has resulted in a change in people's diets especially in Asia to include more animal protein (i.e. meat, dairy products and eggs).
This is a particular problem in both India and China which is creating an ever
increasing demand for this type of protein. So more and more cereals including soya
beans are being used to feed animals, in particular cattle, pigs and chickens, instead
of being eaten directly by people. It is a simple case of counting calories. So 100 cal
of cereals can be either ingested directly to utilize all the calories or be converted via
livestock to 40 cal of milk, 22 cal in the form of eggs, 12 cal worth of chicken, 10 cal
of pork or just 3 cal worth of beef. This can be alleviated by keeping the animals
solely on grass but to meet the escalating demand animals are increasingly raised by
feeding grain. This will mean that even more land will be required to grow crops
than before and it is widely believed that the demand for cereals could double by
2050! Currently 46.5 % of the global ice free land is unused being mountainous,
deserts or permanent forests. However while 14.9 % is taken up by man due to housing, roads, industry including mining and commercial forestry, a massive 38.6 %
(19.4 million square miles) is given over to agriculture. The ever increasing demand
for biomass crops and animal feed shows that the global balance in terms of calories
in crops is moving rapidly from the current 55 % used for human consumption to
where the majority will be used for animal feed and biofuels by the end of the
decade. This is illustrated in Table 10.2 where this is already the case in most developing countries due to intensive livestock farming and the demand for biofuels.

Table 10.2 The relative use of the calories embedded in crops grown in different regions and as a
global average

Area	Calories obtained from crops	
	Human consumption (%)	Animal feed or biofuel (%)
US	27	73
Europe	39	61
Brazil	46	54
India	89	11
Asia	75	25
Africa	72	28
China	58	42
Global	*55*	*45*

To create a sustainable programme to feed the future population and still maintain environmental health then we need to adopt some basic management principles which have also be applied to energy and water (Chap. 15). These are:

- Set a fixed limit to the amount of land available to agriculture (i.e. demand side management) (Sect. 11.2)
- Grow more food on the land already in production
- Use resources more effectively
- Alter diets and the amount of food we eat (Sect. 10.6; 14.2.4)
- Reduce waste (Sect. 10.6.1)

> **More information**: *Johanthan Foley (2014) A five step plan to feed the world. National geographic 225, (May 5), 27–57.*

10.4   The Food We Eat and GHG Emissions

It has proven very difficult to accurately put a reliable footprint on food, especially as it is often unclear where or how our food is produced and how far it has travelled. Also, out of date, unused or rotten foods are all major contributors to landfill methane generation when dumped. We do know that approximately 30 % of European GHG emissions come from the food and drink sector. In the UK this represents 20 % of personal footprints which collectively is equivalent to 170 million tonnes CO_2e per annum. So wasting food is a very expensive way to generate GHG emissions.

10.4.1   Food Miles

Food can travel long distances from the farm to your home. The overall distance can be broken down into the separate journeys from the farm to processing plant, the processing plant to the distribution centre, the distribution centre to the supermarket and finally from the supermarket to your home. We also need to factor in the travel costs of seed, fertilizer, pesticides, animal stock and the whole production cycle. **In the USA, the average distance covered by food increased by about 25 %, from 6760 km in 1997 to 8240 km by 2004**, and has continued to rise annually. Food miles has created a new lifestyle known as locavorism or localism, but surprisingly the least impact in food miles is the transportation from the processer to the supermarket (7 % of overall food GHG emissons), which partially challenges the concept of locally sourced food. For most food it is the transportation of the food from the supermarket to the home which generates most GHGs from transportation (11 %). However, **in terms of food's carbon footprint food miles only represents on average 20 % of the overall carbon footprint**.

It is a common misconception that most GHG emissions are from the transport of food by air. In fact, food that is transported by road produces more carbon emissions than any other form of transportation. Global values for food transport carbon emissions can be broken down as road transport 60 %, air transport 20 %, rail transport 10 % and sea transport 10 %. In UK 90 % of all fruit and 40 % of all vegetables are imported, **with one in four heavy goods lorries on the road carrying food**. So today food in the UK travels 65 % further than it did two decades ago. However, the concept of food miles isn't just about distances it has wider social and ecological implications. So food miles are an inadequate way of assessing environmental damage, although to be fair they were never meant to be proxy for environmental impact.

> **In general food miles associated with air travel are a very small component and are typically associated with expensive perishable items**

10.4.2   Local Is Best?

Is buying local always best? This is a difficult and somewhat emotive question but I think the answer is **not always**, or at least in terms of GHG emissions; **but generally yes** because it has the ability to lower emissions at the community level by creating local employment and stimulating the local economy. It produces potentially fresher foods which have lower food miles in terms of producer to supermarket, and most importantly it educates consumers about the chain of production, who is actually growing our food as well as the importance of food and its relevance within the community.

The transport fraction of the food footprint is really divisive and can often be misleading. For example a study published in the journal *Political Science* showed that imported New Zealand lamb carcasses had a footprint of 688 kg CO_2 per tonne compared to 2849 kg CO_2 per tonne for British lamb (Saunder and Barber 2008). Similarly the concept that buying at the farm gate resulted in lower emissions when compared to buying from a supermarket was also shown to be incorrect when based solely on food miles from the farm gate to the consumer's house (Coley et al. 2008).

Out of season vegetables have to be grown in specially controlled indoor environments. Heated greenhouses require special water systems, heating, specific fertilizers, teams of workers, and road transportation often from one side of Europe to the other. So there are higher energy and transportation emissions for hothouse grown crops compared to vegetables and flowers that are grown out of door using natural sunlight and heat. Even when these products are air freighted from another country, growing crops and flowers in the open where the climate allows higher yields using less energy and producing less GHGs means that sometimes it is better to buy products from overseas. It makes no sense to grow out of season vegetables using fossil fuels for heating and to desalinate water for irrigation, and then shipping these products potentially many hundreds of miles by road. There is another aspect to this and that is creating and supporting sustainable agriculture in developing countries, the very basis of the Fairtrade movement.

For example, Kenya is a major producer of out-of season vegetables and flowers for the UK and Irish markets. Is reducing the average UK carbon footprint from 10.60 to 10.59 t really worth imperilling 1.0–1.5 million livelihoods? We produce a lot more carbon than a Kenyan. If we follow a totally locavore diet this will have a seriously negative impact on many developing economies. So we have to find a compromise in reducing emissions from food but supporting those who are struggling to maintain their communities in developing countries.

10.4.3   Organic Food: Is It the Sustainable Option?

The term organic is often used as an alternative for the concept of sustainable, but is this really true? There is no doubt that organic food is grown by people who care about the environment and work their land to strict conservation principles, but is it really best for the planet overall? Organic farming has made an enormous contribution to altering farm practices and raising public awareness, especially in relation to biodiversity and conservation issues. Indeed many of the underlying principles of organic farming are now seen as best practice in conventional and even intensive farming. But, sales of organic food have been falling in recent years and in 2011 it fell a further 3.7 % in the UK alone with the number of producers also declining by a similar amount to just under 7300. This is in stark contrast to the ethical trading certification products such as Fairtrade which have steadily risen over the same period with a 12 % increase in 2011. In the UK, ethical food and drink made up 8.5 % of household food purchases valued at £7.7 billion. Ethical foods in this context included organic, Fair trade, free range products and other sustainable producers (Fig. 10.6). So why is organic farming having such a bad time at the moment?

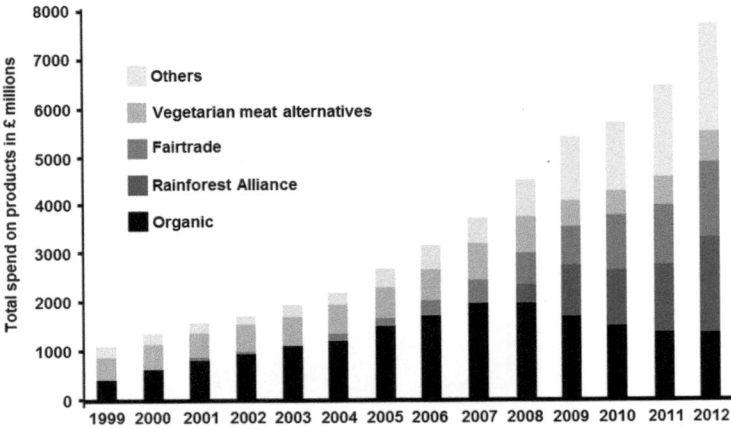

Fig. 10.6 There has been a steady increase in the purchase of ethical food and drink in the UK, although sales of organic food have decreased in recent years. Data from Defra (2014). Reproduced with permission the Department for Environment, Food and Rural Affairs, London

There are three reasons why people buy organic. The majority believe it is healthier (52 %), next come better animal welfare standards (34 %) and a similar number buy organic because they believe it to be a more ethical way of farming (33 %). So what has changed? Quite simply the consumers who are interested in sustainability are using different criteria in buying food such as: Is it in season? Is it local? Does it carry an ethical trading label such as Fairtrade? Have welfare issues been addressed? I must admit to being confused at some organic farmers markets where I have been confronted by out of season vegetables, exotic fruit and vegetables that clearly all have large air and road miles associated with them. Also authenticity is also another problem, especially with the price premium attached to many of these products sold in such markets. A carrot looks very much like another, and as it has now been proven beyond a shadow of a doubt that organically grown is not really better than conventionally grown fruit and vegetables in terms of health or taste. However, as taste is largely linked to freshness then locally grown food could potentially still have an advantage here. But we have to reassess the role of organic farming in terms of sustainability and its ability to supply the growing demand for food. The World Health Organization (WHO) estimate often quoted by organic farmers, 'that 3 million people are hospitalized annually due to pesticide poisoning', has been shown not to be relevant, as the trace pesticide residues found in conventional food, according to the Food Standards Agency in the UK, poses absolutely no risk to health.

Welfare of animals is normally higher with certified organic farmers, however there is sometimes a conflict between using proven chemical intervention (e.g. antibiotics, anti-inflammatory drugs and anthelmintics) and maintaining organic status. In the UK antibiotics are allowed to be used by organic farmers in certain circumstances but largely banned in the US. Farmers may be in a cleft stick, where necessary chemical intervention on welfare grounds could lose them their organic status.

Is organic farming sustainable? Probably not in terms of being able to feed an ever increasing global population. Professor John Reganold in a recent article in *Nature* demonstrated that in developed countries organic farmers are achieving up to 20 % smaller yields compared to conventional farmers which they are able offset financially by charging a premium for organic produce (Reganold 2012). In developing countries most organic fruit and vegetables are exported which brings desperately needed overseas currency into the country, but creates food scarcity within often highly productive areas. Organic certification standards are excellent in Ireland and the UK but do vary widely between countries and their certification bodies, some of which may cause significant confusion to the consumer. So very often the consumer is unaware of exactly what they are buying.

So in terms of sustaining and promoting biodiversity as well as protecting landscapes then organic farming is clearly advantageous over conventional farming, but the majority of farmers are now aware of the importance of these issues and are responding by using a broad range of conservation techniques. In terms of greenhouse gas emissions, going the organic route may not be all that it seems. Certainly soil fertility and quality improves under organic regimes, but research carried at Oxford University suggests that while pollution per unit area of land farmed organically is lower than for conventional farming, it is generally higher per unit of food produced.

But the four tenants on which the Soil Association is based which are health, ecology, fairness and care, are now increasingly at the heart of conventional farming as well. To this effect the Soil Association are now working with non-organic farmers which appears to be a sensible development for all sides of the farming industry. So should we buy organic? If it locally sourced and seasonal then it is preferred by me, but cost will always be a factor as is the need to develop farming to meet the challenges of climate change and increased food demand.

10.5 The Food Footprint

We briefly looked at our food footprint in Sect. 5.2 where we discussed generic values for food. There are many factors that go to create an emissions value for the food footprint. These include food production and type (40 %), cooking (29 %), packaging (5 %), disposal of food waste (3 %) and retail/refrigeration/storage (3 %). So the type of food you eat and how it's produced is the single largest component in your dinner's carbon footprint.

Those items carried by boat are 100 times more efficiently transported than those by planes. For example, bananas have an emissions footprint of 480 g CO_2e per kilogram (g CO_2e kg^{-1}) of fruit transported by sea, while oranges are 500 g CO_2e kg^{-1} when transported by sea compared to 5500 g CO_2e kg^{-1} when air freighted. Strawberries have emissions of 600 g CO_2e kg^{-1} when grown locally in season compared to 7200 g CO_2e kg^{-1} when air freighted. Vegetables tend to fall into two categories bulk and luxury items. So carrots grown locally have a footprint of

250 g CO_2e kg^{-1} compared to 1000 g CO_2e kg^{-1} when brought into the country by international road freight. Asparagus is more of a luxury item and has associated emissions of 500 g CO_2e kg^{-1} when local and seasonal compared to 14,000 g CO_2e kg^{-1} when out of season and air freighted from Peru. Flowers are largely grown and imported from the Netherlands, but a single red rose air freighted from Kenya has emissions of 350 g CO_2e compared to 2100 g CO_2e when transported to the UK or Ireland by international road freight from a heated greenhouse in the Netherlands. So generally speaking in terms of carbon dioxide emissions then transport by sea is best, international road freight is not so good and air freight is worst unless it's an alternative for out of season produce grown in a heated greenhouse.

> **Anything grown and picked in your own garden or allotment in season is equivalent to 0 g CO_2e kg^{-1}**

Cheap non-renewable fossil fuel energy makes intensive agriculture and long-distance transportation economically viable, and has allowed food production and distribution to become global industries. The prices in shops do not reflect the full cradle-to-grave environmental and social costs, so our need to keep food affordable while ensuring carbon emissions are reduced by implementing carbon taxation on fossil fuels are at odds.

More information: Paxton, A. (1994). *The Food Miles Report: The dangers of long-distance food transport*. SAFE Alliance, London. This report is published by the Sustainable Agriculture Food and Environment (SAFE) Alliance which merged with the National Food Alliance in 1999 to become Sustain: the alliance for better food and farming. http://www.sustainweb.org/publications/?id=191 http://www.sustainweb.org/

Red meat is a major source of GHGs as are dairy products, with beef producing 17 kg CO_2e per kg of meat purchased (Fig. 10.7). Beef generally has the highest footprint due to enteric fermentation which produces a rich mix of methane and CO_2 which the animal releases by frequent belching. Compared to other sources of animal protein such as wild fish and chicken, which have the lowest GHG emissions, cattle are very inefficient in the assimilation of energy. Emissions from beef are slowly being reduced by changing the animal's diet, slaughtering animals much younger, by better breed selection and genetic modification. Ireland is the most efficient milk producer in Europe producing less than 1 kg CO_2e for each litre of milk produced and the fifth most efficient beef producer with 20 kg CO_2e produced per kg of beef. **Food Harvest 2020** is a Government initiative to create a more efficient and profitable agricultural sector in Ireland. It aims to increase production significantly while reducing GHG emissions from the sector by 20 %. Run in conjunction with this is a separate scheme **Origin Green**, which are food producers who have committed themselves to reducing their ecological footprint, and currently over 250 companies and farmers have joined the scheme.

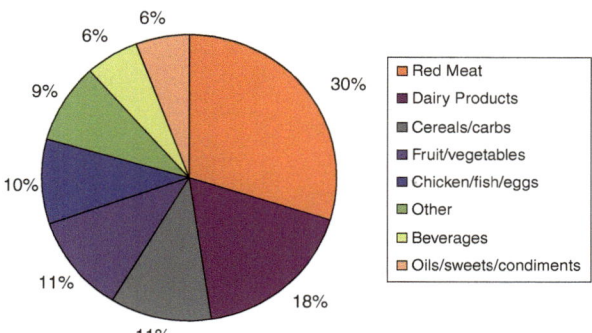

Fig. 10.7 Developed world average greenhouse gas emissions from food. *Source*: USEPA, http://epa.gov/climatechange/ghgemissions/. Reproduced with permission of the US Environmental protection Agency, Washington DC, USA

More information: *Food Harvest 2020* http://www.agriculture.gov.ie/agri-foodindustry/foodharvest2020/; *Origin Green* http://www.origingreen.ie/

Consumption varies from country to country depending on tradition, culture and food availability. So each country tends to have a national diet, although variations can be seen regionally and even locally, and of course we all have individual preferences. Below is a list of the percentage contribution of different food groups to the total emissions (CO_2e) for food production, processing and distribution as related to food consumption in **Sweden**.

28 % Meat and meat products
15 % Milk, milk products (excl. Butter)
11 % Vegetables and root crops
10 % Cereals
7 % Fish and marine products
6 % Fruit and berries
4 % Dietary fats
2 % Potatoes
0 % Legumes

The Swedish Environment Agency carried out a study on personal carbon emission in 2005 and discovered that 25 % of the personal footprint could be directly related to consuming food, equivalent to 2 t per person per year. Tackling food emissions is seen as a major step towards meeting the National GHG reduction targets, as is encouraging farmers to lower emissions on food production. As we saw in Sect. 5.2, food is often excluded from personal footprint calculators.

For the average US citizen red meat is by far the biggest contributor to their food footprint (Fig. 10.8).

Fig. 10.8 Major sources of
GHGs for the average diet of
a US citizen

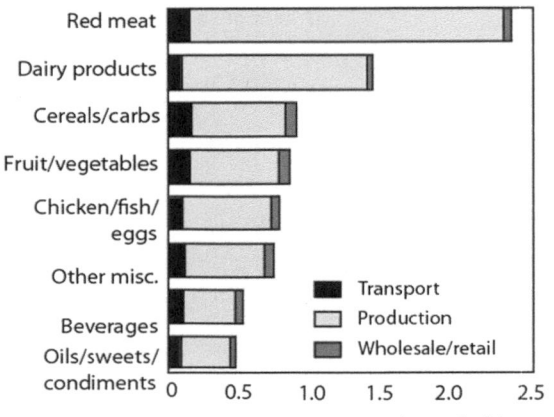

CO$_2$e emissions in tonnes per household per year

Table 10.3 Carbon footprint
in Cardiff study broken down
by food category

33 % Meat and meat products
28 % Milk and dairy
10 % Fruit and vegetables
7 % Cereals
4 % Sugar and confectionary
4 % Oils, fats and spreads
1 % Other
10 % Alcoholic beverages
10 % Non-alcoholic beverages

Source: Cardiff Health Alliance.
Reproduced with permission,
Cardiff Council, Cardiff, Wales

A large study on people's diet and CO$_2$ emissions carried out in Cardiff showed that 23 % of the average personal footprint comes from food and drink when this is incorporated into the overall calculation. They found that the average per capita consumption per year was 72 kg of meat, 143 kg of dairy produce and 196 kg of fruit and vegetables. When this was converted into the carbon footprint it was found that **as a percentage meat and dairy contributed 61 % of the total emissions** (Table 10.3).

By replacing meat with cheese and dairy produce occasionally each week reduced CO$_2$ emissions by 5.9 %, while just replacing a high carbon meat such as beef with a low-carbon meat such as pork reduces the footprint by 26 %. The findings showed that by eating less overall but especially high carbon meat such as beef could significantly reduce personal emissions.

> **Eating less overall but especially high carbon meat such as beef will significantly reduce your personal GHG emissions.**

10.5.1   Calculating the Food Footprint

It is very difficult to accurately calculate your food emissions using carbon foot-printing, simply due to the complexity of the sources and production elements of the foods that we eat, as well as the shear variety of the food we eat on a daily basis (Sect. 10.5.2).

With have no idea about the associated emissions with food once it is brought into our homes. Refrigeration of food is thought to represent about 2 % of the domestic GHG emissions in the UK, with the longer it is stored the higher the foot-print for that item will be. Different preparation and cooking methods also create emissions, with the cooking method in particular an important source of potential emissions. A oven is less efficient than a microwave, although the former also miti-gates its emissions for cooking by also heating the kitchen. Normally these emis-sions are included in the energy usage of the house, but clearly the more you cook at home the greater the energy used and the higher your household footprint will be.

One of the leading companies that has worked in the area of developing carbon footprints for food is Clean Metrics (http://www.cleanmetrics.com/). The company using life cycle analysis and standard protocols, provides a wide range of software that allows you to breakdown your raw products supply chains, the production of the product and your own supply chains in terms of carbon emissions (Sect. 10.5.2). They have produced a very useful basic calculator that gives you the GHG emis-sions for a range of basic food stuffs broken down into production, transportation and waste. While this is an American based calculator, so it uses non-metric inputs (i.e. miles and pounds) and assumes that most food is coming from within the coun-try so that some foods are ignored, it is still give a useful insight of where we can reduce our food related emissions. What is interesting is that often the transport and waste elements of a product are quite small compared to the emissions from its production (Fig. 10.7). Cheese is a highly processed food so it has high emissions from its production. So the manufacture of a single pound weight of cheese pro-duces 4.46 kg of CO_2. If we assume a 10 % wastage by the consumer then this adds another 0.02 kg CO_2 per pound of cheese, but the transport would only be 0.01 kg CO_2 if from a local supplier just 10 miles away compared to 0.07 kg CO_2 per pound when produced 1000 miles away from the retailer. So the overall footprints are 4.49 and 4.55 kg of CO_2 per pound of cheese supplied to the consumer respectively (excluding transport emissions from the supermarket to the consumer's home and emissions arising from storage) that is a difference of just 60 g of CO_2 or 1.3 %. It is worth while exploring the relationship between these three elements and also for your own waste levels.

More information: http://www.foodemissions.com/foodemissions/Calculator.aspx

Bangor University has created a unique carbon footprinting service for farmers called Footprints 4 Food (http://www.footprints4food.co.uk/). They are able to cal-culate the production and transportation footprints using life cycle analysis and then identify where emissions are being produced and how they can be reduced by the creation of a carbon reduction plan. An example of this is shown below in Fig. 10.9.

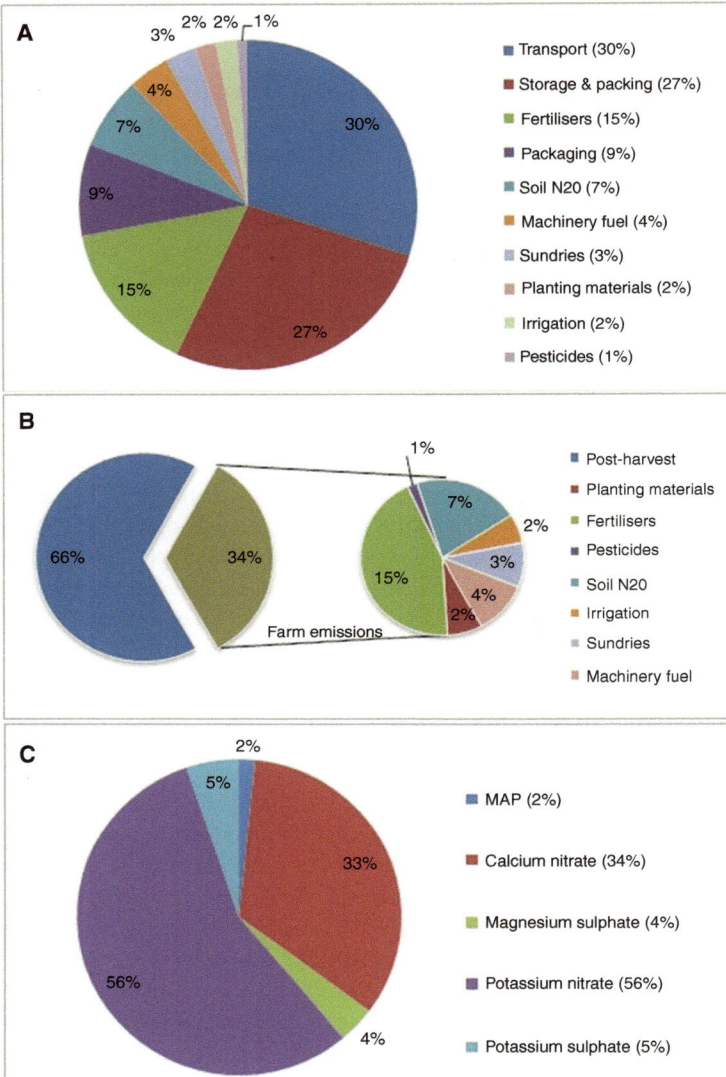

Fig. 10.9 Each carbon intensive area in the life cycle of the product is studied to produce the total carbon footprint (**A**). Next the footprint is divided into farm and post-harvest emissions that can identify where high emissions are being created (**B**). Finally those hotspots can be further interrogated to identify where the emissions are coming from (**C**). From this recommendations can be made to reduce carbon emissions as well as save farmers money, another win-win situation. *Source*: Footprint4Food. Reproduced with permission of Ian Finlayson and Footprints4Food, http://www.footprints4food.co.uk/

Personal food calculators are very variable in their detail and input data. For example, some rely on simply asking you lifestyle or food preference questions such as frequency of eating meat, eating seasonally, recycling packaging or organic produce. Others are far more detailed and ask about weights of different food groups purchased over the week, their source and waste rates (Clean Metrics' Calculator) other rely on your daily calorie intake (Cool Climate Calculator). The calculator from the Marion Institute is based on the monthly expenditure in six food categories…meat, fish and eggs, dairy, eating out, other products (which includes snacks and drinks), fruit and vegetables and finally cereal and bakery products. A draw back with most food calculators is that they tend to be country specific.

10.5.2   Examples of Food Calculators Are

Food Carbon Emissions Calculator (Clean Metrics)
 http://www.foodemissions.com/foodemissions/Calculator.aspx
Carbon Footprint Calculator
 http://calculator.carbonfootprint.com/calculator.aspx
Food Carbon Calculator, Carbon Neutral
 http://www.carbonneutral.com.au/carbon-calculator/food.html
Food Carbon Footprint Calculator
 http://foodcarbon.co.uk/index.html
Carbon Calculator, The Marion Institute
 http://www.marioninstitute.org/serendipity/gaviotas-carbon-offset-initiative/
 carbon-calculator
Carbon Calculator, Friends of the Earth
 http://www.foe.ie/justoneearth/carboncalculator/
Carbon Calculator, Cool Climate
 Http://coolclimate.berkeley.edu/carboncalculator

Of course the easiest way to accurately determine your carbon emissions for the food that you eat is for food to carry a carbon emissions label. These have been around since 2006 when pioneered by the Carbon Trust in the UK. This has been attempted in some countries such as Sweden who began trials in 2008. The Nutrition Department of Sweden's National Food Administration has devised their own labeling system with the overall objectives of reducing carbon emissions and improving health. However, many believe that such a system could be susceptible to abuse by companies claiming offsets during production thereby hiding the true emissions of products. Others prefer a simple colour coding system for foods indicating a range of GHG emissions from very high to low. Increasingly restaurants are giving the calorie count for items on their menus, and likewise the calorie content of food is given on all packaged food. So it is possible to get an estimate of your footprint from your total calorie count. Some general carbon emissions per kg of food is given in Table 10.4.

Table 10.4 Generic conversion factors for some basic foods per kilogram as kg CO_2e

Food group	Food	CO_2e emissions (kg CO_2e per kg food)
Meat	Beef	13.3
	Sausages	8.0
	Bacon/ham	4.8
	Poultry	3.5
	Eggs	2.0
	Pork	3.3
Dairy	Butter	23.8
	Hard cheese	8.5
	Cream	7.6
	Soft cheese	2.0
	Margarine	1.4
	Yogurt	1.3
	Milk	1.0
Bread	Brown bread	0.8
	White bread	0.7

More information: http://www.carbontrust.com/client-services/footprinting/footprint-certification/carbon-reduction-label

So what is our food footprint? Using the values from the Swedish and Cardiff studies then it would appear that we are each emitting 1.5–2.0 t of food associated CO_2e each year. This equates to approximately 340 kg per person per week. This would increase the primary personal footprint for Ireland as measured by Kenny and Gray (2009) from 5.7 to 7.45 t per capita per year (Sect. 5.2).

10.6   Can We Reduce Our CO_2e Emissions in Our Food?

The answer is a resounding yes. All it requires is for us to follow some simple rules which will reduce wastage, improve our diet, improve our health, and also increase our bank balance, all at the same time.

Eat what you buy (Potential Emissions Reduction−25 %)

- Ask people how much they want before you serve them
- Eat the skins of vegetables and fruit
- Plan meals—what needs to be used now?
- Keep vegetables in the fridge
- Rotate contents of your cupboards so as to always eat older stuff first
- Eat a bit less

Eat less red meat and dairy produce (Potential Emissions Reduction—25 %)

- Reduce intake rather than eliminate by reducing portion size
- 50 % reduction as much as 20–25 % reduction in overall food emissions
- Reduce the consumption of red meat and in particular beef
- Eat lower down the food chain

Always read the label (the closer to home produced the less CO$_2$) (Potential Emissions Reduction—10 %)

- Reduce use of hothouse grown fruit and vegetables
- Eat seasonally

  - **October–March**: lettuce, peppers, asparagus, tomatoes, strawberries, cut flowers have come from hothouse or on a plane [eat sparingly]
  - Apples, oranges, bananas all come by boat [eat away]

- Grow your own
- Support local growers/suppliers

Overall buying local is not as important as *what* you eat.

Avoid low-yield varieties (Potential Emissions Reduction—10 %)

- Cherry tomatoes
- Baby sweet corn
- Baby vegetables

Other factors worth considering

- Avoid excessive packaging (Potential Emissions Reduction—3–5 %)

  - Packing is required for hygiene and convenience.
  - However, can be just aesthetic/totally unnecessary
  - A metal dish inside a plastic sleeve inside a plastic bag in a cardboard box seems excessive.

- Recycle all food and packaging waste (Potential Emissions Reduction—5 %)
- Accept misshapen vegetables/fruit (Potential Emissions Reduction—1 %)
- Low-carbon cooking (Potential Emissions Reduction—5 %)

  - Turn down heat to sustain boiling. In fact if you are cooking rice or boiling an egg, once the water has boiled with the rice or egg in place then cover and turn off the heat and it will cook as normal.
  - Always put a lid on pans to save heat
  - Use a microwave whenever possible

- Buy in bulk
- Buy Fairtrade

Eating all that you buy; Reducing meat and dairy; Being more seasonal and avoiding hothouse and air freighted produce will *half* your GHG emissions for food

The meat industry is responsible for about 20 % of all global man-made GHGs and this is such a problem that the Chairperson of the IPCC has asked people to consider limiting their consumption of meat as a step towards reducing emissions. One of the ways this is being done is through not eating meat for 1 day per week. A scheme started in the USA, **Meatless Mondays**, has become a global initiative and has a potentially huge impact not only on GHG emissions but also on personal health as well.

More information: http://thebottomline.as.ucsb.edu/2012/11/meatless-mondays-as-a-program-for-environmental-stability-and-public-health

10.6.1   Food Waste

A study carried out by the Swedish Institute for Food and Biotechnology and pub-lished by the FOA in 2011 calculated that globally approximately one third of all food produced, that is 1.3 billion t per year, is wasted. These losses occur through-out the supply chain from agricultural production to consumption within the home (Fig. 10.10). On average 9 % is actually wasted in the home which works out at

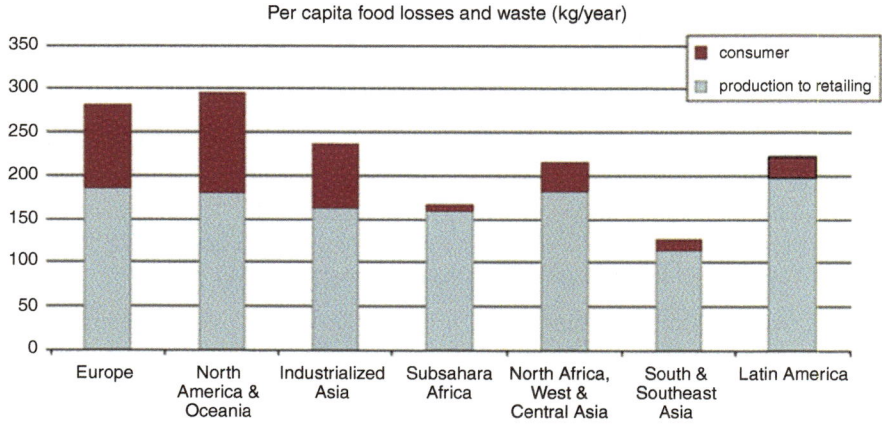

Fig. 10.10 Comparison of food waste by region per capita (kg per annum) showing the relative losses during production and supply and due to consumer wastage. *Source*: Gustavsson et al. (2011). Reproduced with permission of the Food and Agriculture Organization of the United Nations, Rome, Italy

50 kg per capita globally. Wastage rates vary within and between countries, but on average the per capita loss in North America and Europe is between 95 and 115 kg per year with most of this occurring at the household level. This is in stark contrast to developing countries in south and south eastern Asia as well as sub-Saharan Africa where the per capita food wastage is just 6–11 kg per year which are largely attributed to losses during production and storage, with little wasted at the household level.

In developed countries significant losses occur at both ends of the supply chain, with large losses due to quality standards that require specific shape or colour criteria allowing perfectly good fruit and vegetables to be discarded; and also due to restrictive grower-buyer agreements that can often lead to large wastage at farms. So in the developed world our choice of food, who we buy it from and how we manage it in the home has significant effects on the amount of edible food that is wasted.

More information: *Gustavsson, J., Cederberg, C., Sonesson, U., van Otterdijk, R. and Meybeck, A. (2011) Global food losses and food waste: Extent, causes and prevention. FAO, Rome.* http://www.fao.org/docrep/014/mb060e/mb060e.pdf

From this study it has been possible for the FAO to calculate that food waste is equivalent to a staggering 3.3 Gt of CO_2e emissions annually which makes it the single largest source of GHGs after the USA and China. This also means that 30 % of the world's agricultural capacity is being wasted and a staggering 250 km² of water is also being lost unnecessarily. Just like food wastage itself there is huge variability between regions that go to making the global average carbon footprint due to food wastage of 170 kg CO_2e per capita per year (kg ca⁻¹ year⁻¹), varying from just 25 kg ca⁻¹ year⁻¹ in Sub-Saharan Africa to more than 350 kg ca⁻¹ year⁻¹ for the USA. Apart from just simply being more wasteful one of the reasons why the carbon footprint is so much higher in developing countries is that we tend to have a higher proportion of high emissions foods in our diet such as meat and dairy produce or foods that have high post-production emissions such as processed foods or ready meals (Figs. 10.11 and 10.12).

More information: *FAO (2013) Food wastage footprint Impacts on Natural Resources: Summary report. FAO, Rome.* http://www.fao.org/docrep/018/i3347e/i3347e.pdf

A similar situation is seen in the US where food is wasted throughout the supply chain (i.e. farm, processing plants, distribution, storage, at retail outlets, in restaurants and at home), with up to 40 % of food never being consumed. A study by the National Resources Defense Council (NRDC) shows that there are numerous and often complex reasons for these losses but that one area of concern is the retail outlets where food that has short shelf lives such as meat, fish, dairy, fruit, vegetables, eggs, baked products and ready meals, in fact all perishable goods, make up the vast bulk of food that is wasted (Gunders 2012). This has resulted in discarded food becoming the largest single component of solid waste going to landfill. The NRDC make a range of proposals to reduce wastage throughout the supply chain including waste audits and better management of perishable food items at retail outlets. Preventing food waste doesn't just reduce CO_2 emissions it also saves us money

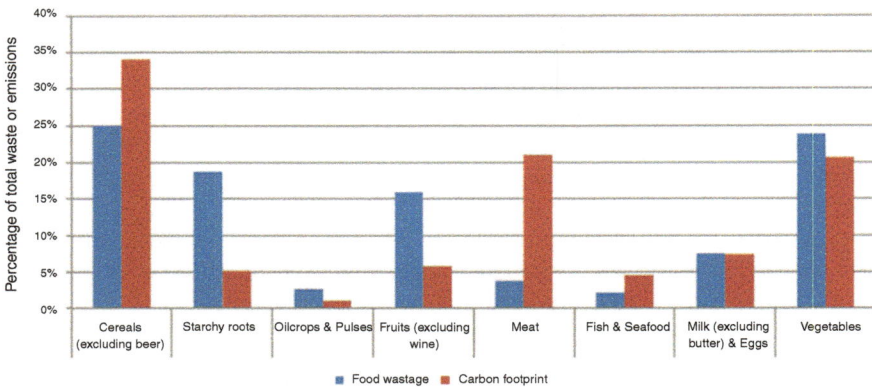

Fig. 10.11 Contribution of each food category to overall food wastage and carbon footprint as global average. *Source*: FAO (FAO 2013). Reproduced with permission of the Food and Agriculture Organization of the United Nations, Rome, Italy

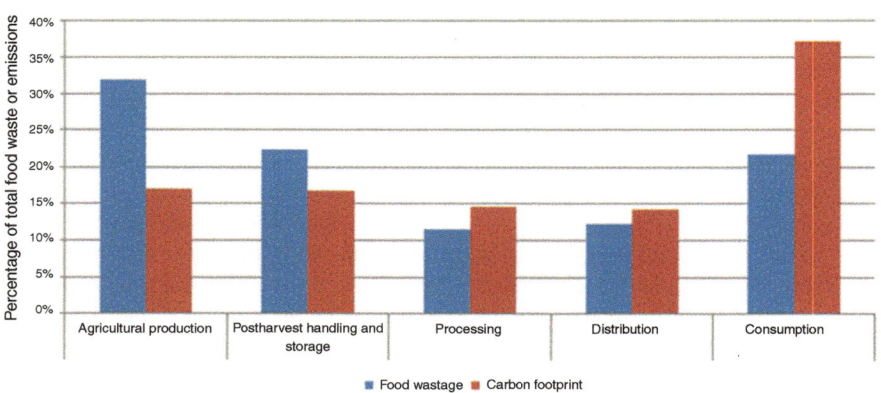

Fig. 10.12 Contribution of each step in the supply chain to overall food wastage and carbon footprint as global average. *Source*: FAO (2013). Reproduced with permission of the Food and Agriculture Organization of the United Nations, Rome, Italy

with the average American family of four currently throwing away $2275 as wasted food each year.

> **A 15 % reduction in food waste in the US would be equivalent to feeding 25 million US citizens each year**

More Information on food waste in the US: *Gunders, D. (2012) Wasted: How America is losing up to 40 percent of its food from farm to fork. NRDC Issue*

Paper: 12-06-B, National Resources Defense Council, New York. http://www.nrdc.org/food/files/wasted-food-IP.pdf

Each year in the UK 6.7 million t of food is wasted by households which is equivalent to every individual throwing away 70 kg of food each and every year. That is a whopping 30.8 % all food purchased simply wasted. Actually that is quite good compared to Ireland where the wastage is four times greater at 280 kg of food each and every year. This wastage is costing UK households on average £420 each per annum, that is £10.2 billion in UK as whole. There is also 1.2 million t of packaging thrown away on food that has never even been unpacked.

Globally less waste means more food to go around and less demand which means lower prices

Food wastage in the UK has been studied by WRAP (Sect. 12.5). They found that 30 % of all food bought is thrown out. The main food wastes are potatoes, bread, apples and meat and fish, with 50 % of all salads bought also thrown out. A staggering 35 % of all bread bought is thrown out, while 25 % of all fruit purchased is thrown out. The study attempted to find out why food was never eaten and so was wasted and among the commonest reasons given were: left on the plate, passed its sell by date, looked, smelt or tasted bad, mouldy or left over from cooking.

Apart from this being just wasteful, this also means a lot of wasted journeys by articulated lorries. To be precise there are 359,000 t of potatoes wasted each equivalent to 51,285 lorry journeys, 328,000 t of bread wasted equivalent to 46,857 lorry journeys, 190,000 t of apples wasted another 27,142 lorry journeys, and 161,000 t of meat and fish meals requiring 23,000 lorry journeys. So **transporting that unwanted food in the UK is equivalent to 15 million t CO$_2$e or equivalent to taking one in four cars off road**.

The food wasted in Europe and the USA could feed the world three times over!

Ireland has one of the highest wastage rates of food in Europe currently at over a million tonnes per each year. To tackle the problem the Irish Environmental Protection Agency has launch a new campaign backed by new legislation. The *Waste Management (Food Waste) Regulations 2009* aims at getting more food waste composted or digested for energy, minimizing waste from catering establishments. It has also set up a new campaign 'Stop food waste' to encourage individuals to reduce food wastage. It offers a wide range of advice and tips on effective shopping, better storage and use of food and cooking (Fig. 10.13).

Examples of help offered by site

TOP SHOPPING TIPS!

Don't go shopping when you are hungry - you'll buy more than you need!

If you are shopping for the week try and plan your meals ahead - check out Sians Plan for a great planning system.

Check your fridge, freezer and cupboards before you go shopping and plan meals around what you find.

Then make a shopping list....and then try to stick to it!!

Beware of special deals - these are great for toilet rolls and shampoo but bad for fruit, veg and salads (anything that can go off quickly). These are the things we buy because of a "good deal" but often does not get eaten.

Try and buy loose fruit and veg - you get what you need and can cut down on packaging wastes in your bin as well.

Check use-by dates to avoid buying food that might get thrown out if not eaten immediately.

Poke around at the back of shelves - you'll often find 'use-by dates' that are further away.

Shop for what you actually eat, not for what you want/wish you would eat (e.g. "I am going to be really healthy this week and eat lots of yogurts!") and then not eat them!

If its an option for you, try shopping on line for the basics - you get what you want and save money - it's like magic!

STORAGE TIPS!

Use your own judgement when it comes to throwing food out - food that can be eaten is worth money. **Use by dates** should be followed, **Best before dates** are a guide. For more information on the various dates used check out our Dates Guide. *But remember that food isn't like Cinderella - it doesn't go off at midnight!*

Labels such as **Sell By** and **Display Until** are used for stock control by shops and are of no interest to householders

Keep all dairy products in the fridge. As the saying goes 'milk left out for an hour is the same as a day in the fridge!'

If you are not going to use meat or fish, freeze it, or cook it and eat it in the following days. Also, if you decide to use just some of it, freeze the rest.

Use your freezer but don't forget what's in there for 3 years!

Make sure your fridge and freezer are maintained properly - this will ensure food is cooled properly and saves on electricity costs too

Supermarkets are smart and use stock control to maximise profits. Try this at home - you'll waste less and save yourself some cash.

Make sure fruit and veg are stored in the correct place - for more information on different fruit and veg click here (PDF).

if you are ever unsure where to store stuff - copy the shops - they try to preserve fruit and veg. for as longs as possible.

Fig. 10.13 Examples of ways of reducing household food wastage on the 'foodwaste.ie' website. *Source*: EPA, http://www.stopfoodwaste.ie/. Reproduced with permission the Environmental Protection Agency, Wexford, Ireland

More information: http://www.epa.ie/whatwedo/advice/waste/foodwaste/; http://www.stopfoodwaste.ie/index.php

A key initiative that is preventing food waste is encouraging producers and suppliers to donate food to charities and food banks. This not only includes farmers and supermarkets, but other retailers such as bakeries, cafes and restaurants. In Ireland one organization **Foodcloud** has managed to bring those with excess food together with charities by using a simple phone app. The suppliers indicate what food is left over at the end of the day and via the app a charity will accept the food and then arrange to pick it up. So the amount of food wasted is minimized, the company saves on disposal costs and the charities are able to get food to those who urgently need it.

More information: http://foodcloud.ie/

Another useful innovation is a new website run by Bord Bia which determines what is in season and available in Ireland thereby avoiding imported out of season fruit and vegetables.

More information: http://www.bestinseason.ie/whats-in-season/

We all have the potential to do more both in terms of using our food supplies better or by growing some of our own food. There are some really useful resources available and two exceptionally inspirational and useful cook books are *Forgotten Skills of Cooking* by Darina Allen and *Frugal Food* by Delia Smith. Both of these books give really useful insights not only how to save money but how to really make most of the food that you buy. Another important aspect of understanding food is growing your own. Not everyone has a garden or can get an allotment, but we can in fact grow food in window boxes, in pots almost anywhere. An interesting book *Square Metre Gardening* by Mel Bartholomew gives some useful tips in how you can intensively use very small spaces to grow a surprising array of plants. There are numerous community groups who are using local open spaces and wasteland to grow food and they are always looking for volunteers. Local schemes include garden sharing, especially with older people who are unable to look after their garden are often delighted to have help to tend a vegetable plot in exchange for a share of the produce, mini allotments, using road side verges and roundabouts and many more.

One of the many examples is the Incredible Edible initiative which was started by a group of volunteers in Todmorden in England who turned plots of unused land into communal vegetable gardens. The idea has gone global and similar schemes can be found all over the world. So can the individual or community make a difference? You bet you can.

More information: http://www.ted.com/talks/view/lang/en//id/1538

> **The tenth step is accepting that food waste is a critical and unnecessary factor in carbon emissions and that by buying only what we need, careful selection of what we eat, and eating everything that we buy, that we can significantly reduce emissions as well as increase global food reserves.**
> **This step also involves you getting involved with growing your own food, even if its fresh herbs in a pot on your windowsill or desk.**

10.7 Conclusions

- Food poverty affects a significant portion of the global population with hunger and starvation linked closely to climate
- Global warming is having a significant impact on agriculture by changing where, what and how we grow food. Climate change will lead to significant migration from rural to urban centres caused primarily by failure of local agriculture, with developing countries not exclusively, but worst affected.
- The supply and demand balance is being critically tested by the removal of so much land from food production to grow biofuels, often exacerbating food shortages. This is making it increasingly difficult to reliably feed the growing global

population as well as making good food deficits caused by national crop failures with resulting in massive rises in food prices that leave people unable to afford their normal food.

• Calculating greenhouse gas emission for food is complex and relies on numerous factors. So reduction in emissions is down to reducing wastage and the frequency of your selection of food items such as red meat and dairy, which have high emissions, in your diet.

Homework!

Start off by keeping a detailed food diary for a week or longer if possible. This will allow you to work out how frequently you eat high emission foods such as beef and dairy produce, or fruit and vegetables from less sustainable sources. Try and reduce the frequency that you eat these items.

Go through your store cupboards and fridge and put all the older dated cans, bottles and packets in the front so that they are used first. If something is getting close to its use by date ensure that you design a meal that uses it before the date runs out.

Then simply eat healthily, waste nothing and always read the label before purchase and put all the extra money you save into your pension plan because you are going to live a lot longer.

Lastly review your options about growing some food of your own. Remember that being wise about the food you eat can significantly reduce your personal carbon footprint as well as save you money.

References and Further Reading

Allen, D. (2009). *Forgotten skills of cooking*. London, England: Kyle Cathie.

Bartholomew, M. (2013). *Square metre gardening*. London, England: Frances Lincoln.

Coley, D., Howard, M., & Winter, M. (2008). Local food, food miles and carbon emissions: A comparison of farm shop and mass distribution approaches. *Food Policy, 34*, 150–155.

Defra. (2014). *Food statistics handbook 2013*. London, England: Department for the Environment, Food and Rural Affairs. Retrieved from https://www.gov.uk/government/uploads/system/uploads/attachment_data/file/315418/foodpocketbook-2013update-29may14.pdf

FAO. (2013). *Food wastage footprint impacts on natural resources*: *Summary report*. Rome, Italy: FAO. Retrieved from http://www.fao.org/docrep/018/i3347e/i3347e.pdf

Gustavsson, J., Cederberg, C., Sonesson, U., van Otterdijk, R., & Meybeck, A. (2011). *Global food losses and food waste*: *Extent, causes and prevention*. Rome, Italy: FAO. Retrieved from http://www.fao.org/docrep/014/mb060e/mb060e.pdf

IPCC. (2007). *IPCC fourth assessment report*. *Working group I report*: *The physical science basis. WG1-AR4*. Geneva, Switzerland: Intergovernmental Panel on Climate Change. Published on behalf of the IPCC by Cambridge University Press, Cambridge, England.

IPCC. (2012). *IPCC fourth assessment report. Working group I: Technical summary (Revised)*. Geneva, Switzerland: Intergovernmental Panel on Climate Change. Published on behalf of the IPCC by Cambridge University Press, Cambridge, England.

Kenny, T., & Gray, N. F. (2009). A preliminary survey of household and personal carbon dioxide emissions in Ireland. *Environmental Impact Assessment Review, 29*, 1–6.

Reganold, J. P. (2012). The fruits of organic farming. *Nature, 485*, 176–177. Retrieved from http://www.nature.com/nature/focus/organicfarming/

Saunder, C., & Barber, A. (2008). Carbon footprints, life cycle analysis, food miles: Global trade trends and market issues. *Political Science, 60*, 73–88.

Smith, D. (2008). *Frugal food*. London, England: Hodder & Stoughton.

Food Poverty and Hunger

Making food poverty history: Labour's Blueprint for Eliminating Food Poverty (2011)
http://www.labour.ie/download/pdf/document.pdf
http://www.worldhunger.org/articles/Learn/world%20hunger%20facts%202002.htm
http://www.cpa.ie/research/foodpoverty.html

Food Scarcity

World Resources Institute. (2014). Retrieved from http://www.wri.org/
FAO. (2008). Retrieved from http://www.fao.org/climatechange/en/
IPCC. (2007). Retrieved from http://www.ipcc.ch/

Food Waste

Gustavsson, J., Cederberg, C., Sonesson, U., van Otterdijk, R., & Meybeck, A. (2011). *Global food losses and food waste: Extent, causes and prevention*. Rome, Italy: FAO. Retrieved from http://www.fao.org/docrep/014/mb060e/mb060e.pdf

Gunders, D. (2012). *Wasted: How America is losing up to 40 percent of its food from farm to fork*. NRDC Issue Paper: 12-06-B. New York, NY: National Resources Defense Council. Retrieved from http://www.nrdc.org/food/files/wasted-food-IP.pdf

FAO. (2013). *Food wastage footprint impacts on natural resources: Summary report*. Rome, Italy: FAO. Retrieved from http://www.fao.org/docrep/018/i3347e/i3347e.pdf

EPA Ireland: http://www.stopfoodwaste.ie/

Food Miles

Paxton, A. (1994). *The food miles report: The dangers of long-distance food transport*. London, England: SAFE Alliance. Retrieved from http://www.sustainweb.org/publications/?id=191

Defra. (2012). *Food transport indicators to 2010*. London, England: Department for the Environment, Food and Rural Affairs. Retrieved from https://www.gov.uk/government/

uploads/system/uploads/attachment_data/file/138104/defra-stats-foodfarm-food-transport-statsnotice-120110.pdf

Impact of Global Warming on Agriculture

http://www.ifpri.org/publication/climate-change-impact-agriculture-and-costs-adaptation

Counter View to Saving Air Miles

http://www.alternet.org/food/145673?page=entire

The American Food Carbon Footprint

http://attachments.brighterplanet.com/press_items/local_copies/52/original/carbon_foodprint_wp.pdf

Diets

http://www.plosone.org/article/info%3Adoi%2F10.1371%2Fjournal.pone.0062228

Chapter 11
Where Does Water Fit in?

Above children in India collect water during drought. Roughly one-third of the world population is estimated to live in areas of water scarcity. In this chapter we examine the problem of both water scarcity and security and how this will be affected by global warming. The idea that water is a peak resource is explored as is the use of water footprinting. Image by A. Ishokon/UNEP, http://unep.org/. Reproduced with permission of the United Nations Environment Programme, Nairobi, Kenya

11.1 Introduction

I remember as a student seeing a slide presentation entitled *Water is life,* which outlined the importance of water in our everyday lives and that without it life itself would not exist. I have seen that title used again and again over the years, but in essence the message behind the title has remained the same, that is water is a precious resource and you pollute it or overuse it at your peril.

© Springer International Publishing Switzerland 2015
N.F. Gray, *Facing Up to Global Warming*, DOI 10.1007/978-3-319-20146-7_11

At a metabolic level water really is the reason for all biological life on our planet. It is critical for all metabolic processes and indeed our survival on a day to day basis. Each and every species is primarily composed of water, from man at about 60–70 % to small creatures such as protozoa and bacteria at 99.99 % water. Water also has a profound effects on our planet's climate. The natural greenhouse effect is caused primarily by water vapour, while the radiative balance at the Earth's surface is modified by snow and ice cover. The distribution of vegetation types is sensitive to the local water balance, with tropical rainforests major drivers of climate including monsoons, while overall regional climate patterns are influenced by ocean currents (Sect. 1.3).

So water in one form or another controls our weather and sustains all life forms so it's pretty important. But what that slide presentation instilled in me was that on a day to day basis it is easy to forget how completely we all depend on water whether it is for drinking, food preparation, washing clothes and dishes, flushing our waste away, and for general hygiene. Water is also critical for our food production, for watering livestock and crop irrigation, then for its processing and supply. Water drives industry and manufacturing, power generation, and is used both directly and indirectly for recreation as well. So water drives the economy as well as sustains communities. Climate change is going to have a major effect on water resources and water availability that will impact on all of us in some way.

The volume of water on the planet remains constant, although its form (i.e. water vapour, precipitation, river or groundwater, seawater, snow, ice) varies as does its chemical quality. Water is constantly being cycled through a system known as the hydrological cycle which is driven by solar energy (Fig. 11.1). This creates water vapour via evaporation or transpiration from vegetation and returns to the surface as rainfall or snow. So when you look at the cycle there seems to be plenty of water about, but in fact only 2.5 % of the water in the cycle is freshwater and of this a third is tied up as ice in the poles or glaciers, while the other two-thirds is found in

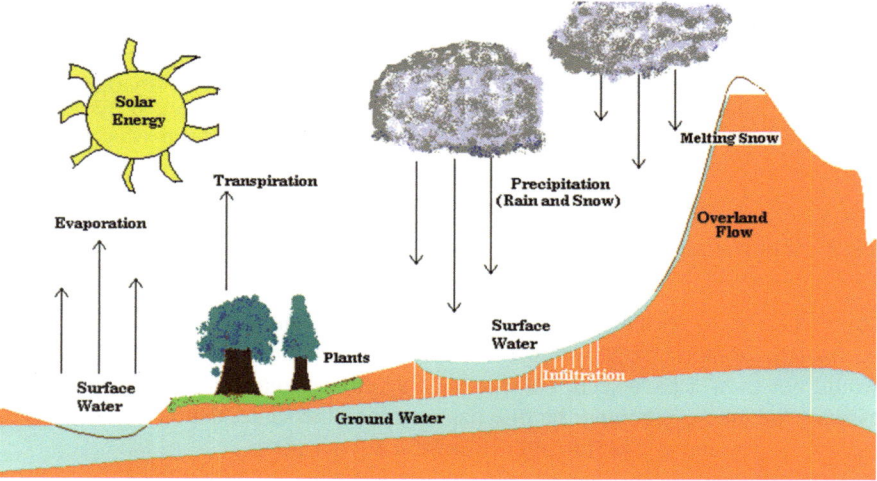

Fig. 11.1 The hydrological cycle showing how water is cycled globally driven by heat from the sun

groundwater, much of which is found in areas of low population or just so highly mineralized or polluted so that it is unsuitable for supply purposes. This leaves <0.2 % present as freshwater in rivers, lakes, and as soil moisture. So in reality water supplies are quite scarce and need to be managed to quite a high degree to ensure continuity of sufficient supplies to satisfy all our needs. A major problem is that as urban areas continue to rapidly grow these supplies become increasing over-subscribed, and urbanization has mainly developed in drier areas so demand frequently exceeds supply.

Replenishment of water resources and soil moisture is from rainfall and as this is not equally distributed then demand and supply are often at odds. In the UK the greatest demand occurs where the largest populations are located, in south-east and north-west England, which does not correspond to the areas where water resources are adequate, so shortages can and do occur (Fig. 11.2).

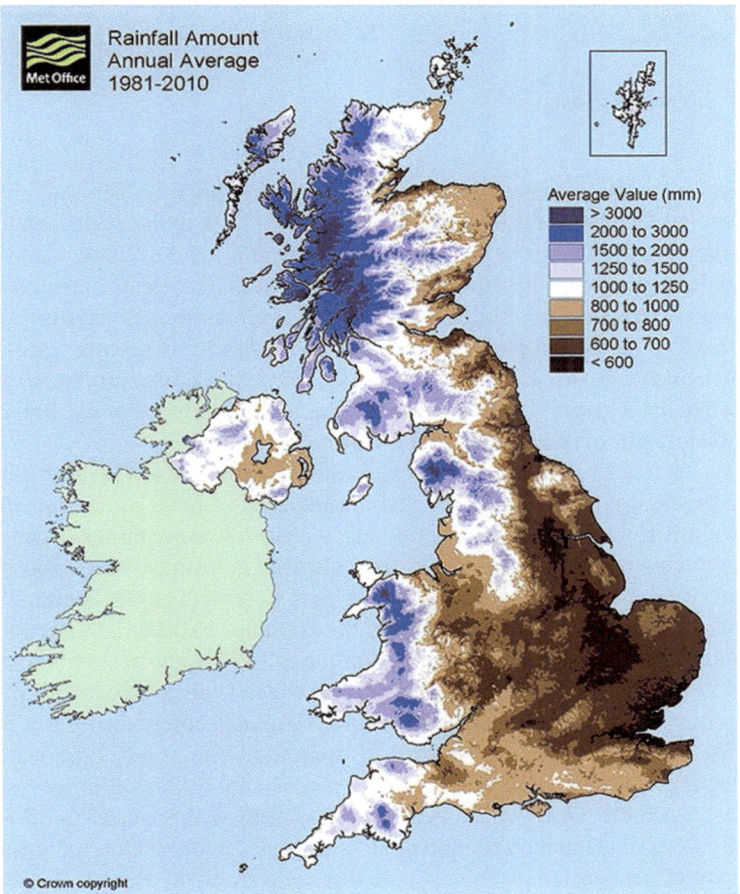

Fig. 11.2 Average rainfall distribution in the UK over the period of 1981–2010. *Source*: The Met Office, http://www.metoffice.gov.uk/learning/rain/how-much-does-it-rain-in-the-uk. Reproduced with permission the Met Office (Crown Copyright), Exeter, UK

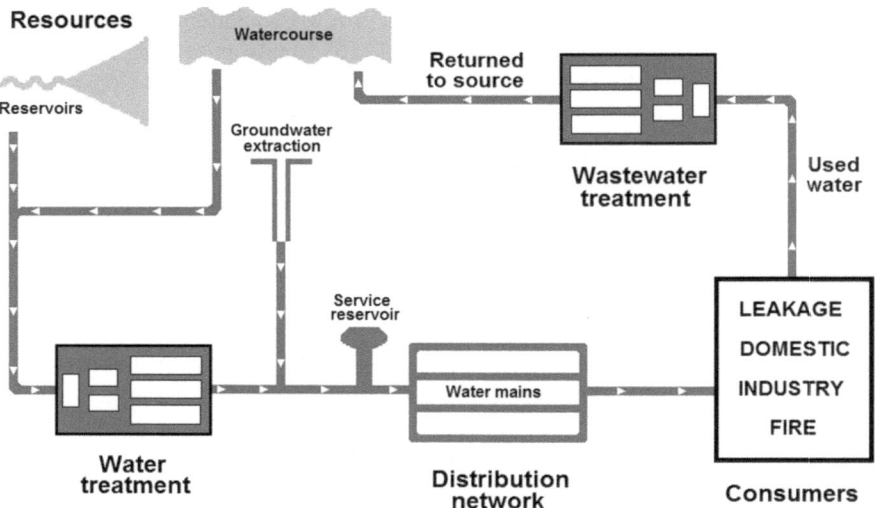

Fig. 11.3 The human water cycle

We borrow water from the hydrological cycle so that householders and industries can utilize it. Once used it is returned after treatment back to the cycle usually via a river; a process known as the **human water cycle** (Fig. 11.3). When water is used for agriculture the water is returned to the cycle through evapo-transpiration and evaporation or groundwater infiltration. Water can be reused over and over again on its way back to the ocean via rivers, but this can result in us leaving traces of our waste such as salt, oestrogens, pharmaceutical and personal care products, caffeine, illegal drugs in fact traces of everything that we use and dispose of via water that is not completely removed by wastewater treatment.

The Water Industry, in conjunction with State Agencies and Government, deals with the processes and mechanisms that are required to manage the human water cycle. It's function is to provide continuous and sufficient quantities of safe palatable drinking water for both domestic and industrial consumers and dispose of the used water to prevent environmental damage and to protect public health. The size of the industry is impressive and in the UK for example, 20,000ML of potable water are supplied to consumers each day which requires 1344 water treatment plants and 326,471 km of distribution mains to bring the water from the treatment plants to your home or place of work. Once used it has to be treated before disposal. Currently in the UK 1,000,000 cubic meters (m^3) of domestic wastewater is produced each day (there are 1000 L in each cubic metre) along with 7,000,000 m^3 of industrial wastewater each day. Wastewater treatment also deals with surface runoff as well, especially from roads, which is equivalent to another 20,000,000 m^3 of wastewater requiring treatment each day. To deal with this there are 9260 wastewater treatment plants located throughout the country serving 95 % of population. The sizes of these plants serve populations ranging from less than a 100 to 1,700,000, so it is big business both in terms of infrastructure and also in terms of money.

Table 11.1 Typical breakdown of water use in a UK household, although these figures vary widely for different families and individuals

Use	%	Use	%
Toilet	35	Wash basin	8
Kitchen sink	15	Outside use	6
Bath	15	Shower	5
Washing machine	12	Dishwasher	4

11.1.1   Water Use

In the past 50 years there have been enormous changes in water use. In the past people would bath and wash clothes once per week, and dishes would have been universally washed by hand. Today it is not unusual to shower or bath several times a day, with clothes changed each day resulting in washing machines used numerous times throughout the week often without a full load. Dishwashers have replaced hand washing and indeed modern kitchen sinks are not really designed to accommodate dish washing either being too small or too deep. Thus the pressure on water supplies is increasing due to changes in our lifestyles and expectations, as well as rising populations, new housing developments, and reducing household size.

A household of two adults and two children consume approximately 510 L per day (L day^{-1}) or half a cubic metre of water. The per capita consumption rate is very variable but averages out at between 150 and 180 L day^{-1}. All water is treated to the same high standard but in practice less than 5 % is used for either drinking or food preparation with the largest percentage used for flushing the toilet (Table 11.1). Laundry washing machines can use on average of 100 L (L) each time they are used while a dishwasher uses about half this volume of water per wash. A bath can use on average 90 L while a shower can use as little as 5 L per minute (L min^{-1}), although power showers use more than 17 L min^{-1}. Outside the house hosepipes can use considerable quantities of water with a garden sprinkler for example using on average 1000 L per hour (L h^{-1}).

11.1.2   Water Scarcity

The demand for water varies significantly between countries due to differences in climate, economic wealth and culture. The USA and Australia have the highest global consumption rates at 580 and 495 L per capita per day (L ca^{-1} day^{-1}) respectively which is 40–60 times more than used by people in water scarce areas (Fig. 11.4). The UN have identified that the minimum requirement for water is 50 L ca^{-1} day^{-1} which is known as the **water poverty threshold** and is based on a daily allocation of:

- 5 L for drinking
- 20 L for sanitation

Fig. 11.4 Water usage per country in litres used per person per day. *Source*: WHO, http://www.who.int/en/. Reproduced with permission of the World Health Organization, Geneva, Switzerland

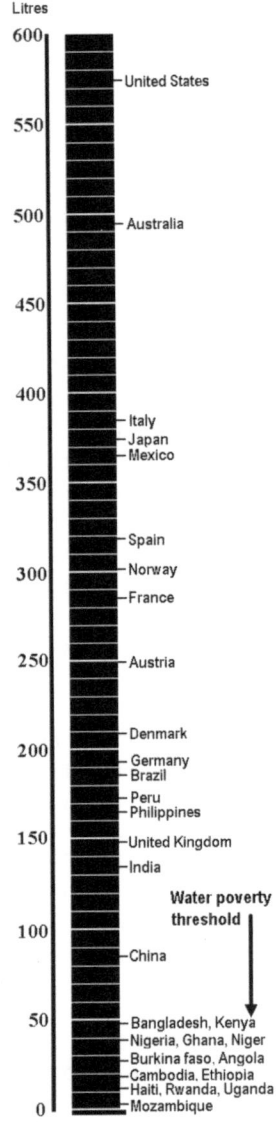

- 15 L for washing
- 10 L for food preparation

In theory the volume of water available for supply is based on the resources available. The threshold between having and not having adequate resources is called the **water stress threshold** and is estimated to be 1700 cubic metres per capita per year (m^3 ca^{-1} $year^{-1}$). Few countries have such high per capita usage rates, for example the average in Europe is just 726 m^3 ca^{-1} $year^{-1}$, while in the USA it is double

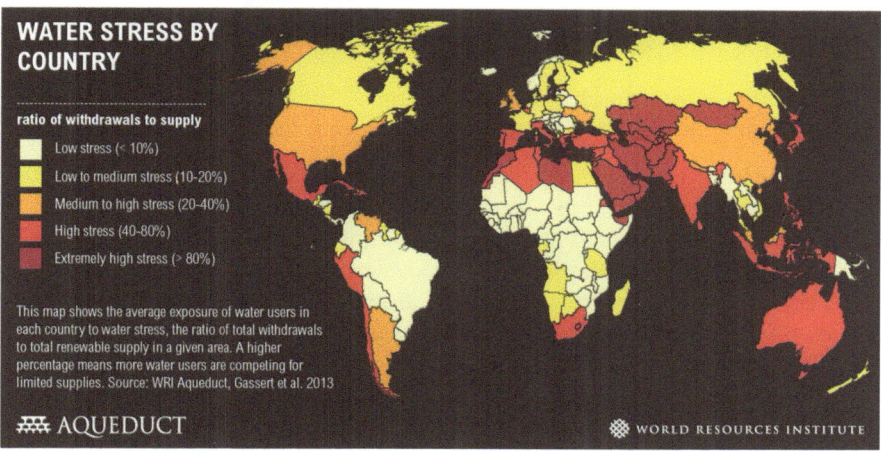

Fig. 11.5 Water scarcity is already common in areas which will be most affected by global warming. *Source*: Water Resources Institute, http://www.wri.org/publication/aqueduct-country-river-basin-rankings. Reproduced with permission the Water Resources Institute, Washington, DC, USA

this at 1693 m^3 ca^{-1} year^{-1}. These figures also include the requirements of industry and agriculture, and so for Europe the average domestic usage is just 10 % of the total figure at 70 m^3 ca^{-1} year^{-1}. While there is sufficient water globally to meet all human consumption, agriculture and industrial needs, including energy generation, over 700 million people live below the water stress threshold, with this figure set to rise to three billion by 2025 (Fig. 11.5). For example, in Brazil over 60 % of their energy comes from hydro-electricity. The falling water levels in the past decade due to over abstraction has led to reduced generating output resulting in a serious energy deficit and a rise in energy prices by 60 % overall.

The World Economic Forum identifies water scarcity as one of the top risks to business. Without water business and industry just can't operate leading to job losses and the breakdown of communities. For example, in 2005 Coca-Cola's Indian subsidiary based in Kerala was accused by surrounding communities of depleting local water resources. Although not proven, it forced the company to shut down. Often where water shortages are exacerbated by the real or perceived actions of others then this can lead to conflict at local regional or International levels.

More information: http://www.weforum.org/issues/global-risks

According to the US Department of Agriculture, in 2012 70 % of US national disaster areas were created by a combination of high temperatures and drought conditions, with widespread failure of soybean and corn crops. So predicting water scarcity is vitally important to both economic and social stability. However, it is difficult to measure and predict water stress with maps like Fig. 11.5 which is based on past records only and with very poor resolution. The World Resources Institute has launched a new online water scarcity mapping system based on 12 different water risk indicators, such as physical water stress, water quality, regulatory and reputational risk, and groundwater stress information. The free model is called *Aqueduct* and is being widely adopted by businesses.

World Resources Institute*:* http://www.wri.org/
Aqaueduct*:* http://www.wri.org/our-work/project/aqueduct

Over one billion people lack access to safe drinking water with a further 2.3 billion living in areas of severe water scarcity, i.e. areas with insufficient rainfall to reliably farm. Many of these are managing on just 5 L a day for all their needs! There are also 2.6 billion people with inadequate sanitation, so seen in conjunction with inadequate drinking water access then as many as 50 % of people in developing countries are suffering from health effects directly associated with these problems. Those living in areas of water scarcity are set to increase to 3.5 billion by 2025 and 4.5 billion by 2050 due to global warming changing weather patterns. These areas are not just in developing countries but include some of the most important cereal growing areas in developed countries in Europe and the USA. This has been explored in detail in a United Nations report ***Beyond Scarcity: Power, Poverty, and the Global Water Crisis***. Published in 2006 by the United Nations Development Programme (UNDP) based in New York.

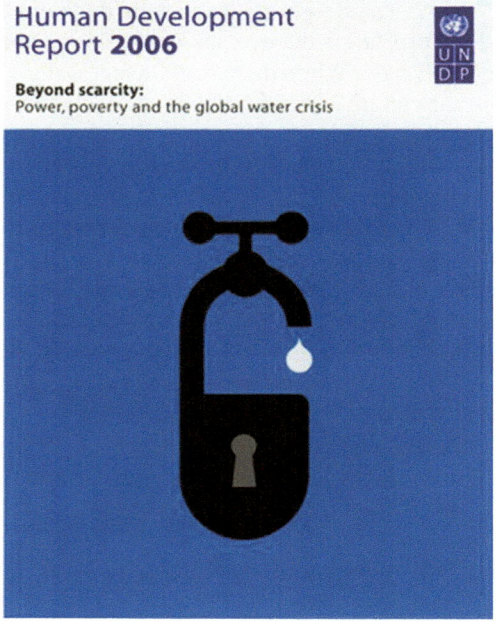

Reproduced with permission the United Nations Development Programme, New York, USA. http://www.undp.org/

The report: http://hdr.undp.org/en/media/HDR06-complete.pdf

...[current] shortages are generally due to poor policy decisions rather
than absolute scarcity, resulting in resources coming increasingly
under threat and over-exploitation.'
UNDP 2006

11.1.3   Water Conflict

Water is not only used for supplying drinking water it is also abstracted for irrigation and to sustain industry, commerce as well as agriculture. It is also used for waste disposal with the degree of pollution arising varying widely between countries. However, water resources have to be shared, but very often rivers, lakes and groundwater cross manmade boundaries. For example, the River Danube which is 2850 km in length passes through Germany, Austria, Slovakia, Hungry, Croatia, Serbia, Romania, Bulgaria, Moldova Ukraine, Switzerland, Slovenia Bosnia, Czech republic, and FR Yugoslavia (Fig. 11.6). So if a river is polluted upstream, then those downstream are unable to use the water. This may especially occur close to international borders where a country may not be able to exploit the resource any further and so may not be incentivized to treat its waste under these circumstances.

During the period 1950–2000 there were 1831 water based conflicts over transboundary river basins. Therefore, in order to avoid conflict the management of water resources must be shared equally between countries and all stakeholders. But conflict may also be local. In some areas of Pakistan water is controlled by landowners who have increasingly exploited water resources for irrigation making drinking water supplies scarce. In terms of availability 1226 m^3 of water used per capita is used each year for irrigation leaving just 26 m^3 per capita per year for domestic supplies.

Water is becoming increasingly scarce not only in poorer arid countries but throughout the developing and developed world. There is an increasingly strong link between water resources and international security leading to conflict. Peter Gleick of the Pacific Institute classifies these conflicts into six categories: (i) *The control of water resources* where water supplies or access to them are the primary origin of tension; (ii) *Use as a military tool* where water resources or water infrastructure are used as a weapon during a military action; (iii) *Use as a military target* where water resources or infrastructure are used as targets for military action; (iv) *Use as a political tool* where resources or supplies are used for a political goal; (v) *Terrorism* where water resources or infrastructure are used as either targets or tools of violence or coercion usually at a local level; and finally (vi) *Development disputes* where water resources or infrastructure used are source of contention and dispute in the context of economic and social development which can be local, regional or international.

Fig. 11.6 The River Danube passed through numerous countries making co-operation in the abstraction of water and the disposal of wastes between all countries imperative if the resource is to be preserved for use by all. Image REGIOgis/European Union, http://ec.europa.eu/regional_policy/index_en.cfm. Reproduced with permission of the European Union, Brussels, Belgium

More information: http://pacinst.org/

Water conflicts are caused or exacerbated by water scarcity

A new water ethic is needed where the water cycle is managed for the benefit of all, and this need is ever more urgent as global warming causes increased desertification, lakes and rivers to dry up and groundwater not to be replenished. The loss of mountain glaciers or snow melting earlier, both classic effects of global warming, results in significant changes in water resources, especially river discharge in the early spring and summer. Conflict over water and the unfolding global water crisis have been explored in detail by two excellent books *Water Wars: Drought, Flood, Folly and the Politics of Thirst* by Diana Ward and *When the Rivers Run Dry* by Fred Pearse.

11.2   Water Demand Management

The problem of dwindling water supplies and escalating demand is a worldwide problem and exacerbated in many areas by climate change and increasing migration within countries from rural areas to urban centres. The traditional engineering approach is to meet this demand by increasing water production by further utilizing dwindling water resources, often resulting in severe ecological damage (i.e. supply-side management). This approach is clearly non-sustainable and has lead to greater public involvement by water companies in trying to reduce water usage as well as attempting to reduce leakage from their distribution networks which in Dublin was as high as 40 % in some areas before large scale mains replacement works began in 2013. Water demand management (WDM) is a more holistic approach to managing water supplies and resources by moving away from expensive and unrestrained infra-structural development associated with increasing water demand by **setting an upper limit on water availability** so that new demands must be satisfied without increasing supplies; a management system that could equally be applied to reducing carbon emissions. This is called demand-side management and uses WDM to achieve its objectives through a range of integrated tools to manage water use including:

- Conservation measures
- Education
- Metering
- Charging
- Building Regulations that include water use minimization
- Increased use of water efficient appliances and fixtures

11.2.1   Water Conservation

This is a key element in managing water resources by ensuring water is not used in a wasteful manner. Water conservation is classed as either *behavioural*, which involves a change in daily water use habits, or *structural*, which involves invest-ment in water efficient technology, rainwater harvesting and the reuse of water.

- **Behavioural changes** include spending less time in the shower or even shower-ing less, washing clothes less frequently, always making sure both the dishwasher and washing machine are only used on full loads, turning off taps when not being used, etc.
- **Structural changes** include mending leaks, dripping taps or cisterns, purchasing water efficient appliances, replacing power showers with energy and water effi-cient shower systems, reusing water, etc. For example reusing the grey water collected from the shower, bath and washbasin to flush the toilet could save 35 % household's water demand, equivalent to approximately 18,000 L per household per year (Table 11.2). Rainwater harvesting can also supply all the water required

Table 11.2 Potential for water reuse

Type of water	Content	Potential use
Black	Urine + faeces	None
Brown	Faeces only	None unless dry composted over several years
Yellow	Urine only	Can be used as fertilizer in garden
Grey	Washing water	Flushing toilets
White	Runoff/rainfall from roofs	Unfiltered but debris free: Flushing toilets: Filtered: Laundry, hot water

for flushing the toilet. Reusing water is far more difficult than it seems in theory Retrofitting can be expensive both in terms of cost and GHG emissions so conservation of supplies and metering appear the simplest approaches.

There is also a huge potential to save water outside the household through a more thoughtful approach whether it is at work or outside work hours. Water is also used outside the home with typical water usages, for example, in offices at 70 L per employee each day, in hospitals at 300–500 L per patient per day and in hotels at 400–500 L per guest per day.

> **It is important that hygiene and public health is not compromised by the introduction of water conservation measures.**

11.2.2   Water Efficiency Labelling

We are all familiar with energy rating labels for household white goods such as washing machines and freezers, but similar labelling schemes for water use efficiency have been developed for all plumbing equipment and water using appliances from taps to shower heads, garden sprinklers to washing machines. This started in Australia which has been trying to cope with a massive water shortage associated in part with extensive droughts which have lasted for over a decade. Part of their response was to introduce a water efficiency scheme in 2005 backed by strict regulations called the Water Efficiency Labelling and Standards (WELS). Every water using product sold in Australia is required by law to be tested by the Government and issued with a WaterMark certificate and a WELS label (Fig. 11.7). Water efficiency for each product is ranked using a simple star system with one star the least efficient and six the most efficient. Details of all equipment covered by the scheme and sold in Australia are listed on an online data base so customers can check the efficiency of all available products. An example of water savings are replacing standard shower heads that use between 18 and 25 L per minute with a three star rated unit that use just 6 L of water per minute which not only reduces water usage by up to 40 % but will also reduce carbon emissions associated with the energy used to heat the water by 40–60 %. However the largest water saving comes from replacing

Fig. 11.7 Example of a water efficiency label from Australia for a dual flush toilet. It only scores fie out of a possible six stars for efficiency as there are even more efficient appliances available. *Source*: Water Efficiency Labelling and Standards (WELS) scheme, http://www.waterrating.gov.au/. Reproduced with permission of the Commonwealth of Australia

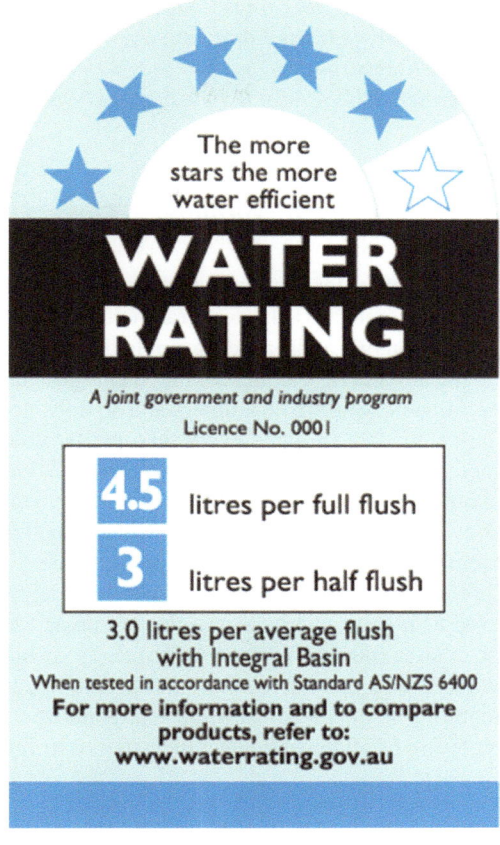

a standard 11 L flush toilet with a high efficiency dual flush system (Fig. 11.7) which can save up to 52 L of water per person per day. Gradually the least efficient products have been removed from sale ensuring water savings each time an appliance is replaced or new plumbing is installed. Also houses that are for sale must be retrofitted with high efficiency taps, toilets and showers before they can be sold. The country is well on target to reach its National target of saving 800,000 ML of water by the year 2021, as well as saving consumers a lot of money in expensive water charges. The scheme has also significantly reduced carbon emissions equivalent to removing 90,000 cars from Australian roads per annum.

More information: waterrating.gov.au

The Water Efficiency Labelling Scheme has been so successful it has been adopted by other countries including Singapore and Hong Kong. A similar labelling system, *Water Sense*, has also been launch in the USA by the Environmental Protection Agency which covers the same range of products as the WELS system.

WELS Scheme Singapore: http://www.pub.gov.sg/wels/Pages/default.aspx
WELS Scheme Hong Kong: http://www.wsd.gov.hk/en/plumbing_and_engineering/
   wels/index.html
Water Sense USA: http://www.epa.gov/WaterSense/about_us/watersense_label.html

11.2.3   Metering Supplies

The use of water meters is the single most effective measure to encourage water efficiency with a 20–45 % overall reduction in use achieved in most homes. Charging is based on two approaches, a single flat or fixed charge or a volumetric charge based on the amount of water used. Fixed charges generally do not get over the importance or value of water to consumers and so does not effectively encourage water conservation. In fact in some cases, especially when first introduced, fixed charges can increase consumption with some customers feeling that if they are paying for it they should get their money's worth. In contrast, volumetric charges have been repeatedly shown to significantly reduce water usage and encourage investment in water efficient products in the home. Charging is based on three concepts: (i) a fixed price per litre irrespective of volume used; (ii) increasing charge per cubic metre (m^3) of water used as higher volume thresholds are passed, or (iii) a fixed free volume per month or quarter (depending on the frequency of billing) and then charge for the excess volume used but at a much high rate than in option (i). In the UK the current level of metering is about 35 % while in Ireland the Government has agreed to universal metering of all households which started in July, 2013.

11.2.4   Household Water Use and CO_2 Emissions

The average annual per capita water usage in the UK and Ireland is 55,121 L ca^{-1} $year^{-1}$. To supply potable water and then subsequently collect and treat wastewater the energy costs are, on average, just 38.6 kg CO_2e ca^{-1} $year^{-1}$. So let's make that clear… your individual carbon footprint for water supply and wastewater treatment is on average just 38.6 kg CO_2e per annum or 0.7 g of CO_2e per litre of water used. However, while the actual supply and treatment of water is not hugely costly in terms of GHG emissions when supplied centrally via a water main and your wastewater is centrally treated, the energy used to heat water in the home can be very large indeed. In some cases heating water can be as high as 5036 kWh ca^{-1} $year^{-1}$ equivalent to 2.83 t CO_2 ca^{-1} $year^{-1}$ if using electricity. Natural gas reduces this by 63 %, so conversion to gas can save up to 4.36 t CO_2e $year^{-1}$ per average household. The optimal scenario is to use solar panels with natural gas as standby, plus water efficient appliances and careful use, which could reduce emissions to as little as 150 kg CO_2e ca^{-1} $year^{-1}$.

> **Saving water reduces GHG emissions very little, but helps conserve supplies in areas where water is scarce.**
> **Major GHG emission reductions can be achieved by selecting low-carbon energy options for heating water for use in the home and by using heated water carefully.**

More information: Hackett, M. and Gray, N.F. (2009) Carbon dioxide emission savings potential of household water use reduction in the UK. *Journal of Sustainable Development*, 2, 36–43. http://ccsenet.org/journal/index.php/jsd/article/viewFile/236/212

11.3   Peak Water

There is no doubt that global warming is going to change the pattern of water availability throughout the World. Some areas will get more rainfall, in some instances significantly higher amounts, or the same as before but as fewer and more intense rainfall events. In contrast other areas will get less rainfall leading to severe and possibly permanent drought. It remains difficult to be precise at this stage how it will affect specific areas and there will be local variations arising from more regional trends. Global warming will also lead to increased evaporation and plant evapotranspiration creating more water movement between the land and the atmosphere as well as melting snow and glaciers releasing more freshwater.

In terms of water resources there will be a continued increase in the loss of snow and ice which are often used as an important water supply resource. Less precipitation will lead to less surface water and less aquifer recharge, and less aquifer recharge means a gradual reduction in ground and possible surface water availability. Intense rainfall events will lead to greater loss of water as surface runoff, leading also to flooding and poorer water quality as we saw in the UK and much of Western Europe in December 2013 to February 2014. Overall demand for water will be driven by the expected increase in temperature, although resources will have been compromised by the more erratic climate. The current trend in increase demand due to urbanization and migration will continue as more people migrate to cities, and there will be an increased water demand for irrigation and livestock. Overall less water results in poorer hygiene and greater risks of disease and disease transfer. In areas where precipitation increases sufficiently, net water supplies may not be affected or they may even increase; however, where precipitation remains the same or decreases, net water supplies will decrease overall (Sect. 14.1).

In areas where snow is an important factor in water availability, the period of maximum river flow may move from late spring to early spring or late winter. Changes in river flow have important implications for water and flood management, irrigation, and planning. If supplies are reduced, off-stream users of water such as irrigated agriculture and in-stream users such as hydropower, fisheries, recreation and navigation could be most directly affected.

> **Global climate change is gradually reducing available water resources but at the same time creating greater demand—this is not a sustainable situation leading to PEAK WATER**

Peak water is reached when the rate of water demand exceeds the rate at which water resources used for supply can be replenished (Bell 2010). Therefore, **all water supplies can be considered finite as they can all be depleted by over exploitation**. So while the total volume of water in the hydrological cycle remains the same, the availability of water does alter. This is particularly true of aquifers (groundwater) and static water bodies such as lakes and reservoirs where the water may take a long time to replenish. So water availability is strongly linked to rainfall and the ability to retain this water within resources which is difficult as increasing intensity of rainfall reduces percentage infiltration.

Due to increasing demand from population growth, migration to urban centres and for agriculture, it is possible that a state of peak water could be reached in many areas if present trends continue. By 2025 it is estimated that 1.8 billion people will be living with absolute water scarcity and in excess of four billion of the world's population may be subject to water stress.

> **Peak water is not about running out of fresh water, but the peaking and subsequent decline of the production rate of the water.**

A question I am often asked is how does a renewable resource become finite? The answer is not as straight forward as first appears. Water availability is governed by a number of possible factors: over-abstraction (i.e. using it before it can be replenish thereby exhausting the supply and causing significant and often permanent ecological damage), not returning water back to hydrological resources after use, saltwater intrusion often caused by over-abstraction, pollution of resources and finally climate change effects (glacier loss, reduced stream flow, evaporation of lakes). Comparatively only a very small amount of water is regularly renewed by rain and snowfall, resulting in only a small volume of water available on a sustainable basis. So all water supplies have an optimal abstraction rate to ensure they are sustainable, but once exceeded then supplies are doomed to failure. As we saw in Sect. 3.2, a modified Hubbert curve applies to any resource that can be harvested faster than it can be replaced (Fig. 3.2). This applies to all water resources but especially to groundwaters.

Peak water is defined in three different ways according to the impact on the resource as: peak renewable, peak non-renewable or peak ecological water.

Peak Renewable Water comes from resources that are quickly replenished such as rivers and streams, shallow aquifers that recharge relatively quickly and rainwater systems. These resources are constantly renewed by rainfall or snow melt; however this does not mean these resources can provide unlimited supplies of water. If demand exceeds 100 % of the renewable supply then the "peak renewable" limit is reached. For many major river catchments globally, the peak renewable water limit has been reached. For example, in excess of 100 % of the average flow of the Colorado River is already allocated through legal agreements with the seven US States and Mexico. So in a typical year the river flow could now theoretically fall to zero before it reaches the sea. Similarly the River Thames can during periods of low flow fall below the volume of water abstracted. The river is prevented from drying up due to over-abstraction by returning wastewater after treatment to the river which is then reused numerous times as it approaches London. Due to the high population within the catchment, the Environment Agency has classified the area as seriously water stressed with towns and cities along the length of the Thames such as Swindon, Oxford and London itself, at risk of water shortages and restrictions during periods of dry weather.

Peak Non-renewable Water comes from resources that are effectively non-renewable aquifers that have very slow recharge rates, or contain ancient water that was captured and stored hundreds or thousands of years ago and is no longer being recharged (a problem that will be exacerbated by climate change), or groundwater systems that have been damaged by compaction or other physical changes.

Abstraction in excess of natural recharge rates becomes increasingly difficult and expensive as the water table drops which results in a peak of production, followed by diminishing abstraction rates and accompanied by a rapid decline in quality as deeper more mineralized waters (i.e. increasingly salty to the taste) are accessed. Worldwide, a significant fraction of current agricultural production depends on non-renewable groundwater (e.g. North China plains, India, Ogallala Aquifer in the Great Plains of the United States) and the loss of these through over-exploitation threatens the reliability of long-term food supplies in these regions.

When the use of water from a groundwater aquifer far exceeds natural recharge rates, this stock of groundwater will be depleted or fall to a level where the cost of extraction exceeds the value of the water when used, very much like oil fields. The problem is that climate change often results in less rainfall creating a greater dependence on aquifers for supply.

Peak Ecological Water is water abstracted for human use which leads to ecological damage greater than the value of the water to humans. The human population already uses almost 50 % of all renewable and accessible freshwater leading to serious ecological effects to both freshwater resources and transitional habitats such as wetlands. Since 1900, half of the world's wetlands have disappeared while approximately 50 % of freshwater species have become extinct since 1970, faster than the decline of species either on land or in the sea (Sect. 7.3). Water supports both man's need and that of its natural flora and fauna. These fragile environments

Fig. 11.8 Spring fed rivers in southern and south east England have been drying up in recent years due to a combination of drought and over-abstraction. Above is the River Frome in Stroud, Gloucestershire. Image: Environment Agency, https://davethroupea.wordpress.com/. Reproduced with permission of the Environment Agency, Bristol, UK

need to be preserved for overall planet health. The simple fact that water supply quality is often a close relationship with the ecosystem, with most water bodies able to self purify its water constantly removing pollutants and improving quality over-all. However, the problem has been in putting an economic value on ecological systems (sometimes referred to as ecological services) and nature as a whole; whereas water used by humans can be easily quantified economically. In the mis-taken assumption that such values are zero has led to them being highly discounted, underappreciated, or ignored in water policy decisions in many areas. Over-abstraction is a major problem in many rivers in southern England that are fed from the aquifer below. As more groundwater is abstracted then the water table falls as does the water level in the river (Fig. 11.8).

It is not only rivers that are drying up due to over abstraction and global warming but some of the largest freshwater lakes in the world such as the Aryl Sea and Lakes Chad and Victoria in Africa (Fig. 11.9).

In the USA, water abstraction and water use peaked during 1975–1980 but has stabilized since (Fig. 11.10). This should have affected economic growth but has been able to continue to grow by implementing better water management strategies to satisfy the new needs of industry. This has been achieved through water conserva-tion, stricter regulations, water efficient and improved technology, education, water pricing, etc. So US citizens are now using less water per capita than ever before. However, many regions of the U.S. face water scarcity (e.g. the arid west) and new areas of water scarcity continue to develop due to climate change (e.g. southeast and Great Lakes region) which all indicate that peak water has been reached (Fig. 11.11). The key question is how long can economic growth be sustained without water becoming a limiting factor?

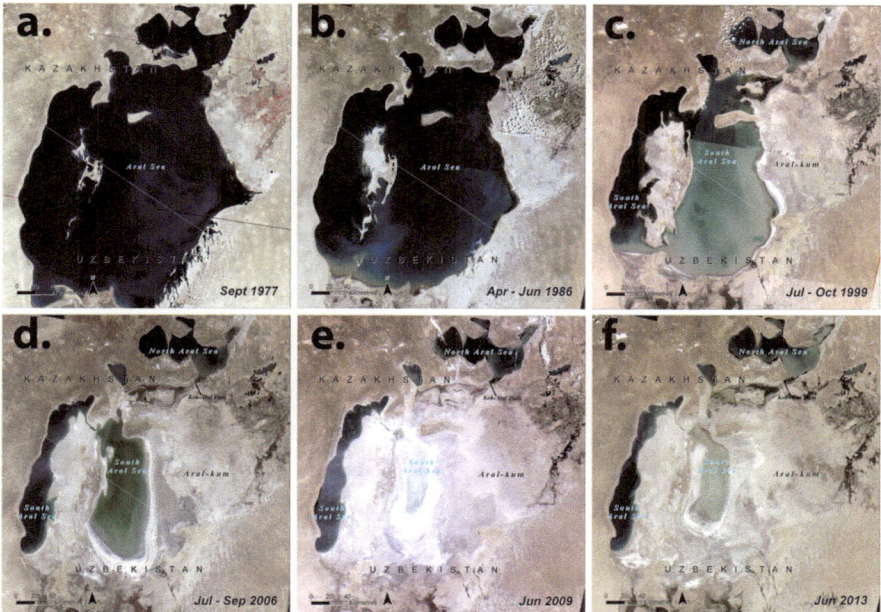

Fig. 11.9 The Aryl Sea was once a massive freshwater lake but is now rapidly shrinking due to excessive abstraction from the rivers that flow into it. The letters **a** to **f** show the time sequence of area since September, 1977 to June, 2013. As abstraction has continued the lake has become increasingly polluted, nutrient enriched and mineralized causing extensive ecological damage. This has happened since the mid 1970s! *Source*: UNEP, http://na.unep.net/geas/getUNEPPageWithArticleIDScript.php?article_id=108. Reproduced with permission of the United Nations Environment Programme, Nairobi, Kenya

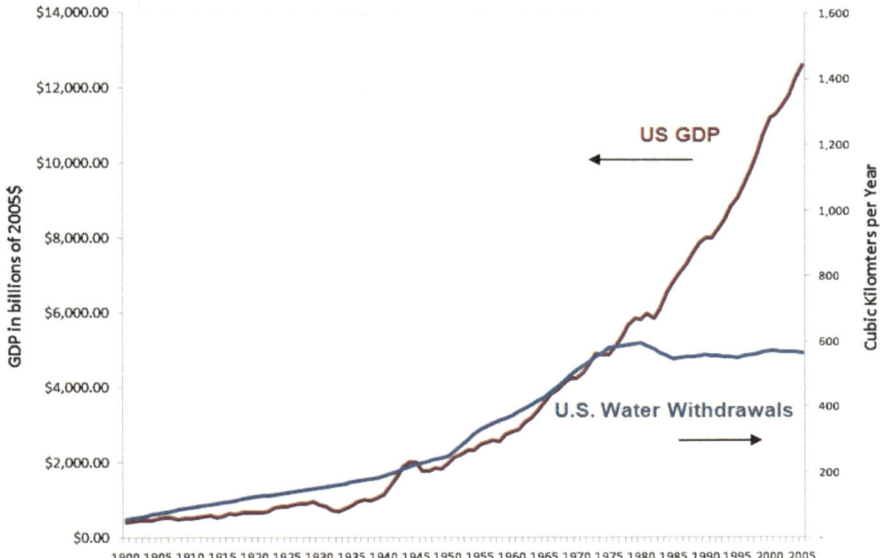

Fig. 11.10 Peak water in the USA has been reached, but continued economic growth has continued by implementing a water demand management approach to the available water supply which is now at peak. *Source*: Gleick and Palaniappan (2010). Reproduced with permission of the National Academy of Sciences, Washington, DC, USA

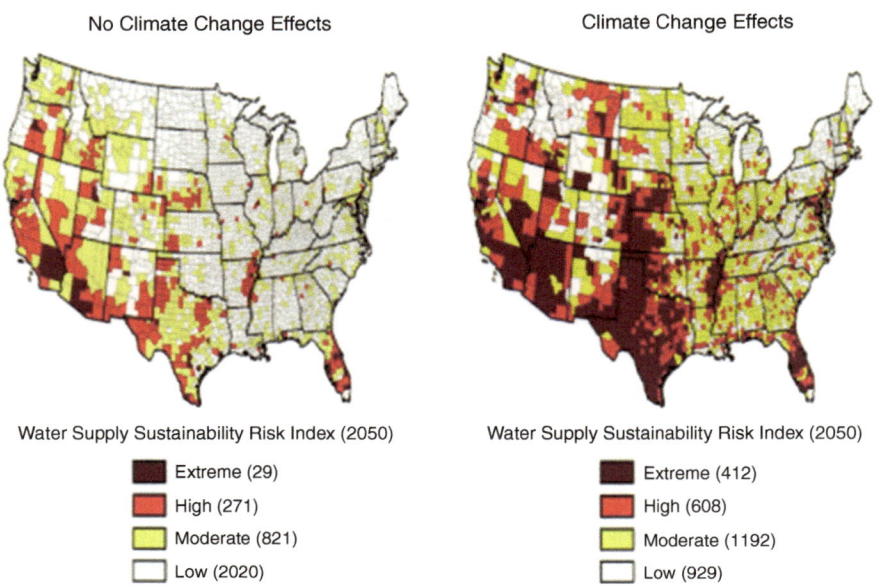

Fig. 11.11 Water supply sustainability index predicted for 2050. In the USA it is estimated that water shortages will become increasingly severe as a consequence of global warming. *Source*: The National Climate Assessment, http://www.globalchange.gov/. Reproduced with permission of the US Global Change Research Program, Washington, DC, USA

More   information: http://nca2014.globalchange.gov/highlights/report-findings/ water-supply)

Will water shortages affect us in Ireland and the UK? The straight answer is yes, and to some extent already is. No one is exempt from the peak water crisis. Due to global warming most arid regions will probably run out of water in less than two decades. In wetter areas, peak water has been reached due to: heavy use of water; pollution of resources (often associated with urbanization); infrastructure not being completed to keep up with demand (China, India) and finally inadequate infrastructure (London, Dublin).

Agriculture represents at least 70 % of freshwater use worldwide and with the demand for food soaring, especially as a result of climate change and increasing crop failure (e.g. China rice failure in 2011) then demand for irrigation and livestock watering will continue to be a major drain on supplies (Sect. 10.3). **Agriculture, industrialization and urbanization all serve to increase water consumption**.

Like peak oil, peak water is inevitable given the rate of extraction

Over-abstraction causes severe ecological damage as lakes dry up and rivers fed by groundwater disappear; a rapid reduction in water quality of groundwater due to mineralization and saltwater intrusion and increased exposure to pollution and pathogens. There are alternative methods of supplying water (i.e. supply-side management solutions) such as river transfer where water is pumped from one catchment to another using natural river systems, extended pipelines carrying water from areas of low demand to areas of high demand, international bulk water transfer using land and ocean going tankers which is already used to supply islands such as Gibraltar; desalination which is creating freshwater from sea water and even fog harvesting collecting water from sea mists and fog using fine nets.

11.3.1 Desalination

Desalination is an attractive alternative for water supply as it uses brackish or salt water which is readily available and so in theory could provide us with exhaustible quantities of freshwater. Currently there are over 14,000 desalination plants globally producing 45 billion litres of water per day, and while they are becoming increasingly efficient with modern designs, most plants are expensive to operate and polluting. The key problems with producing water by desalination is the relatively high capital costs to build the desalination plant and the energy required to desalinate the water both resulting in a high unit cost of the water produced. This cost can be increased by having to transport the freshwater inland from the site of production. For example, after being desalinized at Jubal, Saudi Arabia, water is pumped 200 miles (320 km) inland though a pipeline to the capital city of Riyadh. Desalination also produces concentrated brine which is normally discharged back to coastal areas often causing significant local environmental problems. Apart from these issues, desalination results in high GHG emissions. The commonest method, thermal distillation (multistage flash) (MSF), produces 23.41 kg CO_2 per m^3 of water produced (1 $m^3 = 1000$ L), while the modern multiple effect distillation (MED) produces 18.05 kg CO_2 per m^3. This is in contrast to the pressurized filtration method Reverse Osmosis which produces only 1.78 kg CO_2 per m^3 of freshwater produced.

Desalination produces 78 times more GHG emissions compared to the average emissions associated with the supply of water in Ireland and the UK using conventional technology which is about 0.3 kg CO_2 per m^3 of water supplied rising to 0.7 kg CO_2 per m^3 if wastewater treatment is also included. Therefore, desalination is usually only employed as a last result, when all other supply options have been exhausted being often used as a supply-side management stop gap. For example, Spain is now heavily reliant on desalination in some areas to maintain its fruit and vegetable sector.

Membrane systems are increasingly being adopted in Europe to top supplement supplies during periods of water scarcity as they are cheaper than normal desalination plants and produce less GHG emissions. For example, in 2010 Thames Water plc opened a new reverse osmosis water treatment plant in London which has the capability to produce 150 ML of high quality drinking water per day using the brackish water from the River Thames. The plant, which supplies 900,000 people, runs on renewable energy and will be used only at times of water shortage, which is the most sustainable way of using such technology.

Beckton Water Treatment plant: http://www.thameswater.co.uk/media/10888.htm

11.4   Water Footprints

There are two approaches to measuring water use, either measuring your actual personal use of water as it comes out of the tap, or measuring all the water that has been utilized in the supply of the goods and services that you use. Water footprints can be calculated for an individual, a single business or whole countries being the sum of all direct and indirect water employed in the production of goods and services used (Fig. 11.12).

The direct water usage is the total water entering home which can be easily measured using a water meter or a water diary. The indirect water usage is far more complex to measure and requires the use of life cycle analysis. Indirect water is expressed as the **virtual water** content of product (m^3 of water per tonne of product produced) and is further broken down as **Internal** (water used to manufacture products made in own country); **External** (water used to manufacture products imported into the country) or **Net Imported** which is the difference between the external and internal water footprints.

We use a lot of water in the manufacture of goods and the provision of services, so water use, like carbon emissions, is linked to consumerism and consumer choice.

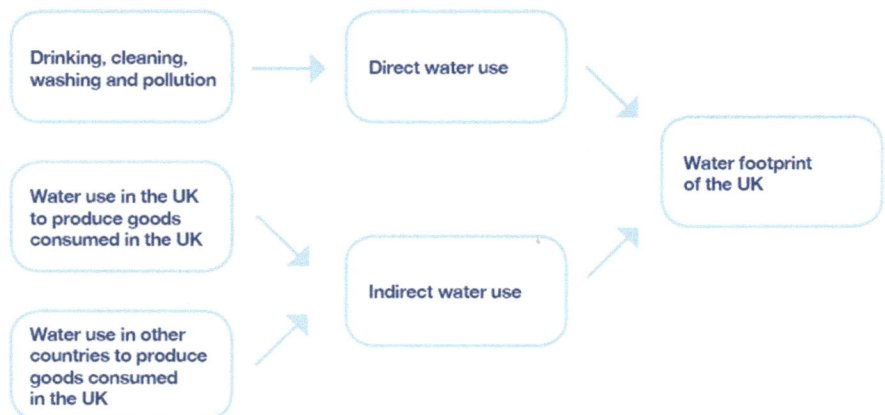

Fig. 11.12 Calculation of the UK water footprint is broken down into direct and indirect water use. *Source*: Chapagain and Orr (2008). Reproduced with permission of the Water Footprint Network, Enschede, The Netherlands. http://www.waterfootprint.org/

Table 11.3 The amount of water used in the production of common household items per unit volume or weight

Per litre	
Water	1 L
Bottled water	4 L
Beer	300 L
Milk	1000 L
Per cup or glass	
Tea	120 L
Orange juice	850 L
Wine	960 L
Coffee	1120 L
Meat per kg	
Lamb	6100 L
Beef	15,000–70,000 L
Eggs	3300 L
Main foodstuffs per kg	
Bread	1300 L
Rice	3400 L
Tea	9200 L
Roasted coffee	21,000 L
Household items (each)	
Pair jeans	10,850 L
Cotton Shirt	4500 L
Cotton sheet (1 kg)	11,000 L
Disposable nappy	810 L
The average car	400,000 L

Examples of the amount of water used in the production of unit volumes or weights of common household items are given in Table 11.3.

The virtual water footprint can also be broken down into percentages of different sources of water used as well as the volume of water polluted during production. These are categorized as BLUE (i.e. the volume of ground or surface water used), GREEN (i.e. the volume of rainwater used) and GREY (i.e. the volume of freshwater polluted). Every crop needs water. The best scenario is where crops are grown solely using rainwater which would be the case with cereal crops grown in the British Isles, but this in contrast to many other crops including vegetables where groundwater and surface water may be required for irrigation. Figure 11.13 gives the total water footprint for the production of a half-litre PET-bottle soft drink which varies according to the type and origin of the sugar. Remember the most sustainable sources are those using most rainwater (green) and causing least pollution (grey).

The UK Water Footprint is summarized in Table 11.4 and is equivalent to 4645 L per person per day (L cap^{-1} day^{-1}) of which 62 % is imported from outside the country. The household (direct) water use, which is based on 180 L per person per day, is equivalent to just 3 % of the overall footprint. The water footprint for China is

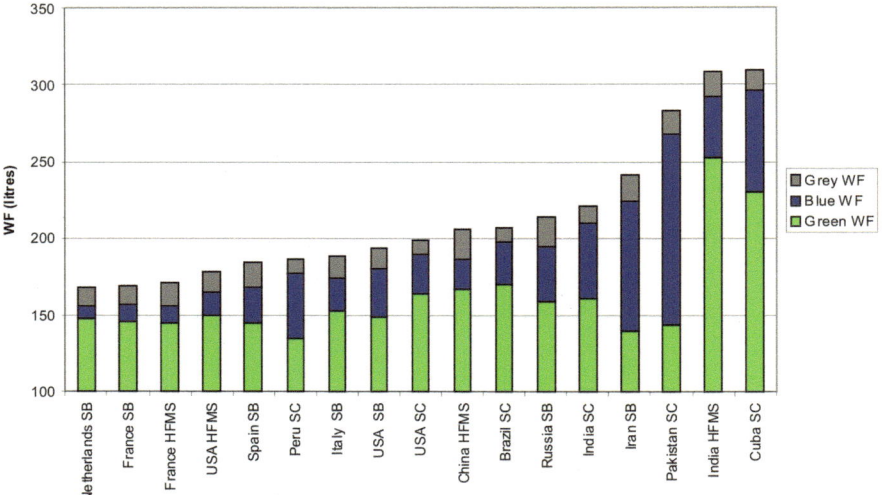

Fig. 11.13 Water volume and sources for the production of a 0.5 L PET-bottle soft drink. Sugar types used are *SB* Sugar Beet, *SC* Sugar Cane, *HFMS* High Fructose Maize Syrup. *Source*: Ercin et al. (2011). Reproduced with permission of Springer, Berlin, Germany

Table 11.4 The UK water footprint showing origin of water used by different sectors in gigacubic metres per year (Gm³ year⁻¹)

	Internal	External	Total	Total footprint
	Gm³ year⁻¹			%
Agricultural products	28.4	46.4	74.8	*73*
Industrial products	6.9	17.2	24.0	*24*
Household water use	3.3	–	–	*3*
Total footprint (Gm³ year⁻¹)	38.6	63.6	102.1	*100 %*
Total footprint (%)	*38 %*	*62 %*	*100 %*	

Source: Chapagain and Orr (2008). Reproduced with permission of the Water Footprint Network, Enschede, The Netherlands. http://www.waterfootprint.org/

2932 L cap⁻¹ day⁻¹ with just 10 % from outside their country while the USA has a water footprint of 7781 L cap⁻¹ day⁻¹ with 20 % from outside country. So compared to other countries the UK is very dependent on external water resources.

The agricultural sector is the single largest user of water and as food is a major import–export commodity then this virtual water is constantly on the move with Brazil, Mexico and Japan the major importers and the USA by far the largest exporter of virtual water (Table 11.5).

This can be further explored in Fig. 11.14 which shows just where the UK agricultural virtual water is coming from.

The major organization involved in water footprinting is the Water Footprint Network (http://www.waterfootprint.org/) which calculates specific product footprints, national footprints, business/corporate footprints and even global water

Table 11.5 The top six agricultural net importers (A) and exporters of water (B) in Gm³ year⁻¹

Country	Export	Import	Net import
A. Net virtual importers of water			
Brazil	91	199	107
Mexico	19	103	84
Japan	4	86	83
China	55	133	78
Italy	38	88	50
UK	15	55	40
B. Net virtual exporters of water			
USA	298	137	161
Australia	71	10	62
Argentina	58	4	54
Canada	70	27	44
Thailand	52	9	43
India	66	24	42

Source: Chapagain and Orr (2008). Reproduced with permission of the Water Footprint Network, Enschede, The Netherlands. http://www.waterfootprint.org/

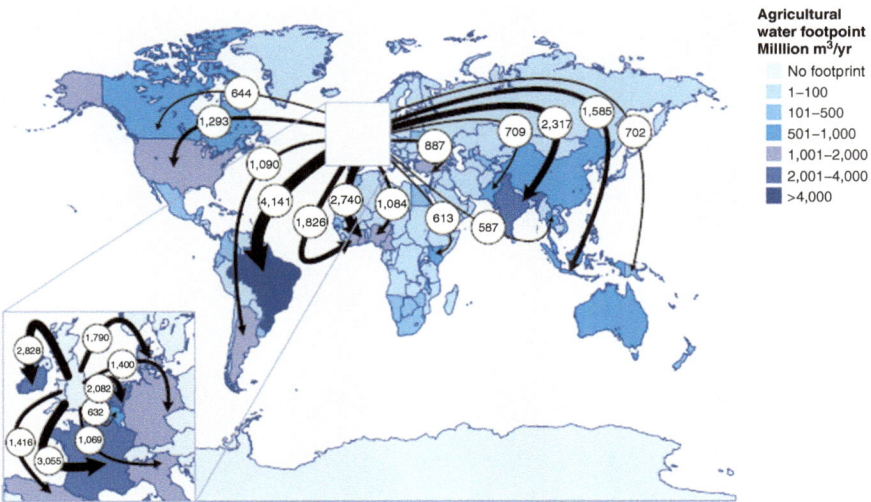

Fig. 11.14 Example of a water movement. The UK External Water Footprint for Agriculture. *Source*: Chapagain and Orr (2008). Reproduced with permission of the Water Footprint Network, Enschede, The Netherlands. http://www.waterfootprint.org/

footprints. While this is a commercial enterprise they do provide a vast amount of fascinating information online including access to personal Water Footprint Calculators which are available as a simple quick version and in-depth evaluation and also as an app.

Personal water footprint calculators:

- **Quick** http://www.waterfootprint.org/?page=cal/waterfootprintcalculator_indv
- **In Depth** http://www.waterfootprint.org/?page=cal/WaterFootprintCalculator
- **App** http://itunes.apple.com/app/water-aflamed-water-footprint/id408976536

Water used by the agricultural sector accounts for nearly 92 % of annual global freshwater consumption

11.4.1   Water Diary

A water diary measures the volume of water used by an individual or household. The concept of water diaries is quite common and has been widely used in Australia in particular, which is facing significant water shortages, to promote more sustainable water use. Water diaries are useful because they help the understanding of our water use and how best to conserve it. Water diaries can be easily developed using your own measurements of flow from taps, showers, toilet flushing, etc. However, generic values are easily obtained (Fig. 11.15) or you can use a

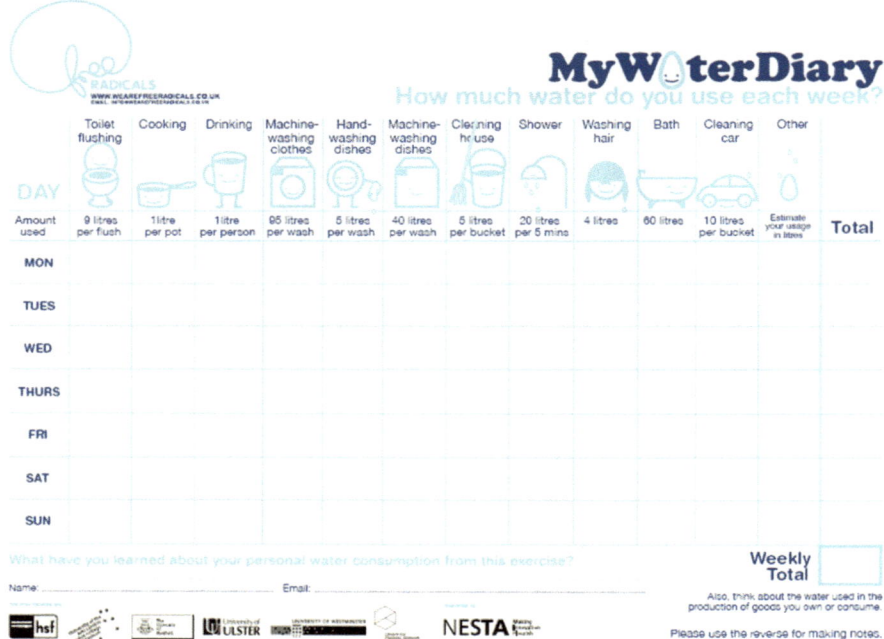

Fig. 11.15 Example of a water diary. This is filled in each day and the generic values used to show not only your daily and weekly water usage but where most water is being used. Water diaries are an important part in personal and household water conservation. *Source*: Free Radicals, http://www.wearefreeradicals.co.uk/FINAL_RESOURCES_DOCS/WATER_DIARY_freerads.pdf. Reproduced with permission of Free Radicals http://wearefreeradicals.co.uk/

simple downloadable water diary. An example of a really good water diary is mywaterdiary which is also available as a free app. Tips and information are shared via its twitter account 'mywaterdiary'

mywaterdiary: http://itunes.apple.com/us/app/my-water-diary/id333197878?mt=8

11.5 Conclusions

- Water is a renewable resource that is now under threat by global warming
- Even where resources and supplies are carefully managed, our insatiable demand for water is leading to increasing environmental degradation
- If we deplete resources then alternative supplies will be required that have larger carbon footprints
- Water shortages lead to migration, conflict, impoverishment, starvation, disease and death
- Increased urbanization, a consequence of climate change, increases risk of water shortages
- All products contain embedded or virtual water. Our direct water use is only approximately 3 % of our total water footprint.
- Water used by the agricultural sector accounts for nearly 92 % of annual global freshwater consumption
- Like peak oil, peak water is inevitable given the rate of extraction
- Our use of water is closely linked to emissions through lifestyle choices
- Water demand management reduces demand by a range of behavioural and structural conservation measures.
- Water demand management and the concept of demand-side management can be directly applied to the use of fossil fuels and be used to help reduce greenhouse gas emissions.

> **The eleventh step is accepting that water is a limited resource and so should be used thoughtfully.**
> **Reduced consumption and its careful use in the home helps to reduce carbon emissions from heating water, and also where water has come from expensive sources such as desalinization, and also reduces the impact of water abstraction on ecosystems.**

Homework?

Well, just how much water do we use individually or as a family? Using the 'mywaterdiary' template above, or the app, calculate your water usage for a full week. This will help you identify what activities use the most water in your household. What are the key things you can do to reduce your water usage? Also complete an online

your water footprint: http://www.waterfootprint.org/?page=cal/WaterFootprint Calculator.

What percentage of your water footprint is your actual direct water use as measured by your water diary?

Finally try and find out how much water your own appliances use and see how they compare to the more efficient models available. Showers and taps are easy to measure using small container. Just time how long in seconds it takes to fill your container then calculate the volume. The easiest way to do this is to use digital kitchen scales. Weight the container empty. Then reweigh when it has the water present in grams. Subtract the weight of the container to give you the weight of water and as 1 L of water weighs exactly 1 kg, then a gram of water is 1 millilitre (mL).

An example is given below where water was collected from my cold tap in the kitchen. The flow was measured when the tap was partially turned on and when fully turned on. Water was collected for exactly 5 s. You can do the same with your shower, with care. If you want to get a really accurate measure you need to collect water in a larger container for a longer period, say 15 s and then take an average of three or four readings.

The weight of the empty container was 104 g
Weight of the container plus water collected for 5 s:

   Tap partially on: **596 g**
   Tap fully on: **1399 g**

Subtract the weight of the empty container and then multiply by 12 to convert the flow into grams per minute.

   Tap partially on : 596 − 104 = 492   492 × 12 = **5904 g**
   Tap fully on: 1399 − 104 = 1235   1235 × 12 = **14,820 g**

As 1000 g is a kilogram and 1 kg is equivalent to 1 L of water, then our water use is:

Tap partially on uses 5.9 L per minute
Tap fully on uses 14.8 L per minute

References and Further Reading

Bell, A. (2010). *Peak water*. Lothian, Scotland: Luath Press.

Chapagain, A. K., & Orr, S. (2008). *UK Water Footprint: The impact of the UK's food and fibre consumption on global water resources, Volume 1*. WWF-UK, Godalming, England. Retrieved from http://www.waterfootprint.org/Reports/Orr%20and%20Chapagain%202008%20UK%20 waterfootprint-vol1.pdf

Ercin, A. E., Martinez-Aldaya, M., & Hoekstra, A. Y. (2011). Corporate water footprint accounting and impact assessment: The Case of the water footprint of a sugar-containing carbonated beverage. *Water Resources Management, 25*, 721–741.

Gleick, P. H., & Palaniappan, M. (2010). Peak water limits to freshwater withdrawal and use. *Proceedings of the National Academy of Sciences, 107*(25), 11155–11162. Retrieved from http://www.pnas.org/content/107/25/11155.full.pdf

Pearse, F. (2007). *When the rivers run dry*. Boston, MA: Beacon.

Ward, D. (2002). *Water wars: Drought, flood, folly and the politics of thirst*. New York, NY: Riverhead Books.

Overview

http://www.ipcc.ch/pdf/technical-papers/climate-change-water-en.pdf
http://nca2014.globalchange.gov/

Water Sustainability in USA

http://www.nrdc.org/globalwarming/watersustainability/files/WaterRisk.pdf

Developing World

UNDP. (2006). *Beyond scarcity: Power, poverty, and the global water crisis*. Human development report. New York, NY: United Nations Development Programme. Retrieved from http://hdr.undp.org/en/media/HDR06-complete.pdf

Virtual Water

http://www.worldwater.org/data20082009/Table19.pdf

Chapter 12
Waste Not Want Not

The symbol above is synonymous with recycling but how effective is recycling and why is it important? In this chapter we explore the concept of recycling, who is doing what and how this equates to reducing your personal carbon footprint. Does it make a difference in terms of global warming and making resources more sustainable?

12.1 Introduction

I have gone through a remarkable transition in relation to stuff. Stuff is what my kids call all the consumable items and things that we use around the house. So when I lament over some broken pottery or the like, the response from my family is usually…*it's just stuff get over it*'. I have to admit that I personally don't like the term as it devalues these items in some way and perhaps it also hides a significant problem…our relationship with stuff.

I was lucky to be brought up in the early 1950s with the post war recession still in place. My parents like everyone else didn't have a lot of stuff. So their attitude

© Springer International Publishing Switzerland 2015
N.F. Gray, *Facing Up to Global Warming*, DOI 10.1007/978-3-319-20146-7_12

was different. Everything that came into the house was expected to last a long time, so it was used carefully always cleaned after use and put away. As a result that same stuff was still around and being used when my mother died in the late 1990s. Don't get me wrong things did wear out and regularly get broken, but she simply tried to buy reasonably good quality things and cared for them. They were exceptionally generous people but they were frugal, and they wasted very little including food. For example, as a family we generally had a large joint of meat for Sunday lunch which was then turned into numerous different meals over the next 2 or 3 days until every bit was used. My mother had the skill and to some extent the time to make these meals both interesting and nutritious. Even the fat (dripping) was carefully saved and reused. Bags, wrapping paper, newspapers, string, in fact anything that could be reused was, often over and over again. They were natural recyclers to whom waste, having lived through the war, was regarded as shocking. Our dustbin was never full, in fact it was so light that even I, as a 5 and 6 year old, could drag the metal bin to the gate for the dustbin (refuse) men to collect. When the 60s came along and things got a little better and choice, affordability, and availability all came together then there was a reaction against being frugal. My mother stopped making jumpers and cardigans for us kids and would buy acrylic jumpers with a sense of pride that she was somehow being extravagant and in being so was doing better for us. Suddenly having stuff, whether you could afford it or not was a sign of success and modernity, and with this transition generations of skill and adaptability were gradually lost. So we have had to reinvent the concept of being frugal and a part of that is relearning to reuse and recycle. Although this has been going on in earnest since the late 1970s, it is only recently that it has become the acceptable norm. The resistance, almost embarrassment associated with recycling in those early years, has thankfully gone, although there is still quite a number of people who see it simply as a way of getting rid of their rubbish. Of course it is much more than this, it is a way for all of us to reduce our CO_2e emissions, preserve resources, as well as save us money. Importantly for the future it starts us thinking about how we can help and sustain ourselves and become more independent people. The idea of living frugally was explored briefly by one of my students, Aoife Beglin, in a very insightful blog post (below).

Recycling is way for all of us to reduce our CO_2e emissions, preserve resources, as well as save us money

12.2   Recycling—The Science of Signs

The Mobius loop (shown at the start of the chapter) was designed in 1970 by Gary Anderson, a graphic designer and architect. It has become the universal sign for recycling and is now found in a huge range of different guises. Apart from the Mobius loop, there are thousands of other symbols, some official others not, that are used in recycling with the aim to encourage and inform us about the product on which they can be found. All these new symbols appear to be concerned with recycling or the sustainability of the product, but what do they all mean? In Europe three symbols in particular are very common, the green dot, tidyman and the Mobius loop itself.

The Green Dot: This is the most widely used symbol used on packaging in many European countries and perhaps the most misleading. To me it looks as though the item is recyclable, however this may not be the case. All the green dot signifies is that the producer has made a contribution towards the recycling of packaging in general. So a green dot can appear on the most surprising objects which are clearly not recyclable and often not good for the environment. So it could be considered as a type of offsetting.

Tidyman: This symbol is again very common but is totally unrelated to recycling and is simply a reminder to dispose of the item carefully and not to litter. Some children who I was talking to during a school visit thought that it meant that this item must be placed in the bin for some reason, whereas another item without the symbol can be safely dropped on the ground. I know that in theory we all know that littering is wrong, but this is an argument I have come across several times especially in rural areas and the people have been quite honest in their confusion thinking that non-marked items will naturally degrade. All litter ends up in the sea eventually and is causing major problems, so littering remains a serious environmental problem.

More information: http://marine-litter.gpa.unep.org/facts/facts.htm

Universal recycle symbol (Mobius loop): This is undoubtedly the most widely used of all packaging symbols and means that the object as a whole or parts of the object are *capable* of being recycled. **It does not mean that that the object has been recycled**. It may be applied to items that are not recyclable in your area and my not refer to all the component parts of the item that carries the symbol, so once again it is of limited value.

Quality vs. Quantity: Mindful Living…or Frugality Often Gets a Bad Rap

Why is it that so often living "frugally" is often linked mentally with want, deprivation and hardship? Mention the word and immediately people conjure up archaic images of teabags drying on radiators, cold showers and penny-pinching, when in reality frugality is essentially a means of achieving a quality lifestyle and fulfilling the needs of the individual in society, whilst leaving minimum negative impact on the community, planet and its resources. Admittedly, the word *frugal* itself is hardly the most onomatopoeically-endearing of the English language, but it certainly does lend a solid air of reality which encompasses the essence of what frugal living is about. For me, being frugal is a means of directing the materials and resources available to me towards a goal which I want to achieve, enabling me to complete such tasks in a less wasteful, more efficient and thoroughly-considered fashion. In no sense do I consider there to be any level of deprivation; in fact I think cutting the clutter and distraction from one's life allows for greater clarity to shine through, enabling one to visualise plainly what are the real essentials in life and how best to go about maintaining these. In the overwhelmingly materialistic, consumer-driven society in which we live, more and more excesses of all has lead to a devaluing of quality in favour of inferior quantity. Here frugality can act as a life raft in the sea of superfluousness, a wake-up call forcing us to examine *what matters*. Giving a brief run-down through this topic does not do it justice as by its very nature greater depth of thought is required, but frugal living can be broadly addressed in three main sections: the why, the how and the outcome.

Frugal living is a sustainable way of life and can be viewed as a way of guaranteeing, to the best of one's ability, that the life one leads is truly meaningful. My own personal definition of sustainability would run something like: "Sustainability is the thought process behind *mindfully living* a considered *life*, endeavouring to maintain an equilibrium between personal development, integrity, and responsibility, such that the needs of future generations are not compromised through mismanagement of the earth's resources by present generations." Lofty ideals you may think, but unless one strives toward

(continued)

an upward goal there would certainly be no motivation to continue on. In the case of any goal that requires longevity, soundly-founded motivations need to be established before anything is ventured and frugal sustainability is no different. We need to ask ourselves *"Why should I do this? What are the benefits to me personally and what are the benefits to the wider community? Am I ready to accept that my personal decisions can and will have a large impact on those around me?"* If these questions can be answered honestly and solidly there will be sufficient motivational undercurrents to sustain one through the practical implementations of frugal living. Starting small and incorporating small changes on a day-to-day basis, great success can be accomplished through the setting of many small, realistic and achievable goals.

Avoiding procrastination is a definite requirement for living the frugal lifestyle, but once this reality is accepted as a hitherto-unrecognised benefit, the sky's the limit for the creative ways in which frugality can be lived out, making your time, money and resources work hardest for you. From practical applications such as reducing waste in the home, buying less consumer goods and increasing productivity by utilising goods to the very end of their product-life, to more fun and unique approaches such as going on self-imposed spending 'diets', enjoying frugal freezer-food meals and trying out 'Meatless Mondays'. Certainly we must realise that there is no one magic-bullet, simplistic route to achieve these goals, but we need to take every small avenue available to us, making numerous small changes.

The outcome of embarking on the frugal living path is that it allows for greater personal freedom. Less "stuff" means a reduction in time and energy consumed whilst absorbed in (or perhaps burdened with) said "stuff". We are afforded more time to reflect and think on our life-goals. We are mindfully conscious of what we are doing and why we are doing it. Savouring experiences and enjoying simple pleasures are incomparably more fulfilling than rushing through a series of tick-the-box life goals set up on a scale belonging to the Joneses. Paring back the excess allows us to appreciate all the more how rich we really are—after all, our attention is no longer divided in multiple different ways but rather we are free to focus and enjoy the delights of a smaller number of better-curated occupations. Each of us are the cultivator of our own person and our personal development, which in turn is amplified and combined with that of others to create the societies in which we live. We have the right to enjoy this, but more importantly we have the responsibility to do so—not only for our own sakes' but for the sake of our fellow man and frugal living is a powerful way to do so.

Aoife Beglin

Source: http://ournewclimate.blogspot.ie/2013/06/quality-v-quantity-mindful-livingor.html

So none of the symbols above are very helpful in terms of giving a clear indication of whether the actual item is made from recycled materials, or whether the whole item can in practice be recycled. Universal recycling symbols appear to be a long way off even though it appears a very straight forward thing to do. Each country has tended to adopt their own system. The UK symbols clearly identify what the objects are made from, which is important in helping people to sort their waste appropriately.

The UK symbols above show that the item to be made of either aluminium, glass, steel or plastic respectively and are all recyclable. However, would these symbols be so easily understood by someone from Japan for example? Certainly the current Japanese recycling symbols below for paper, plastic and aluminium respectively are less readily identifiable:

The only truly International recycling symbol which is universally used by industry and unambiguous is used for the identification of plastics. It employs the Mobius loop and inside it is a number ranging from 1 to 7 which is the plastic type recognition code (Fig. 12.1). All these plastics have the potential to be recycled but it is important that they can be separated into the specific type for maximum benefit to be obtained.

We don't just want to recycle materials, often we want to buy items that have been made from recycled products. This poses the problem of identifying such products and because these often cost more it is important that the authenticity of each product can be assured. I have never really been clear as to why recycled products do cost more but I presume there is a good reason. However, to encourage the use of goods that contain recycled materials many environmentalists have been lobbying to have a lower VAT or purchase tax to encourage us to use recycled materials over new.

Fig. 12.1 Universal plastic
recycling symbols

There are a number of symbols that are designed to help us identify that a product contains recycled material, although the regulation of this is often left to industry bodies. There is nothing wrong in that, except the consumer needs to know simply that this is a recycled product and that they can trust the symbol.

The National Association of Paper Merchants (NAPM) has a quality symbol used by their members to show the percentage genuine waste paper or board fibre in their products. All their paper or board must be made from a minimum of 75 % of waste and no part of which should contain mill produced waste fibre. So this symbol is a really useful guide where available and should perhaps be copied by other manufacturing sectors.

A similar organization is the 100 % Recycled Paperboard Alliance (RPA-100 %) which promotes companies to manufacture 100 % recycled paperboard and then encourages companies to convert to using it for all their packaging. RPA-100 % licences companies and manufactures through a strict quality assurance process resulting in companies able to use the logo on their product packaging. In the US the RPA-100 % now represents all the main manufacturers of recycled paperboard.

The Forest Stewardship Council (FSC) oversees the use of this logo which identifies products which contain wood from well managed forests that have been independently inspected and certified in accordance with the rules of the FSC. So while this does not show the product is made from recycled material it does ensure that the timber has been sourced from sustainable woodlands and forests.

This is a really useful symbol based on the Mobius loop which shows the exact percentage of recycled material contained in the product or packaging. This is widely used within Europe if not globally.

In the US the Mobius loop is also used to indicate the item is made from recycled material. In this case a white loop on a black background indicates 100 % recycled materials, while a black loop on a white background is used for items containing some recycled materials.

So to be successful in promoting the importance of recycling and the reuse of materials we need to have unambiguous symbols that are universally understood and that are carefully regulated to prevent abuse. Likewise we need to understand the importance of ensuring that the materials we recycle are clean and properly sorted. Recycling centres must do their part by being as imaginative as possible and recycle as wide a range of products. Increasingly recycling centres operate exchange initiatives such as book swaps, collection of school books for overseas charities, areas where you can find small building items such a small piece of wood, shelving or screws, and even in some cases a repair areas where items from electrical goods to bicycles can be evaluated and repaired for a small cost if required.

> **The recycling centre is all about extending the life of consumable products, reducing waste and ensuring maximum recovery of resources.**

12.3   Waste Production

In 2008, the total generation of waste in the 27 European Union member States was 2.62 Gt. This means that each EU citizen produced on average about 5.2 t of waste that year of which 196 kg were classified as being hazardous. This figure is a bit misleading as it also contains waste from industry so this represents both your primary and secondary waste footprint. Approximately 500 kg of this average figure can be strictly classed as household waste (Fig. 12.2) and so forms a part of your personal carbon footprint.

Overall the amount of household waste generated in Ireland and the UK has stabilized (Table 12.1), but as the population is steadily growing this actually represents a small net reduction per person. This is partially due the increase in recycling, although waste production figures vary significantly between countries.

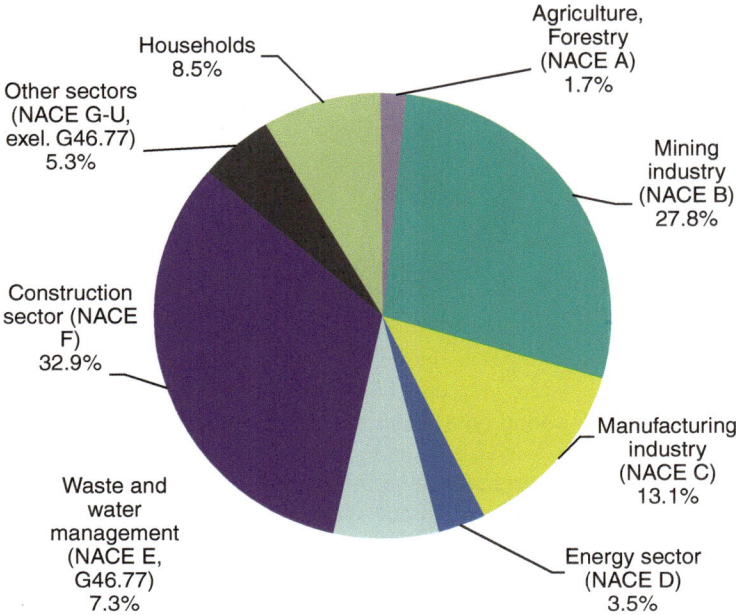

Fig. 12.2 Breakdown of waste generation within the EU by sector. NACE is the EU classification of economic activities or sectors (A–G). *Source*: The European Union, http://ec.europa.eu/eurostat/statistics-explained/index.php/Main_Page. Reproduced with permission of the European Commission, Brussels, Belgium

Table 12.1 Household waste production in tonnes 2004–2008 and 2010

Year	Ireland	UK	EU-27
2004	1,702,345	31,007,480	210,960,000
2006	1,978,711	32,466,328	215,340,000
2008	1,677,338	31,539,338	220,950,000
2010	1,730,000	31,539,000	220,940,000

Source: The European Union, http://ec.europa.eu/eurostat/statistics-explained/index.php/Main_Page. Reproduced with permission of the European Commission, Brussels, Belgium

More Information: http://epp.eurostat.ec.europa.eu/statistics_explained/index.php/Waste_statistics#Total_waste_generation

Within the EU only Demark has higher waste production rates than Ireland producing 801 and 786 kg of waste respectively per person per year (Table 12.2). Ireland also has the highest dependency on landfill mainly due to the national ban on incineration (heat recovery) options. However, these EU figures don't seem to add up because if you divide national household waste production by population you get approximately 410 and 492 kg per person per year of waste in Ireland and the UK respectively. So where is that excess coming from? The answer is commercial

Table 12.2 European recycling rates of municipal waste for the year 2007

2007	Waste per person (kg)	Landfill (%)	Recycle or compost (%)	Incineration (%)
EU-27	522	42	39	20
Germany	564	1	64	35
Greece	488	84	16	0
Ireland	786	61	35	4[a]
UK	572	57	34	9
Denmark	801	5	41	53
Romania	379	99	1	0

Source: The European Union, http://ec.europa.eu/eurostat/statistics-explained/index.php/Main_Page. Reproduced with permission of the European Commission, Brussels, Belgium
[a]Refuse derived fuel

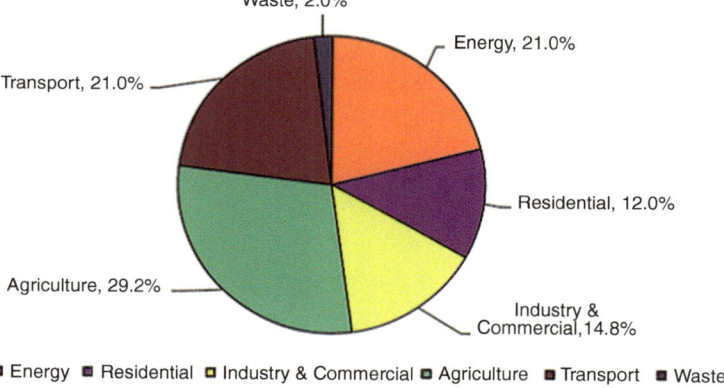

☐ Energy ■ Residential ☐ Industry & Commercial ☐ Agriculture ■ Transport ■ Waste

Fig. 12.3 Greenhouse gas (GHG) emissions from major sectors in the EU including residential. *Source*: The European Union, http://ec.europa.eu/eurostat/statistics-explained/index.php/Main_Page. Reproduced with permission of the European Commission, Brussels, Belgium

waste from offices with most EU countries poor at separating commercial from household waste. While the recycling rate in Ireland is below the EU average in terms of weight, according to the Irish Central Statistics Office 90 % of households do recycle with this rising to 94 % in Dublin, compared to just 45 % in 1999. So while Irish households are actively engaged in recycling they are simply not separating and recycling enough of their waste. In contrast, Eastern European countries are performing badly overall due to a lack of recycling facilities and a market for recyclables (Table 12.2). Irish recycling is reviewed for the year 2009 in an EPA report published in 2011: http://www.epa.ie/downloads/pubs/waste/stats/EPA_NWR_09_web.pdf.

Greenhouse gas (GHG) emissions from post-consumer waste and wastewater are a small contributor (about 3 %) to total global anthropogenic GHG emissions as shown in Fig. 12.3. These emissions totalled 1.4 Gt CO_2e year^{-1} for the period 2004–2005 with the methane (CH_4) from landfills and wastewater collectively accounting for about 90 % of all waste sector emissions, or about 18 % of global anthropogenic methane emissions. So globally the CO_2e emissions from our household waste do not seem to be hugely significant, but it is an important component of our personal carbon footprint and one which we can easily control like food (Sect. 10.6). So this forms an important mechanism in reducing our contribution to GHGs.

12.4   Waste Hierarchy Is Pivotal to Sustainability

Recycling has both positive and negative impacts on carbon emissions and can significantly reduce GHG emissions by reducing the need for raw materials which generally require more energy to produce and transport than when recycled materials are utilized. It also reduces waste ending up in landfills which helps to reduce methane emissions, although not all types of recycled materials are degradable or produce methane when they do breakdown. However, advantages can sometimes be quite small so we must ensure that associated emissions are minimized for example increased emissions due to the transportation by individuals to and from the recycling centre and also due to the processing of the materials.

Overall recycling is not really a very good solution to our waste associated emissions in environmental terms which is why it is the fourth most carbon efficient option. The waste hierarchy (Fig. 12.4) favours waste minimization and waste

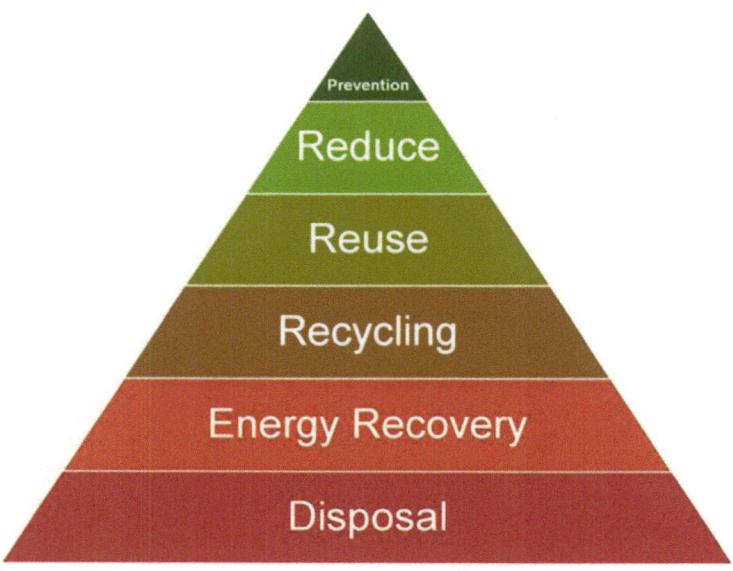

Fig. 12.4 The waste disposal hierarchy. The relative carbon footprint of waste disposal is shown at the base of the pyramid. Recycling, energy recovery and disposal are the least environmental friendly options in that order with the options with the lowest footprint at the top

reuse over recycling, although prevention is always the best environmental option. Reuse and recycling options are more carbon efficient if carried out locally and not transported long distances. However, recycling is more carbon efficient than either landfilling or the incineration of waste with the least environmental option being to landfill waste.

A major problem is that increasingly greater amounts of municipal refuse and recyclables are being incinerated to create heat for electricity generation. In order to make incinerators cost effective a high plastic and paper content of waste is preferred, but does this undermine the ethos of recycling? When mixed household waste is sorted centrally then it goes through a number of very clever steps where paper, plastics, glass, metals both ferrous and non-ferrous, are all separated. The remaining inert material is then removed from residual waste leaving a mixed combustible waste that can then be sold as refuse derived fuel (RDF) that can be used to fuel incinerators and the energy left in that material recovered as heat. Unfortunately it is still common for mixed household waste to be simply burnt in its raw form after screening for ferrous metals which means many valuable recyclable materials are lost. But even mixed waste can generate large amounts of energy with a single bag of household rubbish able to produce enough heat to generate enough electricity to supply electricity to an average household for 3.5 h.

Whether we like it or not our personal and family waste production, including recyclables, is closely linked to our overall lifestyle which includes factors such as consumerism, culture and wealth. The waste hierarchy is a simple mechanism to help us focus on all the components that make up our personal lives.

Another approach derived from the waste hierarchy is the three Rs: Reduce, Reuse, Recycle:

Reduce—buy less and use less is the basic concept here, so use all you buy, wear it out and reduce the amount of material, including food, that is discarded.

Reuse—this is ensuring that all discarded items or components of the item are used again, if possible locally.

Recycle—discarded items are dismantled and sorted so that the key resources can be recovered before disposal and reused.

Although this is very simple, I am not sure if people really understand the philosophy behind the concept. But the three Rs are so important and underpin all sustainable concepts and actions from carbon footprinting to ecofashion. However, the key action must always be **prevention**.

12.5 Facts About Recycling

Currently 40 % of waste from UK households is recycled (2011), compared to just 11 % in 2000/01. This has led to a significant reduction in the amount of residual waste generated per person requiring disposal by 76 kg since 2006/07 to just 275 kg per person each year. Some of this residual waste is incinerated for heat recovery while 55 % of the municipal waste generated in the UK is sent to landfill, compared to an EU-27 average of 40 %. A similar upward trend has been recorded with commercial and industrial waste with 52 % recycled or reused in England in 2009, compared to 42 % in 2002/3. Direct emissions from waste management accounted for 3.2 % of the UK's total GHG emissions in 2009, or 17.9 Mt CO_2e compared to 59 Mt CO_2e in 1990. Of the 2009 total, 89 % arose from landfill, 10 % from wastewater handling and 2 % from waste incineration.

> *Nearly 90 % of GHG emissions from domestic waste come from landfill. So we need to keep our waste out of the general waste stream heading for landfill.*

So has recycling been a success in our fight against climate change? In some respects yes. Recycling is more carbon efficient than either landfilling or incinerating wastes which are our preferred disposal routes at present. The recycling of paper, glass, plastics, aluminum and steel in the UK saves more than 18 Mt CO_2e per year through avoided primary material production. **So recycling in the UK is equivalent to the annual emissions from five million cars or 14 % of UK transport sector emissions in 2006**. Without doubt recycling is making a huge different and helping us reach those carbon reduction targets. The challenge is not only to sustain this level of recycling but to build on it even more. Locating recycling facilities close to where waste is being generated helps to minimize carbon emissions from transporting recyclables to collection points, as does only going to the recycling centre when you can fully load your car with recyclables.

In the US, 251 Mt of municipal waste were produced in 2012 before recycling (Fig. 12.5). Recycling rates in the US are similar to that in Ireland. **In 2012 the U.S. recycled 34.5 % of its waste which saved 168 Mt of CO_2e being generated. This is equivalent amount of GHGs to removing 34 million cars from the road**. Increasing the recycling rate to 38 % would reduce GHG emissions by an additional 17 million metric tonnes of CO_2e annually proving that recycling works. The remainder of the waste, 164 Mt, went to landfill or incineration.

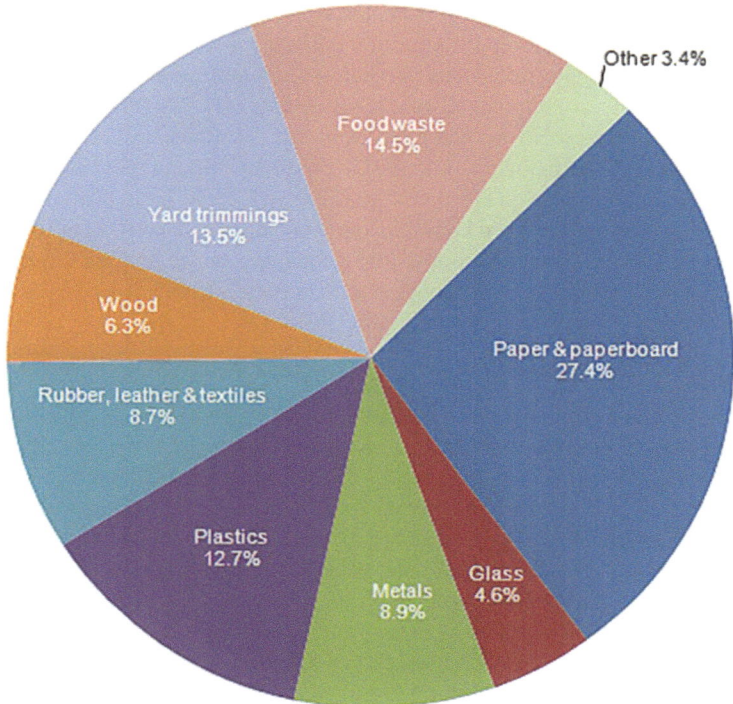

Fig. 12.5 Total municipal solid waste generation in the US by category during 2012. *Source*: The USEPA, http://www2.epa.gov/recycle. Reproduced with permission the US Environmental Protection Agency, Washington, DC, USA

More information on recycling in the US: http://www2.epa.gov/recycle

We are recycling more, reducing CO_2 emissions from our waste, yet our overall emissions do not really reflect this because of lifestyle changes and population increases.

Logo reproduced with permission of WRAP, Banbury, UK http://www.wrap.org.uk/

Many countries have established NGOs to co-ordinate recycling efforts. The Waste and Resources Action Programme (WRAP) was set up to help the UK Government to meet its national and international commitments and build the green economy. Primarily it supports resource efficiency enabling householders, businesses and the public sector to make better use of resources and at the same time save money. Its achievements are impressive. With its help the recycling and reprocessing sectors quadrupled in size between 2000 and 2008. In 2010 over 670,000 t of food was diverted away from landfill, saving consumers over £600 million a year. It has also halted the growth in household packaging waste and developed a new technology for closed-loop recycling of plastic bottles, which has led to the creation of a new market for recycled plastics in the UK.

More information: *WRAP* http://www.wrap.org.uk/

One of the major problems with waste disposal is organic matter. This is difficult to recycle unless composted at home, and even then as it decomposes all the fixed carbon is released as it is broken down by bacteria and fungi. Once organic matter is binned then it is almost impossible to recover it from the mixed waste stream. However, the UK produces vast quantities of organic matter that could realistically be made available for utilization by anaerobic digestion technology, thereby turning the potential CO_2 into methane which can then be used as a fuel. Currently seven million tonnes of food waste is produced each year in the UK (Sect. 10.6) and a further 90 Mt of animal slurry and manure that is already separated and in bulk making collection easy for potential anaerobic digestion to produce methane. The methane can then be used in combined heat and power (CHP) systems to generate electricity. In England alone this could generate at least 3–5 TWh (terawatt hour) of electricity per year by 2020 (a heat equivalent of 6–10 TWh) (see box below). The UK water industry is already treating over two thirds of its sewage sludge by anaerobic digestion, generating in the region of 1 TWh of energy per year in 2010. The great thing about digestion is that all the organic waste can then be used as a replacement for artificial fertilizer as the nitrogen, phosphorous and smaller amount of potassium are still present making the residual stabilized waste from digestion a valuable and sustainable soil conditioner and fertilizer. So the diversion of biodegradable wastes to anaerobic digestion from food processors, distributors, supermarkets and restaurants, as well manure from farms, could reduce greenhouse gas emissions from landfill and from the use of artificial fertilizers as well as reduce our use of fossil fuels to generate electricity…a win-win situation.

---

1 kWh is one kW of power used for 1 h

MWh—megawatt = 1000 kWh (10^3 kWh)
GWh—gigawatt = 1000,000 kWh (10^6 kWh)
TWh—terawatt = 1000,000,000 kWh (10^9 kWh)
PWh—petawatt = 1000,000,000,000 kWh (10^{12} kWh)

The **average household electricity consumption is 4800 kWh per year** while the **total annual electrical energy usage in the UK is around 360 TWh**. So while the 3–5 TWh produced from anaerobic digestion is not going to solve our future energy deficit, it is preventing a huge release of GHG into the atmosphere. Capturing the biogas from digesting 1 t of food waste saves between 0.5 and 1 t of CO_2e, so by just processing this readily available waste food could wipe off up to 5 Mt of CO_2e from the UK national carbon footprint, that's nearly 80 kg CO_2e saved for every person in the UK.

> *Capturing the biogas from the digestion of 1 t of food waste saves between 0.5 and 1 t of CO2e*

12.6 At a Personal Level

The waste hierarchy (Fig. 12.4) provides a key mechanism for personal GHG emissions and potential offsetting, with waste minimization key to preserving resources and an important mechanism in reducing GHG emissions. So before we even think about recycling, we should be exploring ways of preventing the creation of waste, reducing waste and reusing items before deciding to discard or replace them.

Consumerism and waste are closely linked. So we need to:

- Buy less and buy better quality so that it lasts longer.
- Consume all our food and not be taken in by offers such as buy one get one free…if you aren't going to use it then do not take the free one.
- Have the attitude that we don't waste anything
- Wear things out
- Keep electronics and other household or personal equipment longer before replacing (e.g. smart and mobile phones)

One of our problems is that there appears to be no incentives or reward systems for recycling. But this isn't true. By reducing the amount you buy and using it all then you save money. By keeping things longer you save money, by buying better quality then they can be repaired, reused or sold again saving money in the long term. Nearly all of us now pay waste charges for our rubbish that goes to landfill creating an incentive to reduce waste at doorstep. The average Irish household only produces one sack of rubbish per fortnight costing between 6 and 8 euro for disposal. They also produce another 2–3 sacks of recyclables that can be taken to a recycling centre for free, saving them between 12 and 24 euro a fortnight in disposal charges. So an annual waste disposal charge creates a disincentive to recycle whereas a charge per bag creates an incentive. Many recyclables are very light, so charging by weight does little to reduce the volume going to landfill or to encourage recycling. Other incentives include a proposal to reduce VAT on second-hand or goods made from recycled materials, although one of the problems in its implementation is the ensuring the provenance of recycled materials and goods.

When I was a kid nearly all drink bottles were returnable; milk, beer and soft drink bottles could and were expected to be returned. Apart from milk bottles they often had a returnable deposit with the amount usually printed or cast into the bottle. So a Corona soda bottle for example had a returnable deposit of sixpence. Today Carlsburg, who brew a wide range of beers, are producing 35 billion bottled items a year. They have been involved in a successful initiative in Denmark where all drink retailers have to register their products so that a returnable deposit can be levied. With 15,000 items now registered, cans and bottles can be returned to automatic reverse vending machines which accept your returnable containers for reuse in the case of bottles or to be recycled in the case of aluminium cans, and pays you your deposit back. This has been a huge success with 90 % of all the registered items being returned which is equivalent to over three million items each day. By placing a returnable deposit on the bottles they are kept out of the waste stream avoiding contamination and so are much cleaner and mainly undamaged and therefore easier to reuse. This reduces costs for the drinks industry who are able to reuse the bottles.

While recycling is very worthwhile there is a problem, and that is the cost and emissions associated with individuals taking their recyclables to a recycling centre that may be many miles away. It is very demoralizing to see people arriving at these centres with just a single carrier bag of material. It is important that as few trips to recycle are made as possible each year, so ensure your car is packed to the roof with bags of clean materials to recycle. If you don't have sufficient storage space, then try to create a rota with neighbours and friends. Perhaps a better alternative is commingling. This is recycling waste not via the recycling centre but by having the waste collected from your home. This has the added advantage that all your recyclables can be put into a single bag except for glass and left out for collection. This actually has been shown to be more efficient with 20 % more being recycled. The insistence of separating recyclables into different bins for collection is often about maximizing profits for collecting agencies as some recyclables are worth much more than others. It has a negative effect on recycling with people often confused and even afraid to separate items. Recycling has to be made as easy as possible for householders and sorting centrally is in my opinion the best option for both householder and the processor.

The problem with mixed recycling is contamination which has resulted in much of our exported recyclables being returned due to high levels of impurities, usually glass shards and organic contamination. Mixed waste is taken to material recovery facilities (MRFs) where the waste is both mechanically and manually sorted. The problem is that up to 32 % of the contents of the green bin, is contaminated or non-recyclable materials that should not have been included, and this can often lead to the whole lorry load of recycleables having to be landfilled (Table 12.3). The problem is getting worse, so coupled with a downturn in the price for recyclables on the open market, that instead of collection companies being paid for this waste, they now being charged between €20 and €30 per tonne for MRF plants to take it, although this varies from month to month. So poor recycling practices are undermining the efforts of the majority, so much so that we are dangerously close to having to pay to recycle items which would be a huge step backwards. So contamination

Table 12.3 Average contents of the Dublin Green Bin

Waste category	%	Waste category	%
Mixed paper	51	Mixed plastic cartons/trays	1.5
Cardboard	8	Opaque plastic (e.g. milk containers)	0.8
Soft plastic	2.5	Aluminium cans	0.7
Steel cans	1.8	Colour plastic (e.g. shampoo bottles)	0.2
Clear plastic bottles	1.5	**Contamination**	**32**

Contamination includes nappies, food, garden waste and glass which should be recycled separately via a bottle bank. *Source*: Panda Waste. Reproduced with permission Panda Waste, Dublin, Ireland. https://www.panda.ie/

is a big problem to the whole recycling industry and for that reason the EU has stepped in and introduced new legislation to deal with it.

From 2015 new EU regulations concerning recycling will mean that householders will require at least four different recycling bins if they have a centralized collection service. Currently in the UK householders have three bins one for household waste, one for mix recyclables and a third for garden waste. However councils are under pressure for paper, metal, plastic and glass to be collected separately. This has been prompted by the difficulty of removing glass shards from paper in particular resulting in much paper waste being rejected by recycling companies and ending up in landfill. Under the Revised Waste Framework Directive that comes into effect on January 1st 2015, it is a requirement to keep these four main recyclable waste streams separate in order to prevent contamination and generate higher quality recyclable material and at the same time ensure maximum recycling is achieved. So it will be a legal requirement to collect paper, metal, plastic and glass separately from 2015 which effectively means the end of mixed recyclables collection and processing.

12.6.1   Electronic Items

The EU Waste from Electrical and Electronic Equipment (WEEE) Directive became law throughout Europe in 2005 and requires all suppliers to build in a recycling charge on all new electrical items. This pays for these items to be subsequently collected and recycled. Under the legislation suppliers must also accept equivalent goods for recycling when a new replacement item is purchased, although all electrical goods can be recycled free at any recycling collection centre.

White goods make up 5 % of household waste with the average person consuming 3.3 t of electronic waste in their lifetime, or on average around 0.016 t (16 kg) per person each year. What is good news is that most of the components of electronic waste are recyclable. A fridge, for example, has up to 95 % recoverable material, much of which can be classified as non-renewable resources. In 2008, 30,000 t of electrical goods were collected in Ireland equivalent to 9 kg per person which is double the EU average making Ireland the best EU Member State at recycling WEEE items.

The life span of electrical goods is generally much longer than the replacement period, therefore the failure of an item is usually not the reason for replacement. For example, computers are typically replaced every 2 years and mobile phones every 18 months. The recent transfer to digital broadcasting led to 800,000 perfectly good television sets being scrapped in Ireland and replaced with digital sets. Household goods often have surprisingly large embedded emissions associated with their manufacture. A desktop computer and CRT screen (weighing 24 kg), for example, uses ten times its weight in fossil fuels during its manufacture compared to a car or fridge which may only use 1–2 times their weight in fossil fuels during manufacture.

12.6.2 Someone Somewhere Wants It

While some items are worn out, soiled, or broken and beyond repair, the majority of items we throw out are simply no longer wanted. This may be due to a variety of different reasons such as the item is no longer fashionable, or doesn't have the capability for your needs anymore and you need to upgrade, you no longer use it, or perhaps you have been given a better model. Regardless of the reason the product still works or perhaps just needs a simple repair. So many of the items that are discarded are still useful and this can include electrical and mechanical goods, cooking and gardening equipment, clothes, building materials and even food.

Reuse is a major factor in sustainability and good quality manufactured goods can have significantly extended life. So before you throw anything out consider that perhaps someone else may have a use for it and so be able to extend its life. There are many local, regional and national organizations that are engaged in reusing unwanted items and they play an important role in reducing GHG emissions. Reuse options include bring and buy sales, swap shops, charity shops or online free trade sites. There are also charity restoration groups which repair and restore furniture, electronics, white goods, bicycles and much more for redistribution also there are an increasing number of sustainable fashion groups who take old clothes and repair or redesign.

Examples of online free trade sites:

http://uk.freecycle.org/
http://www.freetradeireland.ie/
http://www.jumbletown.ie/forums/index.php

Table 12.4 Average waste production per household in the UK by weight

3.8 kg	Cardboard and paper
0.7 kg	Dense plastic packaging
0.4 kg	Ferrous packaging (steel and tin cans)
0.2 kg	Aluminium packaging
0.6 kg	Miscellaneous metal (ferrous and non-ferrous)
1.7 kg	Glass packaging
0.3 kg	Textiles
3.1 kg	Putrescible kitchen waste
2.9 kg	Garden waste
3.3 kg	Misc. combustible waste (DIY combustibles)
0.6 kg	*Miscellaneous plastic (e.g. plastic coat-hangers, plastic film)*
0.3 kg	*Sanitary wastes*
0.6 kg	*Misc. non-combustible waste (brick, rubble)*
0.1 kg	*Dust and ash*

Items in italics were considered not recyclable. *Source*: Defra (2013). Reproduced with permission of the Department for Environment, Food and Rural affairs, London, UK

12.6.3 What Is in Your Bin?

The Open University carried out a detail study of household waste in the UK for DEFRA and found that the **average household produces 18.6 kg of waste per week** (Table 12.4). Only the items in red in the list were unrecyclable. However even the miscellaneous plastic and non-combustible wastes are recyclable but need careful separation. So in theory nearly everything we throw away can be recycled. The reason why items that are recyclable are not being recycled appears to be mainly due to contamination and the time it takes to clean and separate our waste. Of course having the space to store the separated items is also the problem as well as the limitation of recycling centres to take all items that can be recycled. A note of caution, the values in Table 12.4 are weight not volume which can be a bit misleading especially when dealing with soft and rigid plastics that can quickly fill up your refuse sack but weigh very little. Remember that generally more energy is saved by recycling plastics than is gained by burning them, which is especially true of denser plastic items.

So is it worthwhile recycling? Well look at the table below which gives the GHG emissions saved by each kilogram of a material recycled (Table 12.5).

Table 12.5 GHG emissions saved by recycling (per kg) common waste materials

Cardboard and paper	1.5 kg CO_2e
Dense plastic packaging	2.0 kg CO_2e
Ferrous packaging	1.5 kg CO_2e
Aluminium packaging	10 .0 kg CO_2e
Glass packaging	0.5 kg CO_2e
Textiles	8.0 kg CO_2e
Putrescible kitchen waste	4.5 kg CO_2e
Garden waste	1.0 kg CO_2e
Miscellaneous combustible waste	1.5 kg CO_2e

Maximum recycling of household material is equivalent to saving GHG emissions equivalent to 38.2 kg CO2e per household per week, or 16.6 kg CO2e per person per week

Of course the bottom line is that we should really look at ways of reducing all of our waste including the recycled fraction. So just because you are taking loads of stuff to the recycling centre doesn't mean that you are doing as much as you could be. Most of the recyclable waste in my household comes from packaging and this problem has attracted quite a lot of attention in the concept of creating sustainable packaging. To be classed as sustainable, packaging should meet the following criteria:

- be designed holistically with the product in order to optimise overall environmental performance;
- be made from responsibly sourced materials;
- be designed to be effective and safe throughout its life cycle;
- meet market criteria for performance and cost;
- meet consumer choice and expectations;
- be recovered efficiently after use.

While 73 % of packaging can currently be recycled in the UK only 33 % actually gets recycled. A key problem area is food packaging with every tonne of food associated with a quarter of a tonne of packaging. **So our starting point should be to buy food with less packaging**.

More information: http://www.europen.be/index.php; http://www.recyclenow.com/

12.6.4   The Way Forward

Recycling saves GHG emissions and is the only sustainable way of prolonging life of non-renewables. As demand grows through population growth and lifestyle enhancement then recycling may be the only way to manufacture some consumables, and we may have to free up metals by replacing them with more common materials (e.g. replacement of copper pipes with plastic ones). The problem of recycling rare Earth metals is that they are used in such small amounts (Sect. 3.2). However, the University of Cardiff is developing ways to extract platinum, a vital component in catalytic converters and fuel cells, from the dust in urban areas which contains just 1.5 parts per million of platinum. This may lead us to find new innovative ways to recover these rare metals such as indium and gallium. In Denmark 90,000 vehicles are recycled every year. After removing any oil, fuel or other fluids from the vehicle it is literally shredded into small pieces with 85 % of the elements recovered and reused. The value of these recycled materials is about €500 per car so there is money to be made for the companies who have invested in the technology required to do it. However, that type of technology is not cheap. The challenge is to recover even more and as metals in particular become scarcer then more innovative solutions like those being developed at Cardiff University will be needed to separate and recover them.

To become sustainable we need to able to recover and reuse every last fragment of potentially useful material that we discard and create a closed loop.

In 2000 the average lifespan of a car in the UK was 13.5 years which means that in excess of two million cars are being scraped each year. The amount of steel in a car has been steadily declining being replaced with plastics and lighter metal alloys to reduce vehicle weight making them more efficient in terms of emissions and fuel economy (Fig. 12.6).

Recovery and reuse of parts is still an important element in recycling vehicles, with most items serviceable so that they can be resold back into the market place as parts which can include body parts such as doors, bonnet wings, headlamps, starter motors, alternators, gearboxes, engines, drive shafts almost anything that can be removed can be salvaged and be reused saving customers money as well as preventing these items being recycled and/or landfilled (Fig. 12.7). Reusing parts is a very important part of achieving the 85 and 95 % recovery targets which have been introduced under the EU End of Life Vehicles Directive and also prolongs the life of older cars reducing their annual embedded lifetime emissions (Table 9.11).

Incorporating accurate GHG emission values for waste or incorporating carbon credits for recycling is difficult and for that reason most calculators simply ignore them. At best they may incorporate generic values based on 100 % recycling, for example incorporating a simple question such as *do you recycle*? (yes/no). There are

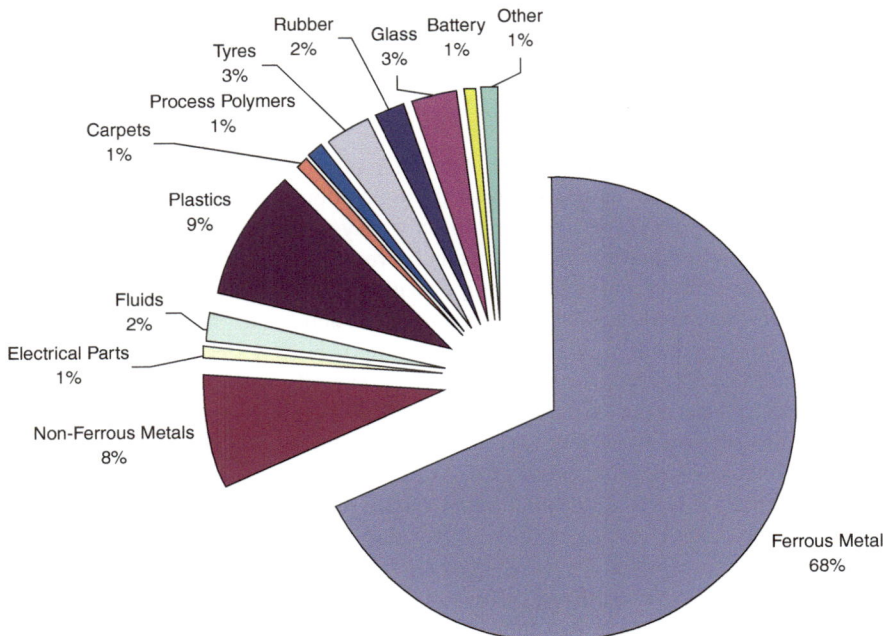

Fig. 12.6 The recyclable components of a car in the year 2000 by weight. Since that time the percentage of ferrous metal has decreased and plastics increased. *Source*: The European Union, http://ec.europa.eu/eurostat/statistics-explained/index.php/Main_Page. Reproduced with permission of the European Commission, Brussels, Belgium

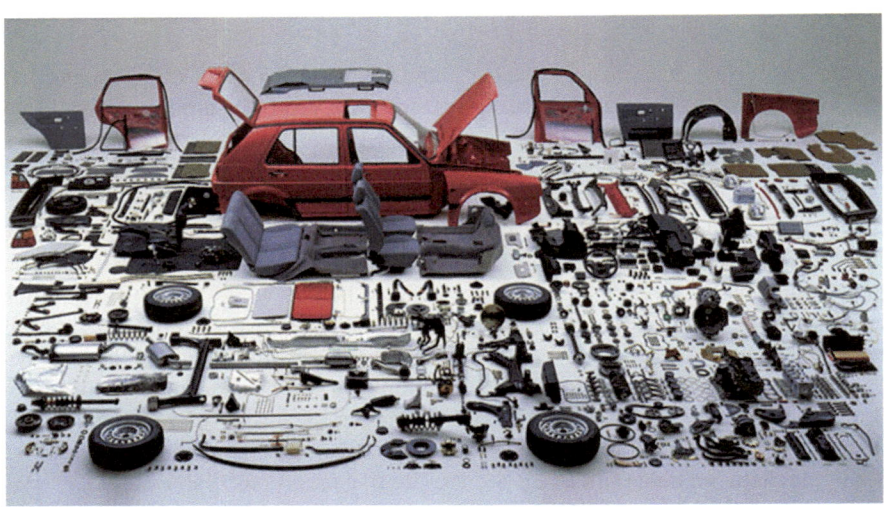

Fig. 12.7 A car is made up a hundreds of separate parts the majority of which can be recovered, serviced or reconditioned and then sold as spare parts. Parts can be salvaged from both end of life and crashed vehicles. *Source*: Ford Motor Company, https://corporate.ford.com/homepage.html. Reproduced with permission of the Ford Motor Company, Dearborn, USA

a number of calculators that give you an idea of just how much energy you can save through recycling but be warned conversion is difficult, so it is best to make own calculations using the values in Table 12.5.

One of the most powerful personal waste calculators is the Waste Reduction Model (WARM) produced by the US EPA. The calculator lists 46 material types and allows you to fill in a simple spread sheet of items recycled to give you the amount of energy in kWh or CO_2e in tonnes that you have saved. It is regularly updated and the calculator can also be used to compare different disposal options.

USEPA WARM waste calculator: http://www.epa.gov/warm

There is a lot of negativity about recycling based on where materials ultimately end up and are reprocessed. In Ireland the bulk of recyclables go to the UK for reprocessing (336,197 t per annum), with the remainder going primarily to China (161,579 t per annum) and to a lesser extent to Belgium and other EU countries (160,739 t per annum). But even exporting waste saves GHG emissions. For example the UK exports 4.7 Mt waste paper and 500,000 t plastic mainly to China. Due to the trade imbalance ships, which had brought manufactured goods from China, would return empty. So shipping waste to China, according to a WRAP study, costs only 10 % of carbon saved. However, scandals about recyclables being dumped into third world economies or increasingly sorted waste being incinerated rather than reused are sadly largely true and have caused a crisis of confidence in recycling. The problem is that the margins in terms of viability on recyclables is very low and so if it is contaminated then generally it is not worth processing and so ends up being incinerated. So we have to adhere strictly to that waste hierarchy, and reduce the volume of waste we generate whether recyclable or not, and then ensure that what we recycle is clean and genuinely recyclable.

> *We have to minimize the volume of waste we generate whether recyclable or not, and then ensure that what we recycle is clean and genuinely recyclable.*

12.7   Conclusions

- Waste is generated from a wide number of sectors with household waste representing <10 % overall by weight. Yet this can represent an important contribution to our personal carbon footprint.
- Recycling has the potential to reduce personal and business carbon emissions significantly. For example, manufacturing an aluminium can from recycled aluminium uses only 5 % of the energy to make a new one which is an energy saving equivalent to a 20 W light bulb on for 20 h.
- Reuse or prevention can have enormous environmental benefits in terms of GHG emission reductions as consumerism inevitably leads to emissions.

- There is an urgent need to minimize landfill and associated methane emissions and this can be achieved by taking waste food out of the waste stream for digestion.
- It is best to recycle and reprocess as close to source as possible as transport is a major avoidable carbon penalty with often the carbon saved by recycling lost by the energy used to take it to the recycling centre.
- Incineration should be the bottom line but necessary due to contamination of some materials and those non-recyclable items.

The twelfth step is acknowledging that recycling can make a significant difference both to GHG emissions and preserving non-renewable resources. This involves you changing your lifestyle by seriously adopting the three Rs and to:

- **Buy less and buy better quality**
- **Consume all your food**
- **Use consumables carefully**
- **Wear things out**
- **Keep electronics longer before replacing (e.g. mobile phones)**
- **Become an active and predatory recycler**

Homework!

I want you to have a quick look inside your own refuse sack and critically revaluated what could have been reused or recycled. Why was it discarded and how could this be avoided in the future? Identify what constraints there are on you to recycle even more.

Now look at all your recyclables. How could this have been reduced? Locate your closest recycling centres and try and see where you can recycle all the materials in your refuse sack, as some centres may take only a limited number of materials. Remember they have to be clean.

If you can't recycle something can you replace it with something that can?

References and Further Reading

Defra. (2013). *Waste prevention programme for England: Call for evidence.* London, England: Department for Environment, Food and Rural affairs. Retrieved from https://www.gov.uk/government/uploads/system/uploads/attachment_data/file/221130/wpp-consult-doc-20130311.pdf

Environmental Benefits

http://www.wrap.org.uk/downloads/Environmental_benefits_of_recycling_2010_
update.08688e75.8816.pdf

Reuse

http://www.wrap.org.uk/downloads/Final_Reuse_Method.2f4770bf.11443.pdf
http://www.rpa100.com/recycled/

Ireland Waste Sites

http://www.epa.ie/whatwedo/resource/
http://www.repak.ie/index.html
http://www.managewaste.ie/e_guides/index.asp
http://www.managewaste.ie/

UK and Europe Waste Sites

http://www.idea.gov.uk/idk/core/page.do?pageId=9499350
http://randd.defra.gov.uk/Default.aspx?Menu=Menu&Module=More&Location=None&Complet
ed=0&ProjectID=14644
http://ec.europa.eu/environment/waste/index.htm

US Waste Sites/Calculation Models

http://www.epa.gov/waste/nonhaz/municipal/index.htm
http://www.epa.gov/epawaste/conserve/tools/iwarm/index.htm
http://www.epa.gov/region10/pdf/climate/wccmmf/Reducing_GHGs_through_Recycling_and_
Composting.pdf
http://www.epa.gov/climatechange/wycd/waste/tools.html#warm

Recycling

http://www2.epa.gov/recycle
http://www.recyclenow.com/
http://www.earthodyssey.com/symbols.html

http://www.all-recycling-facts.com/recycling-symbols.html

Part IV
Responding to the Impact of Global Warming

Chapter 13
The Planet's Health

Since that very first picture of planet Earth taken on December 22nd 1968 by the astronauts of Apollo 8 (Fig. 2.1) we now are able to produce stunning high resolution images of our planet using satellite data. In this chapter we explore the processes that control the planet's ecosystems and how global warming is altering them. Are we reaching irreversible tipping points in the planet's health? *Image*: Created by Reto Stöckli, Nazmi El Saleous, and Marit Jentoft-Nilsen using the resolution Imaging Spectroradiometer (MODIS) instrument aboard NASA's Terra satellite and the NOAA's Geostationary Operational Environmental Satellite (GOES). http://earthobservatory.nasa.gov/IOTD/view.php?id=885. Reproduced with permission of NASA, Washington, DC, USA

It has been a rough ride so far, and up to now it's been all pretty gloomy in places and I have given you some harsh facts…so I feel that I should be saying:
Don't Panic!

© Springer International Publishing Switzerland 2015
N.F. Gray, *Facing Up to Global Warming*, DOI 10.1007/978-3-319-20146-7_13

Doom and Gloom. Image by Rupert Besley. Reproduced under licence.

It is inevitable when talking seriously and honestly about global warming and climate change that you are not going to feel great, in fact I would be surprised if you didn't feel pretty depressed. However, the whole point of this book is about giving you the facts and exploring how we should be responding personally. To tackle global warming effectively we need to understand where we stand right now, we also need to appreciate the real urgency of the situation and that the message has not been lost by constant repetition. That is why so often actions really do speak louder than words. Hopefully by now you are beginning to see what you can start doing about it.

13.1   Whose Planet Is It Anyway?

We have looked at many issues so far but in this chapter I want to examine some of the effects these are having on the health of the planet. I am not really discussing pollution here, but more importantly the process that keeps the biosphere habitable for humankind.

The planet is made up of two entities the geological (inorganic) and the biological (organic). Our perception of the planet is based largely on the organic entity which has evolved into a complex highly diverse and to our eyes a rather beautiful place. For most of us the ideal of a healthy planet is some idyllic eighteenth century landscape which of course is highly artificial and created entirely by man. The planet is made up of numerous niches with species adapted to each. Unlike any other species, man has managed to reach, explore, colonize and exploit every one of these niches and has modified or completely altered every one of them. He has become so successful in using the planet, as no other creature has before, that he has started to modify it inorganically; so that the natural inorganic processes are no longer able to maintain the balance/equilibrium that has produced this period of sustained organic development allowing this highly diverse and biologically rich planet to evolve. We see the *status quo*, or some form of idealized status quo, as the norm. However, in reality it is very difficult to honestly define a healthy planet as it has been evolving for millions of years and will continue to evolve with or without humans present.

> **Until we stabilize the atmospheric CO_2 concentration we simply do not know what kind of planet we have to deal with, or if CO_2 concentration can be stabilized, or whether there is a place for us in the future, or if we can survive these changes.**

13.1.1 Biodiversity

The world has evolved over millions of years into a very stable physical environment which has allowed a hugely complex and diverse organic society to emerge. This complexity and diversity is a function of environmental (climate) stability, and the organic society (mainly vegetation) collectively maintains this physical environment. Over just a few hundred years humankind has increasingly impacted on the environment through terrestrial, atmospheric and aquatic pollution; global warming; deforestation; urbanization; agriculture and over fishing. Humankind has critically altered this organic-inorganic balance leading to the gradual collapse of this equilibrium which we now know is quite fragile leading to increasing instability especially in relation to our climate.

We are very familiar with normal predator-prey relationships in ecology, but humankind is a super predator. Collectively we have eliminated food and non-food species by over hunting, destroyed habitats, poisoned other species through deliberate or accidental release of chemicals, and destroyed natural cycles and processes that control the climate and water resources. We have even poisoned the seas through nutrient enrichment, toxic chemicals, physical particulates, overfishing (largely local and regional) and now through acidification and warming (global). The result is a massive and continuous extinction of species (plant and animals, terrestrial and aquatic). This has established itself as a continuum and **it is inevitable that unless we take positive action now humankind will simply become a part of this sequence of extinction**. Humankind has developed into some all consuming destructive force which is no longer living in equilibrium with the planet. We are no longer living in harmony with nature, rather we are actually destroying the very processes that sustains it as an organic living entity. We are simply destroying ourselves and the majority of other species through our sheer numbers and our lifestyles…**it is time to take a step back and really decide if we want to leave a planet for our children that they may not be able to save for future generations**. We have to act now, each and every one of us.

13.2   Maintaining Earth's Current Organic Balance

There are three key cycles that control Earth's inorganic balance on which the organic is totally dependent. These are the carbon, nitrogen and hydrological (water) cycles. Apart from man there are natural sources of GHGs (e.g. volcanoes, permafrost, wildfires, and methane hydrate) that affect these cycles and within them feedback lops that often cause the rate of change to speed up. Global warming is causing these planet life support systems to gradually break down creating a modified and increasingly unstable environment.

Where does humankind fit into this new global environment?

13.2.1   The Carbon Sink

The Earth has a natural carbon cycle in which carbon flows from one reservoir to another over time scales ranging from days to decades and even millennia and longer. The major reservoirs of carbon are the oceans, terrestrial vegetation, soils, and the atmosphere. The **carbon cycle helps regulate the amount of carbon dioxide (CO_2) present in our atmosphere**, and is therefore a major component of the climate system. Over the millennium prior to the Industrial Revolution, atmospheric concentrations of CO_2 were relatively stable. This is because the two major carbon fluxes (i.e. between terrestrial vegetation and the atmosphere; and between the ocean and the atmosphere) were generally in equilibrium. This is summarized in Fig. 13.1. The CO_2 concentration in the atmosphere was balanced by the natural carbon fluxes shown by the large arrows in the figure. The problem is that now the cycle must also absorb the massive extra inputs of carbon from burning fossil fuels. Every year some 3.3 gigatonnes of carbon per year (Gt C $year^{-1}$) shown by the small arrows in Fig. 13.1 are added to the atmosphere from anthropogenic sources that cannot be absorbed either by the land or sea. This is causing the mean atmospheric CO_2 concentration to steadily rise.

Approximately 20 % of the world's annual CO_2 emissions result from land-use change, primarily deforestation in the tropical regions of Central and South America, Africa, and Asia. These lands are shifting from relatively high-carbon stock natural forests to generally lower-carbon stock crop agroforestry, grazing, or biomass fuel plantations and urban areas (Fig. 13.2). While this transformation by land clearing, forest harvest, and fire provides short-term economic benefits and rural livelihoods, it is also a major source of GHG emissions and other social and environmental problems. The key climate factors from deforestation are carbon release, loss of biodiversity and change in weather patterns. However, significant carbon sequestration and greenhouse gas mitigation potential still exists in the tropics, and other regions outside the US and Central Europe through reforestation.

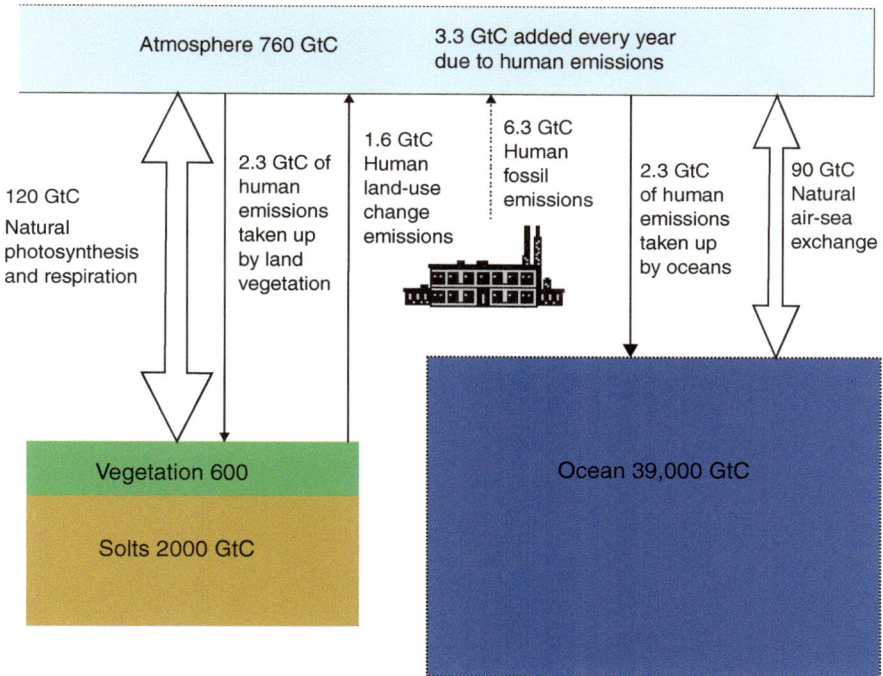

Fig. 13.1 Natural carbon cycle and the effect of man-induced carbon emissions. The *large arrows* show natural carbon fluxes and *small arrows* anthropogenic induced carbon fluxes. Values are in gigatonnes of carbon (GtC—equivalent to a billion tonnes of carbon) with values against *arrows* showing annual fluxes as GtC per year. *Source*: IPCC (2007). Reproduced with permission of the Intergovernmental Panel on Climate Change, Geneva

Fig. 13.2 Gold mining operation in the Amazon rainforest. Image by Rhett Butler. Reproduced with permission of Rhett A. Butler/Mongabay. http://www.mongabay.com

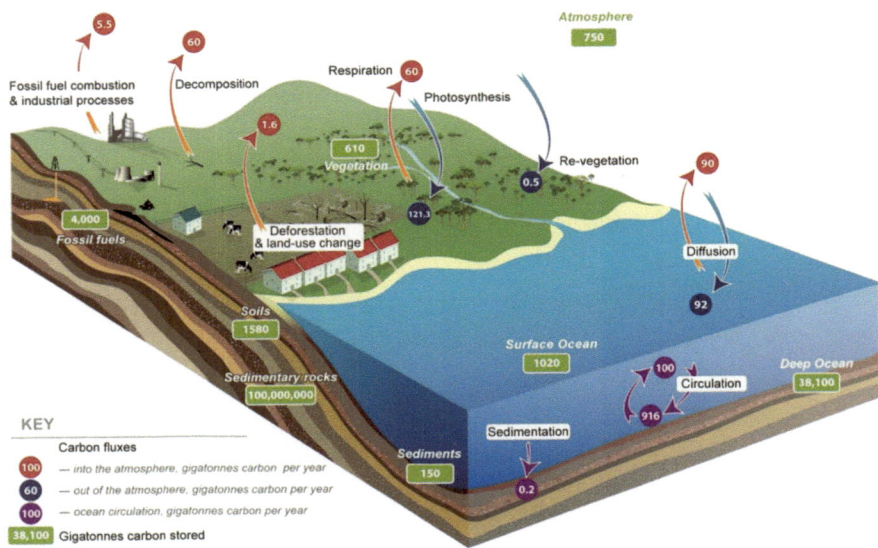

Fig. 13.3 The carbon cycle and the carbon reserves. *Image Source*: co2crc.com, http://www. co2crc.com.au/. Reproduced with permission of CO2CRC Limited, Carlton, Australia

The atmosphere of the planet has gone from being anaerobic to aerobic during its lifetime. Early organic forms followed by algae and more complex plants stripped the CO_2 from the atmosphere and polluted it by releasing oxygen causing the concentration in the atmosphere to rise and that of CO_2 to gradually fall over millions of years. This has created the organic planet we all know and are a part of today. This excess CO_2 has been permanently removed from the atmosphere as plant material converted into oil, gas, lignite, peat, and methyl hydrate. When left alone this stored CO_2 remains intact. There is also a vast amount of carbon stored in sedimentary rocks formed by dead crustaceans, shell fish and corals whose shells and exoskeletons are made from carbonate a form of CO_2 originally from the atmosphere (Fig. 13.3). There are a number of different inputs of carbon to the atmosphere. Decomposition and respiration are balanced by photosynthesis and creation of new plant material, with the excess being absorbed by the oceans. It is the extra input of CO_2, taken from the deep reserves in the form of fossil fuels, that the carbon cycle is unable to deal with, so that the excess is stored in the atmosphere causing a gradual increase in CO_2 concentrations resulting global warming.

13.2.2   Volcanoes

Why bother about our emissions when tonnes of GHG are released by volcanoes? I am asked that question almost every day and with 22 active volcanoes on Iceland alone, it is quite an important question to ask. It is true that volcanoes can have massive carbon footprints when they erupt. For example, Mount Pinatubo in the Philippines emitted 42 million tonnes (Mt) of CO_2e when it erupted in 1991 **Collectively all the world's volcanic activity produces 300 Mt CO_2e per year** and while this seems a lot it is still less than 1 % of human emissions. These relatively small amounts of naturally produced CO_2 can be assimilated by long-term processes in the oceans by laying down carbonate sediments in the form of corals and shells, but this mitigation pathway has been affected by excessive CO_2 assimilation causing acidification in the oceans that slow or prevent these processes (Sect. 13.8).

Volcanoes also counter these warming effects by cooling with ash and emitted sulphur dioxide (SO_2) thrown into the stratosphere where they reflect sunlight back to space.

So on balance the Mount Pinatubo eruption caused a net cooling of 0.5 °C the following year, however all these cooling effects are only temporary.

13.2.3   Other Warming and Cooling Effects

We have already seen that nitrous oxide and CFC's warm the planet. Carbon black which are particles released from combustion…remember I wondered where all the smoke and car fumes used to go?… well they increase heat uptake from the sun by reducing the albedo effect on the surface by making snow darker so it absorbs more heat, but it also cools the earth by shading (Table 4.2). Diesel fuel usage is also rapidly increasing and the small particulates from diesel engines also cools the planet.

Sulphur dioxide (SO_2) in the atmosphere, which is also released from fuel combustion, forms an aerosol of tiny droplets that reflects solar radiation back into space causing cooling. Between 1940 and 1970 high levels of SO_2 balanced the warming effect of other greenhouse gases. However, through intense environmental pressure the release of SO_2 was eventually reduced to counteract the effects of acid rain which destroyed so much forestry and acidified lakes in northern areas. The result of this reduction in SO_2 was that the masking effect was then lost which caused a rapid rise in heating of the surface of the planet. The effect of SO_2 aerosols is relatively short lived so effects are not global but regional. In 2000, global SO_2 levels began to rise once more largely due to the proliferation of coal fired power stations in China, again causing a masking affect and cooling. However, these emissions are now being scrubbed which will again accelerate warming. While climate models attempt to incorporate these effects, SO_2 emissions may be masking the true rate of global warming.

13.3   Wildfires

Natural fires are part of the natural carbon cycle, but the frequency and extent of wildfires is rapidly increasing as a consequence of longer periods of dry weather due to global warming making vegetation far more susceptible to fire (Fig. 13.4).

Fig. 13.4 Wildfires cause massive and increasingly permanent loss of forests and biodiversity, as well as causing significant damage to housing and other infrastructure. This example is a forest fire in Greece. *Source*: Wild Forest Fires, http://www.wildforestfires.com/. Reproduced with permission Wild Forest Fires

Because of the increased frequency and scale of wildfires the carbon released from these fires is no longer being sequestered back at an equal rate, resulting in a net generation of CO_2.

Global warming is turning wildfires into a positive feedback loop which is speeding up average temperatures by increasing CO_2e emissions. Fires cover between 3 and 4 million km^2 of the globe on average each year and are responsible for the release of 2–3 Gt of carbon to the atmosphere. Fire emissions impact climate by the direct emission of greenhouse gases and globally is the source of 10 % of methane and 10–20 % of nitrous oxide released annually. Secondary effects include altering both aerosol and ozone concentrations. Wildfires also result in both biological and physical changes to the land surface that affects carbon exchange in subsequent years and alters the surface radiative balance for several decades. Also, areas are not recovering as has been the case previously due to changing in rainfall patterns and higher temperatures, altering the regeneration process, resulting in desertification or at best significantly altered habitats. Unlike SO_2 the impacts of fire on CO_2 emissions to the atmosphere can be large at both the regional and global scales. Also there is significant uncertainty regarding the magnitude, timing, and variability in CO_2 emissions from fires making predictions on global atmospheric CO_2 concentrations difficult.

Arson involving crops, scrubland and forestry is not taken very seriously unless it causes significant financial loss in relation to buildings or loss of life.
However, such actions significantly damage the environment and add to GHG emissions.

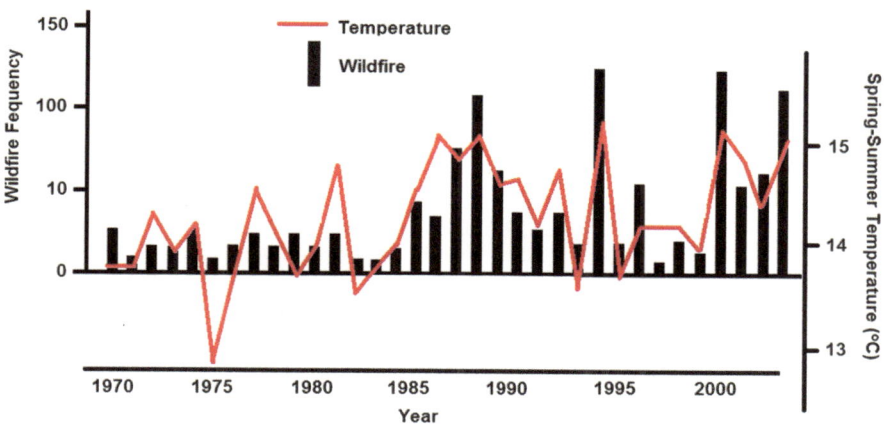

Fig. 13.5 Increasing incidence of wildfires in western USA is blamed on rising temperatures

Wildfires are incredibly important in terms of our fight to stabilize global CO_2 levels. They are a major contributor both in terms of CO_2 emissions and the creation of carbon black particles. Approximately 100 t of carbon is released per hectare from a wildfire which is equivalent to 367 t CO_2 per hectare (i.e. 3.67 times its weight in $CO_2 = 367$ t CO_2). The major bushfires in Australia in 2009 released a 165 Mt of CO_2e and due to changes in climate much of that area will not fully regenerate and so will be unable to reabsorb the equivalent weight of CO_2 as new vegetation. Globally **wildfires contribute on average** 2.1 Gt of CO_2e per year based on 1997–1998 estimates. So today this figure is well in excess of **3 Gt of CO_2e per year**. It is generally accepted that the frequency of wildfires is directly linked to temperature (Fig. 13.5).

The major problem for us today is that as the climate changes, forests are becoming drier and wildfires more common. In Australia some people feel that leaving brush material on the floor to encourage wildlife has also increased the risk and intensity of fires. Climate change has also changed the nature of the soil, which again may not be able to recover as before, so these fires are frequently not carbon neutral (i.e. carbon emitted is removed by new growth). Each year fires in Federal US forests release 90 Mt of CO_2 and this is expected to rise by 50 % by 2050 and double by 2100 (Fig. 13.6). Fires in the contiguous United States and Alaska

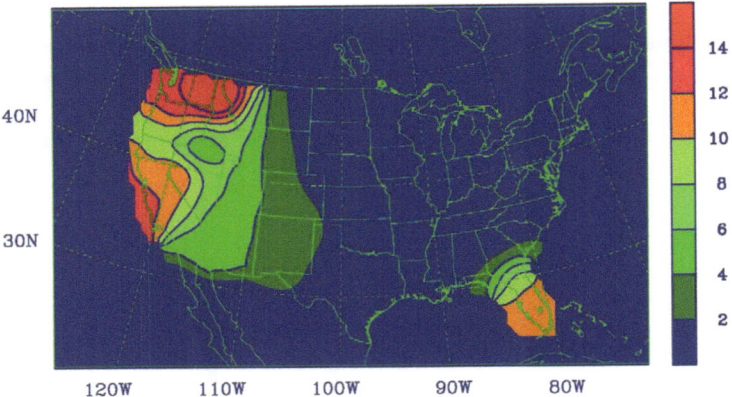

Fig. 13.6 Spatial distribution of annual CO_2 emissions from wildfires in the USA (t km^{-2}). *Source*: http://www.oregon.gov/ODF/BOARD/docs/FFAC_LiuYong_CO2.pdf?ga=t. Reproduced with permission of Professor Yongqiang Liu, Research Meteorologist, USDA Forest Service/Forestry

release about 290 Mt of CO_2 per year, which is about 4–6 % of the amount of the greenhouse gas that the country releases through fossil fuel burning. We could possibly prevent large amounts of GHG emissions by trying to reduce wildfires through better forest management, regional and international investment in fire fighting equipment, a more co-ordinated international response in dealing with large wildfires, and making wildfire arson a very serious offense to try and discourage deliberate setting of fires as seen during 2014 in Spain and Greece.

More information: Wiedinmyer, C. and Neff, J.C. (2007) Estimates of CO_2 from fires in the United States: implications for carbon management. *Carbon Balance and Management*, 2:10. doi:10.1186/1750-0680-2-10

Global fire maps: http://earthobservatory.nasa.gov/GlobalMaps/view.php?d1= MOD14A1_M_FIRE&eocn=home&eoci=globalmaps

13.4   Ice Cover

Everyone is aware that ice cover is getting less. In particular mountain glaciers are getting smaller with many now disappeared altogether However, it's not just glaciers that are in decline, snow cover in Northern hemisphere, arctic ice cover, frozen ground (permafrost) are all declining during the summer months. All models predicted that snow will start to accumulate later and start melting earlier, with less accumulation. Permafrost will continue to thaw at ever increasing rates in the Northern hemisphere, including Alaska (Sect. 13.6). Melting sea-ice floats and so displaces the same amount of water so when these melt they do not affect sea levels, which is in contrast to land ice and or snow which increases sea level when they melt. All major ice sheets (e.g. Greenland) are getting smaller, even though in some cases snow cover is getting thicker on top. This massive influx of freshwater will also reduce salinity, especially in the Northern Atlantic, which will affect fisheries and marine wildlife. Another less familiar problem is that the reduction in sea ice reduces the reflection of sunlight resulting in more heat being absorbed by the sea which in turns speeds up the heating of the water increasing the rate of ice melting. This is an example of a positive feedback mechanisms which are common phenomenon in global warming (Sect. 4.3).

The Arctic Ice sheet has been gradually retreating since accurate records began in the late 1970s, with the minimum ice extent measured at the end of each summer in September giving a retreat rate of 12.0 % per decade (Fig. 13.7). Some summers start earlier forcing greater than expect retreats such as in 2007 when a new record was set of just 4,140,000 km^2 of ice remaining. In 2012 this recorded was again broken leaving just a minimum of 3,500,000 km^2 of ice. Most modellers predict the Arctic will be ice free during the summer by 2050, although Prof Peter Wadhams who is head of the Polar Ocean Physics Group at Cambridge University has suggested in a recent paper in *Nature* that this could happen much sooner and within this decade. He also suggests that the ice free window will slowly increase and that by 2035 that the Arctic Ocean of could be ice free for 6 months each year.

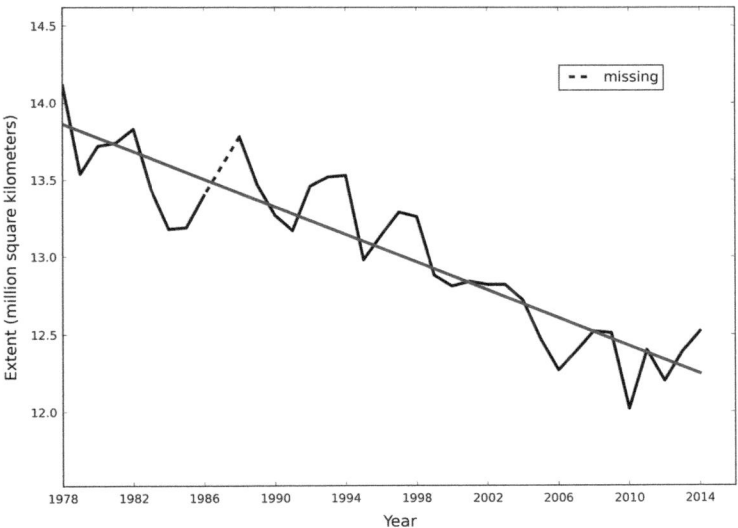

Fig. 13.7 Seasonal extent of the Arctic sea ice (in millions of square kilometres) based on satellite records. The monthly December ice extent showed a decline of 3.4 % per decade over the period 1979–2014 relative to the 1981–2010 average. The *dashed line* indicates a period of missing data. *Source*: The National Snow and Ice Data Center, http://nsidc.org/. Reproduced with permission of the National Snow and Ice Data Center, Boulder, CO, USA

The projected temperature rise in the Arctic will be significantly higher than the global average due to a thinner atmosphere which means less air to heat before the heat is transferred to the surface, also less energy is lost by evaporation.

More information: http://nsidc.org/arcticseaicenews/

13.5   Sea Level

The sea level has been stable for the past 1600 years. However, since 1900 it has begun rising at a rate of 1.7 mm per annum until the early 1990s when the rate doubled to 3 mm per annum due to global warming. Sea level rises is a result of glaciers and small ice caps melting, land ice melting (coastal sections of Antarctic and Greenland ice sheets), portions of ice sheets sliding into the sea, and importantly sea water expansion due to rising water temperature. Approximately half of the current rate of rise is from melting land ice and the same amount again from expansion of sea water. Sea level rise is not the same everywhere due to several factors such as changes in ocean currents, differences in ocean temperatures and salinity. The IPCC estimates that on average, sea levels will rise using a range of different scenarios, by between 0.18 and 0.59 m by 2100 relative to the level in 1980–1999. There will be a small offset as higher temperatures will increase the amount of snowfall over central Greenland and Antarctica. Interestingly land is also rising and falling which has localized effects on relative sea levels. For example, the sea level rise is relatively higher in the southeast than in other parts of England due to the land sinking.

The historical, current and projected change in sea level is shown in Fig. 13.8. This is based on the loss of land based ice from Greenland and Antarctica continuing at the same rates as observed from 1993 to 2003. Of course these rates could increase or decrease in the future we just don't know. **If melting increased linearly with global average temperature, the upper range of projected sea level rise by the year 2100 would be between 0.48 and 0.79 m**.

The Ice sheets present serious problems in relation to sea level rise. The West Antarctic and Greenland Ice Sheets are very vulnerable to increase in temperature. So this presents another layer of uncertainty about sea level rise. **The West Antarctic Ice Sheet alone contains enough water to raise the sea level by 5–6 m**. This ice sheet or parts of it could slide into the sea if warming continues which would cause a devastating global tsunami as well as raise the sea level by several metres immediately and increasing as it slowly melts. **The Greenland Ice Sheet contains enough water to raise sea level by 7 m**. This is not unstable like the West Antarctic Ice Sheet but will continue to slowly melt (Fig. 13.9).

Although the West Antarctic ice sheet has always been considered more vulnerable to melting recent research indicates that the East Antarctic ice sheet, which holds ten times more ice than Western Antarctica is also unstable and is at risk of

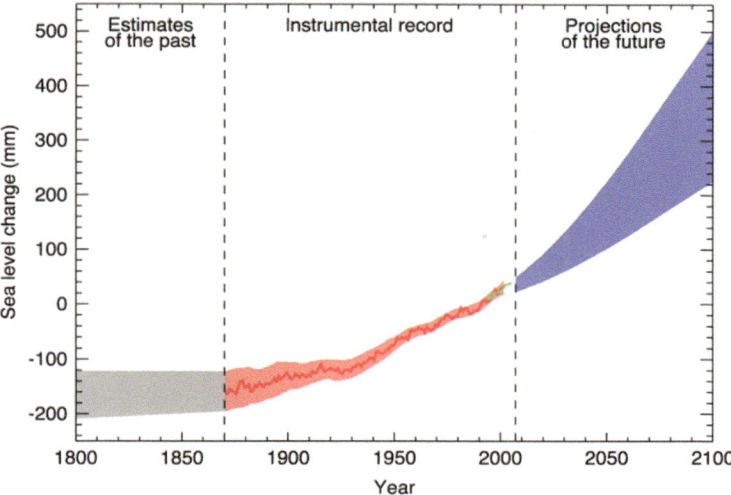

Fig. 13.8 Past and projected global average sea level. The *shaded area* is uncertainty, the *green line* most accurate as measured by satellite. The *purple shaded area* represents the range of model projections for a medium growth emissions scenario (IPCC A1B) (Sect. 4.4.2). *Source*: IPPC (2007). Reproduced with permission the Intergovernmental Panel on Climate Change, Geneva, Switzerland

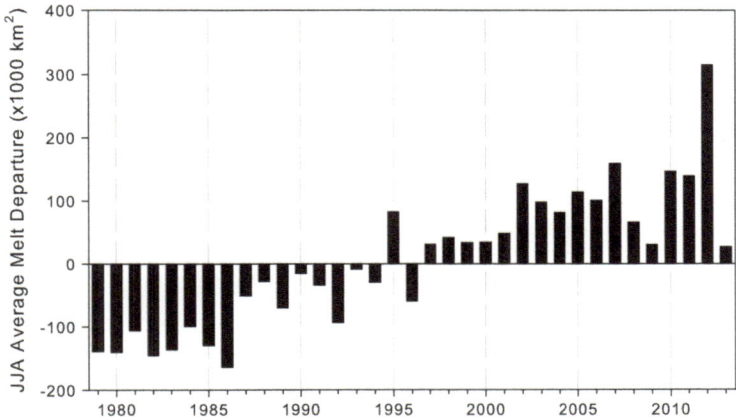

Fig. 13.9 Variation in the extent of the annual melt of the Greenland ice sheet compared to the average for the period 1978–2013. *Bars* show the sum of the extent of the melt during the summer (June to August—JJA) with the average subtracted. A clear increase in the area melting each year is seen in recent times. *Source*: Thomas Mote of the University of Georgia and the National Snow and Ice Data Center. http://nsidc.org/greenland-today/. Reproduced with permission of the National Snow and Ice Data Center, Boulder, CO, USA

melting with the possibly that sections of the sheet could eventually slide into the sea. The West Antarctic ice sheet largely slopes inland but sections of East Antarctic slope seaward, including the huge Wilkes ice sheet which is very sensitive to changes in temperature. The Wilkes ice sheet is held in place by a layer of coastal ice, rather like a wedge and if this ice is removed the entire basin will melt leading to a global sea level rise of 3–4 m. It is estimated that the wedge will be removed by a global sea level rise of 80 mm which will trigger the ice sheet to melt with the possibility that the whole ice sheet will become instable and slowly start to slide into the sea and melt accelerating sea level rise. This will possibly occur early in the next century.

More information: *Mengel, M. and Levermann, A. (2014) Ice plug prevents irreversible discharge from East Antarctica. Nature Climate Change.* doi:10.1038/nclimate2226 http://www.nature.com/nclimate/journal/vaop/ncur-rent/full/nclimate2226.html

There will be substantial variability in future sea level rise between different locations due to variation in winds, atmospheric pressure and ocean currents. These factors already cause the sea level to rise more rapidly along the Mid-Atlantic and Gulf Coasts, and less rapidly in parts of the Pacific Northwest. Therefore, some locations will experience sea level rise above the global predicted average, others less.

Even small increases in sea level can have devastating effects on coastal habitats. As seawater reaches farther inland other effects include destructive erosion, flooding of wetlands, contamination of aquifers and agricultural soils, and lost habitat for fish, birds, and plants. When large storms hit land, higher sea levels mean bigger, more powerful and destructive storm surges. In addition, hundreds of millions of people live in areas that will become increasingly vulnerable to flooding and higher sea levels will eventually force them to abandon their homes and relocate. Low-lying islands could be submerged completely. Most at risk are the coastlines of Bangladesh and Vietnam, the Philippines, small pacific and Caribbean islands; and also large coastal cities or those affected by tidal range such as Tokyo, New York, London, Cairo, Dublin and all low lying coastal and estuarine settlements.

Figure 13.10 shows a more-or-less steady increase in global mean sea level of 3.2 ± 0.4 mm per year during the period of 1992–2014. While there is not enough evidence to show that the rate of increase will continue to rise; most recent estimates agree that the rise will be between 0.8 and 2.0 m by 2100 enough to swamp many of the cities along the U.S. East Coast. The worst case scenario which includes a complete meltdown of the Greenland ice sheet, would raise sea levels by 7 m (23 ft) which would submerge London and Los Angeles as well as most of central Dublin.

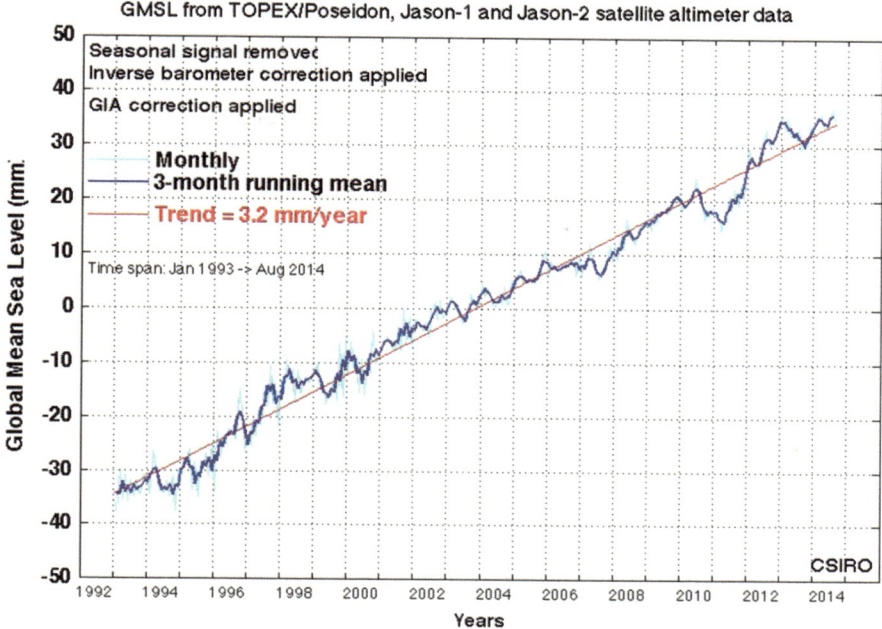

Fig. 13.10 Actual rise in sea level based on the most accurate global satellite data for the period 1992–2014. This shows an average rate of sea level rise of 3.2 mm per year. *Source*: Church and White (2011) and CSIRO, http://www.cmar.csiro.au/sealevel/. Reproduced with permission of the Commonwealth Scientific and Industrial Research Organisation (CSIRO), Australia and the authors

13.6 Permafrost

Permafrost is frozen soil which forms the Tundra region of the Arctic. Billions of tonnes of organic matter has been permanently frozen in the ground since the last ice age which occurred thousands of years ago. The land is frozen all year around so the organic matter is literally in a deep freeze and unable to undergo decomposition. Approximately 5 % of the land area of the Northern hemisphere is permafrost representing a total area of 13 million km² in Canada, Siberia and northern Europe (Fig. 13.11).

Permafrost can be categorised into four types. **Isolated** permafrost are frost hollows where the mean annual temperature is 0 °C. **Sporadic** is where less than 50 % of the landscape is affected and the mean annual temperature is 0 to −2 °C. **Discontinuous** permafrost is formed in patches covering 50–90 % of the land area with the mean annual temperature is 0 to −5 °C, but as the air temperature is just below zero it is formed only in sheltered areas. **Continuous** permafrost forms

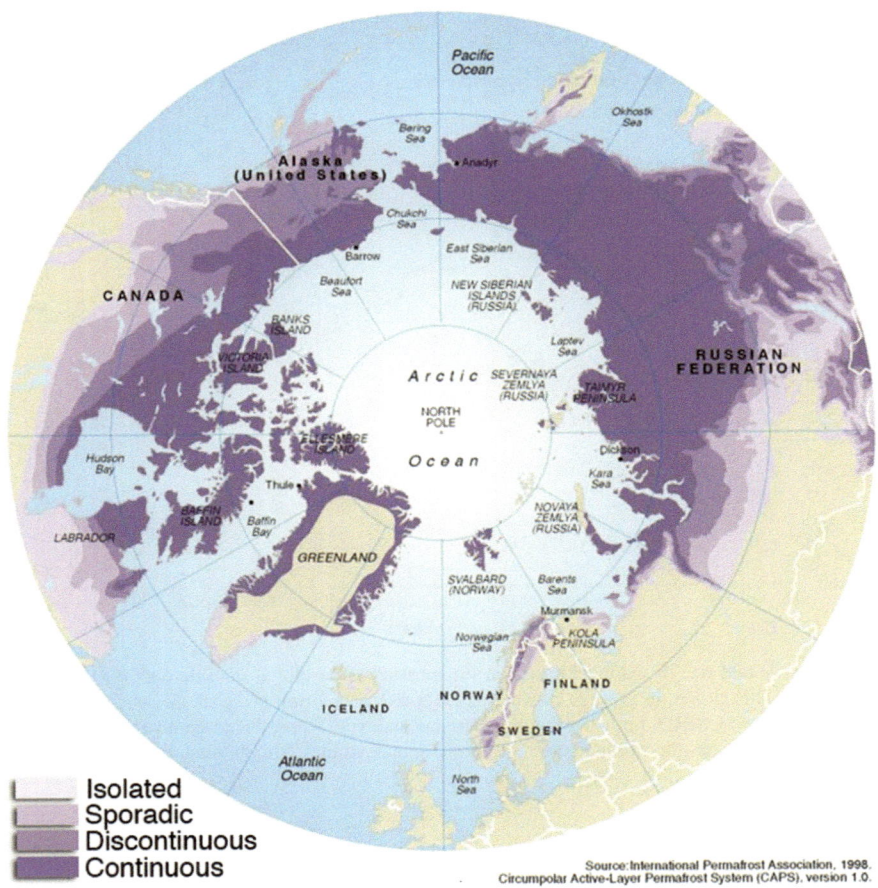

Fig. 13.11 The permafrost extends right around the globe of the Northern hemisphere with the area covered decreasing the further south you travel from the pole. *Source*: The IPA, http://ipa. arcticportal.org/. Reproduced with permission of the International Permafrost Association, Potsdam, Germany

everywhere (100 %) as the air temperature remains constantly below 0 °C. In these areas the frozen layer can be up to 1500 m in depth. **Within the permafrost both peat and methane are stored which represent between 1400 to 17,000 Gt of carbon which is double the amount of CO_2 in the atmosphere at the moment!**

Between 2020 and 2030 the permafrost will stop becoming a net sink for carbon and become a major source of CO_2 and methane as it starts to defrost and then decay. Permafrost is already melting at the southern tip of the Tundra. In the summer of 2007 the average temperature in the Tundra regions was 3 °C above normal. The soil should get colder as you go towards the surface however, now the opposite is true and the soil gets warmer towards the surface. Vast areas of permafrost are already close to −1 °C and so near to melting. As it melts two actions occur.

Fig. 13.12 Examples of ice-wedge polygons in the Russian permafrost landscape. After the soil has completely frozen the extreme cold can cause the ground to shrink and crack. Water then flows into these cracks in spring only to refreeze and expand next winter creating these characteristic ice wedges. Image by Konstanze Pie of the Alfred Wegener Institute. Reproduced with permission the Alfred Wegener Institute, Potsdam, Germany; and the United Nations Environment Programme, Nairobi, Kenya

First the stored organic matter will decay releasing CO_2 and secondly the lakes and waterlogged areas that are formed will start to produce massive quantities of methane as the organic matter decays anaerobically. It is not only in the tundra that we will see this effect (Fig. 13.12). As regions become drier then natural peat bogs will dry up and start to decay releasing CO_2.

Many scientists believe we could reach an irreversible tipping point in terms of carbon emissions if the permafrost melts. This point could occur by 2030.

We have to prevent the permafrost from melting, as the effects on the living planet will be similar to a massive meteor strike in terms of damage to biodiversity and humankind. This is the single most important problem that we face in relation to climate change. All the signs are that the process of thawing is

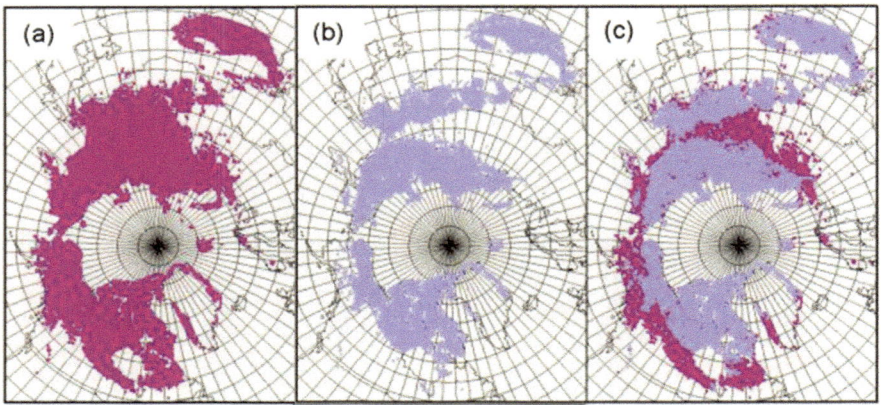

Fig. 13.13 Extent of permafrost in the Arctic region, (**a**) 1990–2000 and (**b**) predicted for 2090–2100. (**c**) Predicted areas of permafrost lost by 2100 shown in *pink*. *Source*: USGS, http://www.usgs.gov/. Reproduced with permission U.S. Geological Survey, Reston, VA, USA

underway although the timescale in terms of how quickly this will occur is still speculative (Fig. 13.13).

> **Some predictive models suggest that 60 % of the permafrost will melt by the year 2200 releasing 190 Gt of carbon.**
> **This will be cataclysmic in terms of human survival**

Look at these videos: http://link.brightcove.com/services/player/bcpid1786720821 ?bctid=786834489001 http://www.youtube.com/watch?v=yN4OdKPy9rM&feature=player_embedded#!

13.7   Methane Hydrate (Methane Clathrate)

There are vast amounts of methane frozen underground within an ice lattice called methane hydrate. It originates from the microbial breakdown of plankton under anaerobic conditions. The amount of methane hydrate could easily be in excess of all the fossil fuels that have so far been discovered so they represent a huge reserve of sequestered carbon that was once present as CO_2 in the atmosphere. Most of reserves are located in ocean sediments although some are associated with permafrost soils. In both cases warming of the ocean floor or of frozen soil will release methane to the atmosphere. Also where these sedimentary deposits are close to the surface of the ocean bed and are eroded , blocks and sheets of frozen methane hydrate float to the surface where the gas is released on warming. **The risks are unquantifiable and have been ignored by IPCC models**, but if triggered then CO_2 levels could soar.

13.8   Sea Acidification

Up to 50 % of the CO_2 released by burning fossil fuels over the past 200 years has been absorbed by world's oceans. Ocean acidity has increased 30 % since the Industrial Revolution, the fastest change in ocean chemistry for at least 65 million years. Currently about a third of our CO_2 emissions are removed from the atmosphere primarily by the sea, however, this buffering effect will gradually reduce as the solubility of CO_2 is reduced in warmer water and the ability to absorb CO_2 at the same rate declines. Also biological sequestration of CO_2 will also decline with an increase in temperature. **This means that as warming increases, the rate of rise in seawater acidity will also increase due to natural sequestration declining**.

The mechanism of seawater acidification is well understood. The absorbed CO_2 in seawater (H_2O) forms carbonic acid (H_2CO_3) (rainfall is in fact carbonic acid and has a natural pH of 5.2), lowering the water's pH level and making it more acidic. Hydrogen ion concentration in the water is raised thereby increasing acidity (Fig. 13.14) and it is the acidity which limits an organisms' access to carbonate ions needed for shell and skeleton construction, etc. The pH scale is a logarithmic one which means that each pH unit is an order of magnitude larger or smaller than the

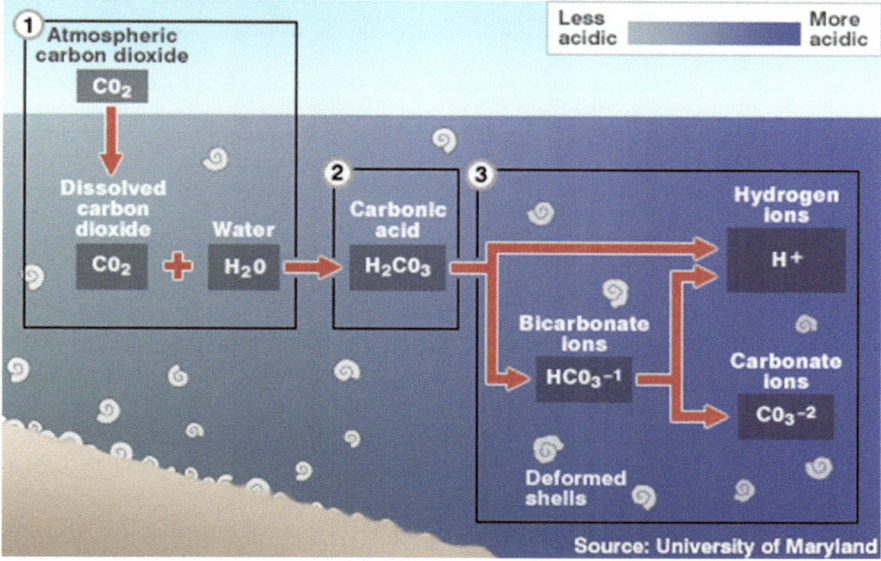

Fig. 13.14 The chemical steps that is leading to the acidification of the oceans. There are three critical steps. (*1*) CO_2 is absorbed from the atmosphere by the sea which (*2*) forms carbonic acid which is a key driver in ocean acidity. The carbonic acid dissociates into bicarbonate and hydrogen ions (H^+), with the bicarbonate breaking down to carbonate, which is used for shell construction which will finally fall to the sea floor forming sedimentary rock layers, while more hydrogen ions are also released. However, as more H^+ are released the water becomes more acidic. Reproduced with permission of the University of Maryland

Changes in Aragonite Saturation of the World's Oceans, 1880–2013

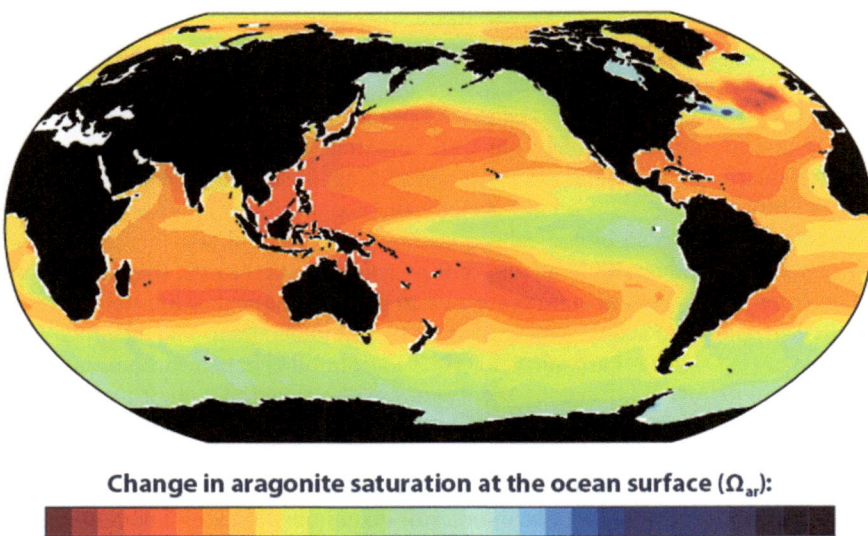

Change in aragonite saturation at the ocean surface (Ω_{ar}):

-0.8	-0.6	-0.4	-0.2	0	0.2	0.4

Data source: Woods Hole Oceanographic Institution. 2014 update to data originally published in: Feely, R.A., S.C. Doney, and S.R. Cooley. 2009. Ocean acidification: Present conditions and future changes in a high-CO₂ world. Oceanography 22(4):36–47.

For more information, visit U.S. EPA's "Climate Change Indicators in the United States" at www.epa.gov/climatechange/indicators.

Fig. 13.15 Change in global aragonite saturation from 1880 to 2013 which can be seen as an analogue for pH. Argonite is a form of calcium used for shell, skeleton and coral development. Negative changes show a fall in aragonite saturation making it more difficult for animals to maintain their skeletons and shells. Data from Woods Hole Oceanographic Institution and image by USEPA http://www.epa.gov/climatechange/science/indicators/oceans/acidity.html. Reproduced with permission of the US Environmental protection Agency, Washington, DC, USA

last. So pH 7 is ten times more acidic than pH 8, or pH 6 is 100 times more acidic than pH 8, which means that even small changes in the pH can be very significant to aquatic animals.

The pH of the world's oceans is not consistent (Fig. 13.15). Some areas have a relatively low pH (the purple areas) which is the result of the upwelling of deeper colder water which is rich in CO_2. However, no region is expected to escape the impact of falling pH with the IPCC forecasting that ocean pH will fall by *"between 0.14 and 0.35 units over the 21st century, adding to the present decrease of 0.1 units since pre-industrial times"*. Researchers warn that it could eventually result in a massive extinction of life in the seas. Creatures that form alkaline shells are likely to be particularly affected and coral reefs may begin to crumble before the end of the century, with many coral reefs already showing signs of damage from acidification.

As global warming increases, the rate of rise in seawater acidity will also
increase due to natural sequestration declining.

13.9 Tipping Points in Planet Health

Models have tried to identify the so called tipping points in planet health. These are
the rise in global surface temperature temperatures at which key climate systems
will irrevocably break down. Except for possibly the loss of the Arctic Sea ice, tipping points are irreversible. It is generally now accepted that in 2007 the cycle of
free-thaw of the Arctic Ice Sheet moved into an unsteady state. This in turn will
affect the Siberian Yedoma permafrost accelerating heating which will become irreversible at 1.5 °C, so that a thaw could occur before 2030. Increase in global temperature will change weather patterns to warm Asia more quickly and speed up the
thaw, which will further drive global warming accelerating the rate of melt of the
Greenland Ice sheet (Sect. 1.3; Fig. 13.16). This destabilizes the Atlantic thermoha-
line circulation which pumps water around global seas, and could shut it down
altogether changing world oceanic currents leading to severe weather changes and
possibly the increased heating of the southern oceans. This cascade or domino effect
where one tipping point acts to increase the likelihood of the next has been extensively studied and is supported by a large body of evidence (Levermann et al. 2012).
Global temperatures have already risen by 0.8 °C over the past century, so even if
all GHG emissions were stopped immediately, global temperatures would continue
to rise another 0.3 °C, so it would appear that **the Yedoma permafrost tipping
point is inevitable**. Our current GHG emissions have already locked us into a predicted rise in global temperature of 1.1 °C, with 2 °C our target limit under Kyoto.
But even if the Kyoto limit is achieved, which is currently highly unlikely, then
further tipping points could be triggered. So the outlook doesn't look too good if we
allow global warming to continue unabated. Like so many predictions in climate
science the concept of tipping points and how they might interact remains contentious, but they appear to be increasingly likely.

13.9.1 Should We Care?

But here is an interesting fact that you may not have considered. **The reality of all
this is that planet Earth doesn't care**! It has probably not even noticed this small
blip (i.e. us) in its continuous inorganic evolution. We tend to think of the planet as
a living entity, Lovelock's Gaia. In reality it is an inorganic system and the biological component is just a thin surface layer. The planet will continue to evolve and to
change and new organic diversity will evolve again and will create its own new

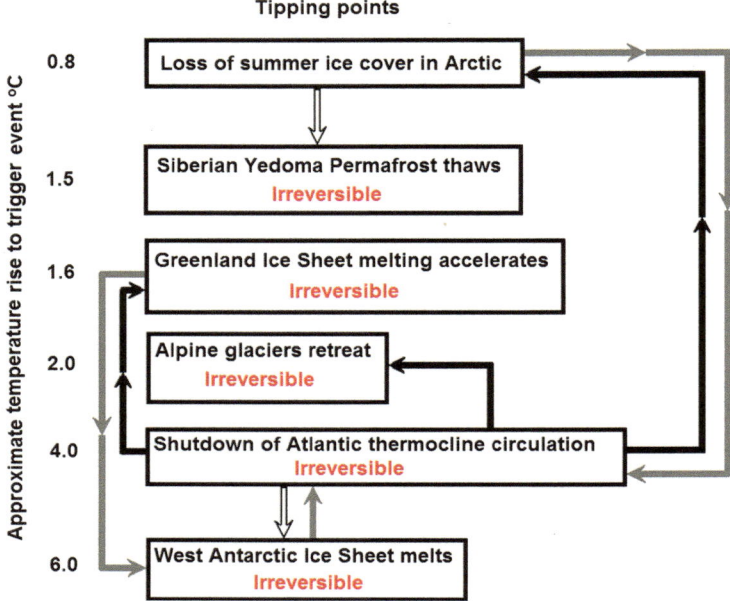

Fig. 13.16 Critical tipping points caused by temperature rise that trigger a whole range of inter-related events that are probably irreversible

utopia which may last for just thousands or even perhaps millions of years until it too becomes extinct by perhaps some cataclysmic event such as being hit by a meteor.

Global warming is a cataclysmic event inflicted by humankind on itself. Our concerns over global warming are not about saving the planet or even saving biodiversity. This is all about us, humankind, saving ourselves as a species, maintaining us as the dominant species who has managed to manipulate the geological side of the earth like no other organic form has done before. **Tackling global warming should be seen as a lifeboat exercise as we move into a new era of whole planet engineering and management**.

> **Controlling GHGs to control global temperature rise is vital to prevent the uncontrolled release of fossil carbon that will lead to a cataclysm.**
> **It is well within our power to prevent this even now…we have just got to get on with it.**

> **The thirteenth step is realizing that planet Earth is at a tipping point inorganically and that our existence as a species is dependent on all of us realizing that we have exploited the planet too far and that we must act now to prevent these fundamental changes occurring.**
> **It is also realizing that we can still prevent the worse scenarios occurring if each of us chooses to act right now.**

13.10 Conclusions

- The health of the planet is in crisis with its natural feedback systems under threat
- The carbon balance is collapsing and is unable to cope with the vast amounts of previously stored GHGs now being released back into the atmosphere
- Excess CO_2 is causing the oceans to become more acidic
- Wildfires are a major cause of GHG emissions which are no longer being sequestered back into new vegetation
- Ice cover is melting resulting in sea level rise, a function of additional freshwater and expansion due to temperature rise
- As the climate warms more carbon is released from sinks and emitted into the atmosphere: the permafrost is melting releasing GHGs, peat bogs are drying up releasing GHG, methane hydrate is releasing GHGs. Nearly all these potential sources of GHGs are in addition to our warming predictions based on atmospheric CO_2 levels. These sources of emissions are driven by numerous feedback loops
- We are facing a series of tipping points, most of which are irreversible that will change our global society for ever

Homework!

The time has come to become more proactive in exploring ways that you can make a real difference. First you need to start formulating a personal plan as to how you can live a more sustainable lifestyle and start to reduce your own GHG emissions.

It is also time to think about whether you feel able to make a personal commitment to living sustainably. Up to now you have been putting your homework into a physical portfolio or into a virtual file on your computer. Now you need to start organizing this into a personal plan of action. So start by reviewing all the information you have gathered about yourself and your family's emissions. Personal action also needs to influence your family, friends, workmates and the community.

If you don't feel you can make any kind of commitment right now then that is fine but try and justify why in words.

References and Further Reading

Church, J. A., & White, N. J. (2011). Sea-level rise from the late 19th to the early 21st century. *Surveys in Geophysics, 32*, 585–602.

IPPC. (2007). *The AR4 synthesis report*. Geneva, Switzerland: Intergovernmental Panel on Climate Change. Retrieved from http://www.ipcc.ch/publications_and_data/ar4/syr/en/contents.html

Levermann, A., Bamber, J., Drijfhout, S., Ganopolski, A., Haeberli, W., Harris, N. R. P., … , Weber S. (2012). Potential climatic transitions with profound impact on Europe: Review of the current state of six 'tipping elements of the climate system.' *Climate Change, 110*, 848–878. Retrieved from http://www.pik-potsdam.de/~anders/publications/levermann_bamber12.pdf

Carbon Cycle and Fluxes

http://carboncyclesandsinks.org/
http://www.esrl.noaa.gov/research/themes/carbon/

Climate Change in General

http://epa.gov/climatechange/index.html

Forest Fires

http://www.globalforestwatch.org/

Deforestation

http://travel.mongabay.com/deforestation.html

Ocean Acidification

http://oceanacidification.wordpress.com/
www.mcbi.org/what/ocean_acidification.htm

Chapter 14
Your Health and Wellbeing

In this chapter we explore the effect that global warming will have on our health. How we tackle this and also sustain wellbeing is examined

14.1 Health

The effect of climate change on health is quite complex with a range of positive but mostly negative effects linked to existing problems relating to poverty, food and water scarcity. Predictions suggest that most mortalities will arise due to the frequency or severity of familiar health problems which will be made worse by global warming affecting those already at risk. The effects will be both direct and indirect as shown in Fig. 14.1 and can be broken down into a number of key health risks:

- Temperature-related illness and death
- Extreme weather-related health effects

© Springer International Publishing Switzerland 2015   357
N.F. Gray, *Facing Up to Global Warming*, DOI 10.1007/978-3-319-20146-7_14

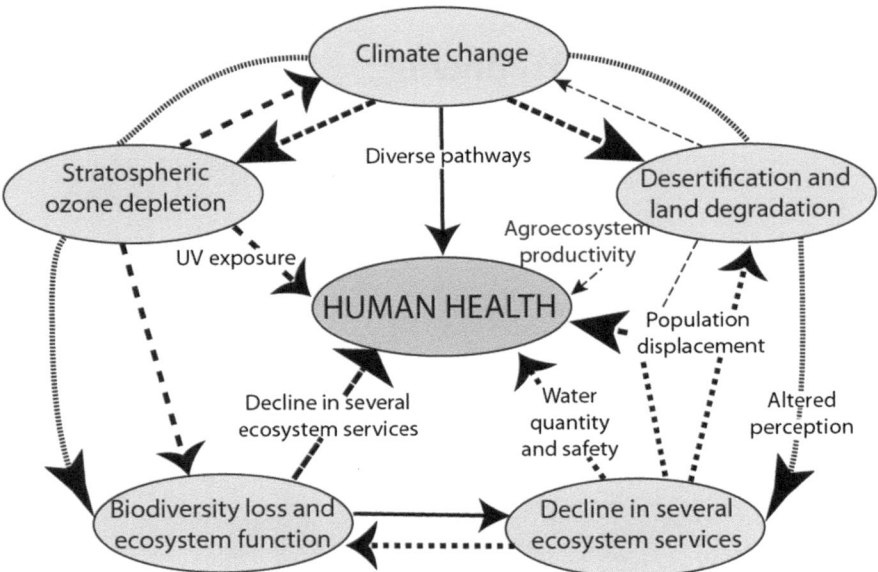

Fig. 14.1 The indirect pathways in which climate change can affect human health. *Source*: http://www.who.int/globalchange/ecosystems/en/. Reproduced with permission: The World Health Organization

- Air pollution-related health effects
- Water and food-borne diseases
- Vector-borne and rodent-borne diseases
- Effects of food and water shortages
- Effects of population displacement

People are already dying globally from a wide range of poverty-related causes, and each year 3,500,000 individuals die from malnutrition, 2,200,000 die from diarrhoea and a further 900,000 die from malaria…that's over six million people each year! In contrast, relatively small numbers die from extreme weather events, about 60,000 individuals each year. Unfortunately the former doesn't make the news and latter does, so our understanding of the effects of global warming is a little over emphasised on weather related incidents. The problem is that some of the largest disease burdens are climate-sensitive so it is those at most risk to these diseases, due to malnutrition and poverty, who will suffer the most as temperatures rise. The World Health Organization (WHO) has estimated that climate change is responsible for the annual loss of about 160,000 lives and the loss of 5.5 million years of healthy life. This is expected to rise to 300,000 deaths by 2020 and the loss of ten million years of healthy life. Climate change will affect overall health and longevity (chronic) rather than cause direct (acute) mortality. It will trigger migration leading to increased pressure in surrounding areas, especially urban centres resulting in a significant reduction in both wellbeing and health.

It is not only the effects of climate change that affect health, we have to be very careful that the adaptation and mitigation policies implemented by Governments and organizations do not have any negative implications for health and wellbeing. The protection of health and wellbeing is central to the United Nations Framework Convention on Climate Change which requires that all countries give due consideration to health when introducing climate mitigation strategies.

> *'Human health and wellbeing is a basic human right and contributes to economic and social development. It is fundamentally dependent on stable, functioning ecosystems and a healthy biosphere. These foundations for health are at risk from climate change and ecological degradation.'*
>
> The Doha Declaration on Climate health and wellbeing

14.1.1   Temperature-Related Illness and Death

Extremes of both heat and cold can cause potentially fatal illnesses such as heat stress or hypothermia, as well as increasing death rates from heart and respiratory diseases. As global warming alters weather patterns, these extremes can occur almost anywhere, often unexpectedly. Statistics on mortality and hospital admissions show that death rates increase during extremely hot periods, particularly among very old and very young people living in cities. Abnormally high temperatures in Europe in the summer (August) of 2003 were associated with at least 27,000 more deaths than the equivalent period in previous years. Temperatures were 10 °C above the 30-year average, with no relief at night with Belgium, the Czech Republic, Germany, Italy, Portugal, Spain, Switzerland, the Netherlands, and the UK all reporting excess mortalities during the same period, In France over 60 % of the reported deaths during this period were people aged 75 and over. Deaths in Paris over that summer heat wave are shown in Fig. 14.2. The lines represent the maximum (green) and minimum (yellow) daily temperature.

More information: http://www.eea.europa.eu//publications/climate-impacts-and-vulnerability-2012

14.1.2   Vector-Borne and Rodent-Borne Diseases

Increased temperature and precipitation create more disease-friendly conditions in regions that did not previously host diseases or disease carriers. Climate change accelerates the spread of disease primarily because warmer global temperatures enlarge the geographic range in which disease-carrying animals, insects and microorganisms (as well as the bacteria, protozoa and viruses they carry) can survive. In addition to changing weather patterns, global warming affects disease transmitted via insect vectors such as mosquitoes (vector-borne disease) or through rodents

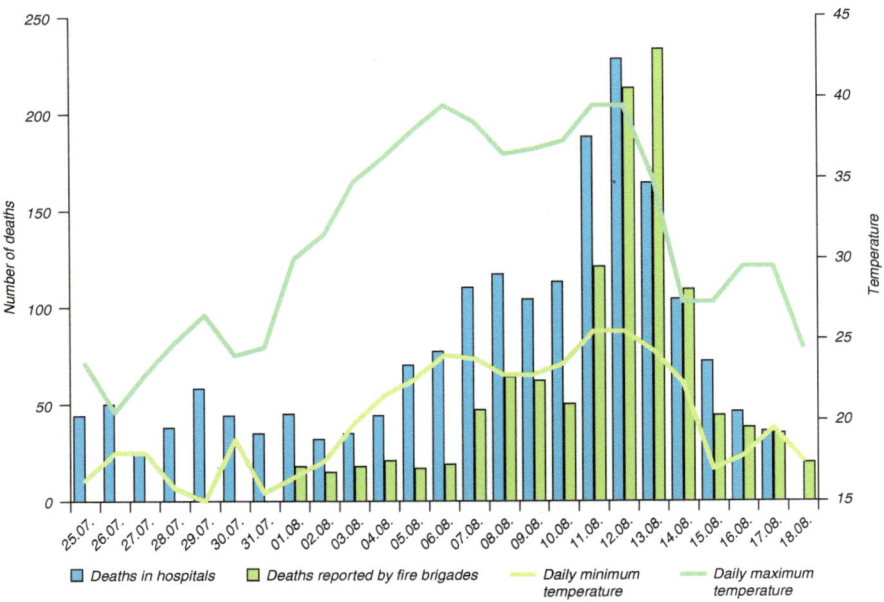

Fig. 14.2 Reported deaths during the Paris heat wave in the summer of 2000. *Source*: EEA, http://www.eea.europa.eu/publications/climate_report_2_2004. Reproduce with permission: The European Environment Agency, Copenhagen, Denmark

(rodent-borne disease). Deadly diseases often associated with hot weather, like the West Nile virus, Cholera and Lyme disease, are spreading rapidly throughout North America and Europe due to the increased temperatures in those areas allowing disease carriers like mosquitoes, ticks, and mice to thrive.

Malaria is the world's most important vector-borne disease, resulting in 216 million clinical cases and 655,000 deaths in 2010 alone. The vast majority of these cases are in sub-Saharan Africa in children under the age of 5 years. However, increasing temperature and moisture for breeding grounds (especially irrigation areas) and also migration of infected people will extend its range with global warming (Fig. 14.3).

Figure 14.4 shows the current distribution of malaria in yellow. New climate conditions are going to favour the distribution of the *Anopheles* mosquito that carries malaria so it is inevitable that it will spread as climatic conditions change. The red areas on the map predicts how the disease will be extended by 2050, while the grey areas show where it will be excluded due to conditions no longer being favourable. Malaria has been in Ireland and UK previously and could easily return, so this map may underestimate the true extent of the disease which could engulf the whole of western Europe by 2100 given current climate change predictions.

Although malaria is under control in the USA, there were 1691 cases reported during 2010 according to the Centres for Disease Control and Prevention (CDC),

Fig. 14.3 Female *Anopheles gambiae* mosquito is a vector for malaria. Here it is obtaining a blood meal thorough through its needle-like proboscis. *Source*: Centers for Disease Control and Prevention, http://phil.cdc.gov/phil/home. asp. Photograph is by James Gathany. Reproduced with permission of the Centers for Disease Control and Prevention, Atlanta, GA, USA

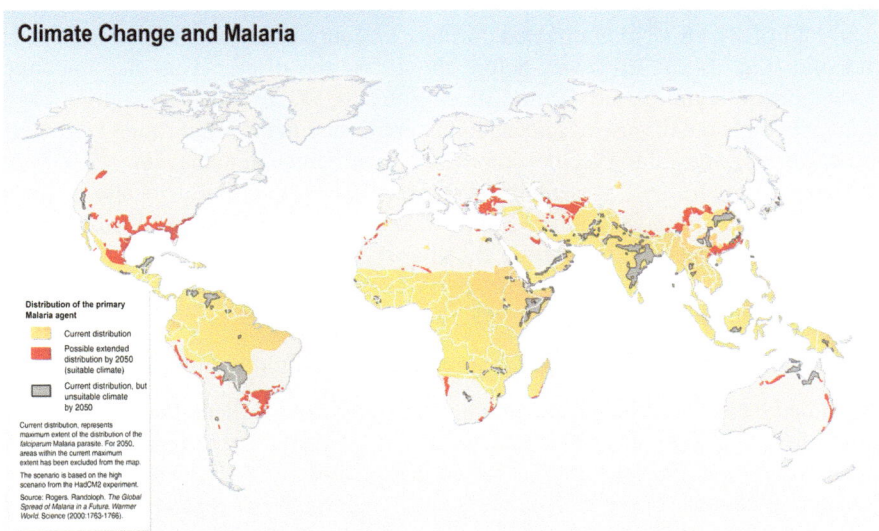

Fig. 14.4 Current and predicted global distribution of malaria. Image by Hugo Ahlenius, UNEP/ GRID-Arendal from data from Rogers and Randolph (2000). Reproduced with permission of UNEP/GRID-Arendal. http://www.grida.no/publications/et/ep4

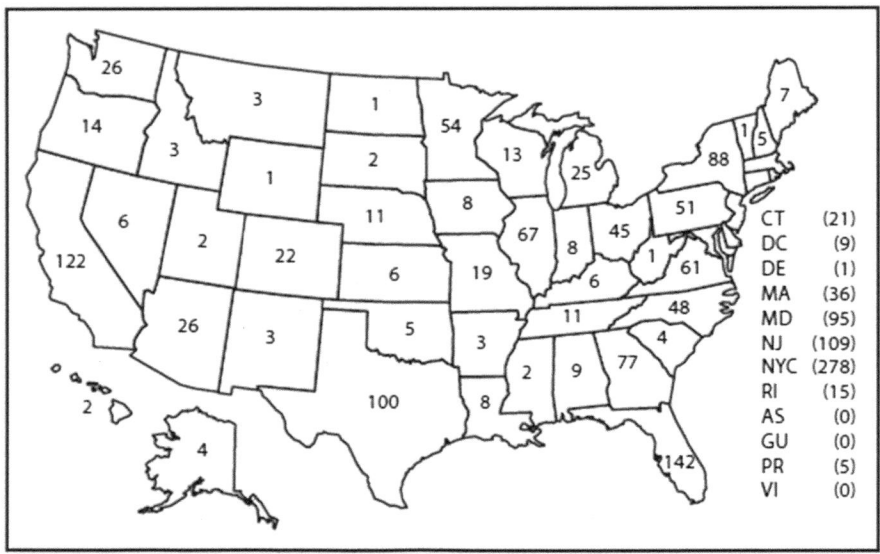

Fig. 14.5 Map of the United States, showing the number of malaria cases, by state in which the disease was diagnosed, in 2010. Other territories are AS = American Samoa; GU = Guam; PR = Puerto Rico; VI = U.S. Virgin Islands. *Source*: The US Centres for Disease Control and Prevention, http://www.cdc.gov/malaria/. Reproduced with permission of the US Centres for Disease Control and Prevention, Atlanta, GA, USA

nearly all of which were contracted during visits abroad to areas where malaria is endemic (Fig. 14.5). However, before the 1950s the disease was also endemic throughout the south-eastern United States with 600,000 reported cases during 1914 alone. So while the disease has been largely eradicated in the US through better living conditions, mosquito (vector) control and environmental management; the vector mosquitoes are still present in these areas and occasionally locally derived malaria cases are reported. However, with changes in climate then the conditions for the vector mosquitoes will improve allowing it to expand its range and presenting an ever increasingly difficult challenge to controlling the disease in the future throughout the Southern States of the US (Fig. 14.4).

More information: http://www.cdc.gov/mmwr/preview/mmwrhtml/ss6102a1.htm

Climate-sensitive diseases are among the largest global killers. Plant and animal pests and diseases are also associated with climate change. For example, there are a number of animal diseases that are in Northern Europe and UK and so are on the doorstep of Ireland. These include the Schmallenberg virus a disease of sheep and to lesser extent cattle. It is primarily transmitted by insect vectors (e.g. midges, mosquitoes). There is no direct transmission from animal to animal, other than maternal transmission from mother to offspring during pregnancy, so this is similar to malaria in requiring an insect vector. The first case in Ireland was reported in October, 2012.

Blue tongue, is a virus that is spread amongst sheep, goats and cattle by biting midges (*Culicoides* spp.). While African Horse Sickness is also a virus that is transmitted by insect vectors (primarily by *Culicoides* spp. but also by other midges, mosquitoes). These two diseases are rapidly moving into Northern Europe. Rift Valley Fever, is primarily a livestock disease that can also be picked up by people handling infected meat (still mainly Africa) and West Nile Virus, which is transmitted by mosquito from infected birds to both animals and humans (now in eastern Europe). All these new diseases are thought to have arrived in North Western Europe due to global warming creating warmer, wetter conditions from April to November which favours their spread.

More information: http://www.agriculture.gov.ie/animalhealthwelfare/diseasecontrol/

14.1.3 Waterborne Diseases

Global climate change will affect the hydrological (water) cycle in many different ways that will alter the pattern of waterborne pathogen survival, infectivity and distribution. These changes include the frequency and intensity of heavy rainfall events, flooding, droughts, increased temperatures and sea level rise. Indirectly, climate induced changes will impact on the effectiveness of the traditional infrastructural and the barrier approaches to pathogen control on which water industry has evolved and now relies. Climate change will also result in an alteration in the behaviour of populations and test their ability to cope with changes in risk from infection and possible failures in the barriers that protect them from pathogens. These barriers include waste water treatment, natural elimination in water resources (i.e. rivers and lakes), water treatment and continued chlorination of drinking water within the distribution system leading to your home. Diarrhoea is closely related to temperature and precipitation (Fig. 14.6). In Lima, Peru, this relationship was studied and it was discovered that diarrhoea cases increased by 8 % for every 1 °C temperature increase.

14.1.4 Extreme Weather-Related Health Effects

Weather extremes, such as droughts, heavy rains, floods, and hurricanes, can also have severe impacts on health. Over the period of 1995–2004, a total of 2500 million people were affected by disasters, with losses of 890,000 dead and costs of US$ 570 billion with 95 % of these in poor countries. Most disasters (75 %) are related to weather extremes that climate change is expected to exacerbate. For example, in 1998, Hurricane Mitch stalled over Central America and released six feet of rain,

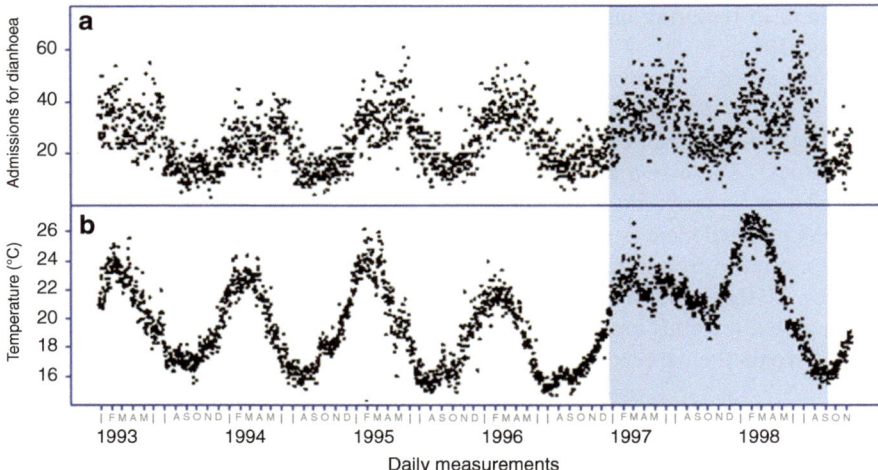

Fig. 14.6 Cyclic relationship between temperature and number of serious cases of diarrhoea requiring hospital admission in Peru. *Source*: Checkley et al. (2000). Reproduced with permission of Elsevier Science, Oxford, UK

causing massive mudslides and claiming 11,000 lives. In October 1999, a cyclone in Orissa, India, caused 10,000 deaths. Hurricane Katrina, in 2005 resulted in excess of 4000 deaths (Sect. 1.3). The total number of people affected was estimated at 10–15 million. In 2006, Sichuan Province, China experienced its worst drought in modern times, with nearly eight million people and over seven million cattle facing water shortages.

14.1.5 Air Pollution-Related Health Effects

In cities, stagnant weather conditions can trap both warm air and air pollutants leading to smog episodes with significant health impacts. The prevalence of asthma in the USA has quadrupled in past 20 years and has been linked to climate-related factors. In the Caribbean asthma has increased due respiratory irritants in the form of dust clouds from Africa's expanding deserts being swept across the Atlantic by trade winds due to warmer ocean temperatures. Altered seasonal distribution and increased levels of allergenic plant pollen and soil fungi may also be involved. Global warming, through the modification of the climate, will have very complex indirect effects on our health, with migration a key factor in putting existing communities and systems under pressure (Fig. 14.7).

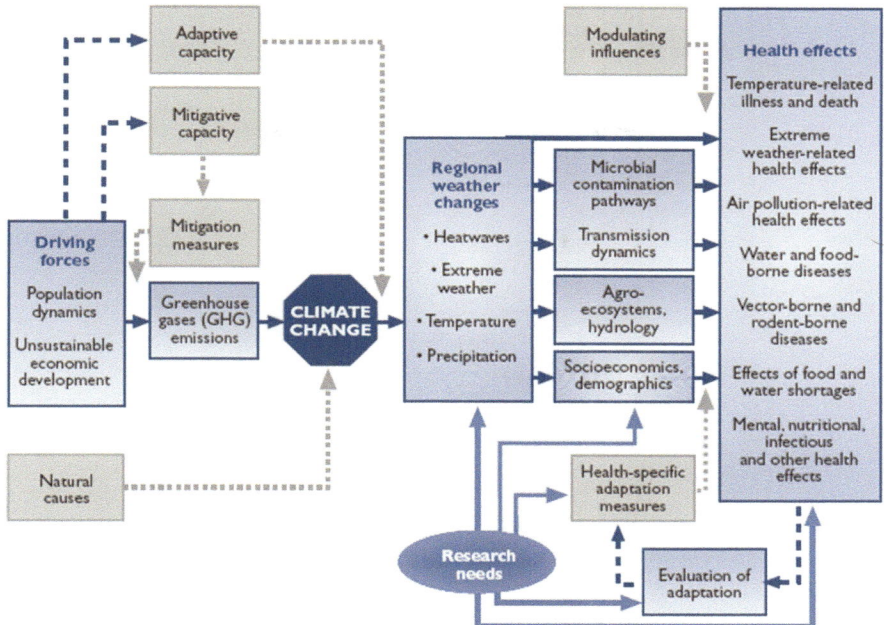

Fig. 14.7 Summary of global warming induced health effects. *Source*: WHO based on the study by Patz et al. (2000). http://www.who.int/globalchange/climate/summary/en/index12.html. Reproduced with permission of the World Health Organization, Geneva, Switzerland

14.1.6   Who and How Many Are at Risk?

Global warming is estimated to be killing between 140,000 and 160,000 people per year (1970–2005) and is steadily rising with temperature induced climate change. Climate change deaths are primarily due to increases in malnutrition (77,000 deaths), diarrhoea (47,000 deaths) and malaria (27,000 deaths). This is a relatively small number when compared to 6,000,000 deaths per year mainly from childhood and maternal malnutrition and 109,000 deaths per year from carcinogen exposure. Health impacts of climate change vary greatly across the world (Fig. 14.8). In general the areas least responsible for changing the climate, are suffering the most deaths from climate change, with about half of such deaths occurring South and Southeast Asia (e.g. Bangladesh, Bhutan, Democratic People's Republic of Korea, India, Maldives, Myanmar, Nepal), which are home to 1.2 billion people (Fig. 14.8).

Very few countries are well prepared to deal with extreme weather events or have adaptive polices to deal with the effects of climate change on health. Although poorer countries are most vulnerable, certain groups are at greater risk regardless of the wealth of the country. These include the urban poor, children, the elderly, subsistence farmers and those living in low lying coastal and flood plain areas (IPCC 2007).

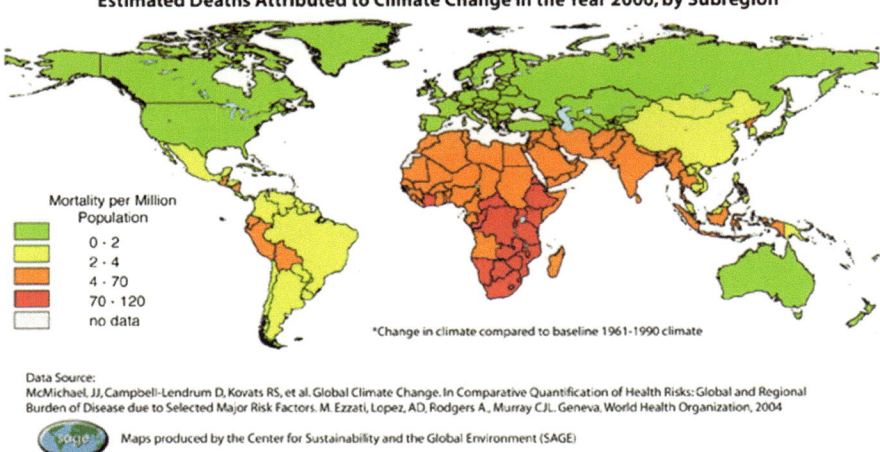

Estimated Deaths Attributed to Climate Change in the Year 2000, by Subregion*

Mortality per Million
Population

- 0 - 2
- 2 - 4
- 4 - 70
- 70 - 120
- no data

*Change in climate compared to baseline 1961-1990 climate

Data Source:
McMichael, JJ, Campbell-Lendrum D, Kovats RS, et al. Global Climate Change. In Comparative Quantification of Health Risks: Global and Regional Burden of Disease due to Selected Major Risk Factors. M. Ezzati, Lopez, AD, Rodgers A., Murray CJL. Geneva, World Health Organization, 2004

Maps produced by the Center for Sustainability and the Global Environment (SAGE)

Fig. 14.8 Estimated deaths attributed of climate-sensitive deaths due to diarrhoea, malaria and malnutrition in 2000. This is expected to increase rapidly as global temperatures rise. The map was produced by Sarah Olson if the Center for Sustainability and the Global Environment (SAGE) at the University of Wisconsin-Madison based on data produced by WHO. http://www.sage.wisc. edu/people/patz/UWmedpubhealth_article.pdf. Reproduced with permission of SAGE and the Board of Regents of the University of Wisconsin System and the World Health Organization, Geneva, Switzerland

At each COP meeting a declaration is made in relation to health and wellbeing. In December 2012, The Doha Declaration on Climate Health and Wellbeing was published and world health organizations were asked to endorse its aims (see box below).

The Doha Declaration on Climate Health and Wellbeing

What we seek from climate action
Recognising health in all policies and strengthening health systems globally can advance human rights and help create safe, resilient, adaptable, and sustainable communities.
  We call for:

1. **The health impacts of climate change to be taken into account domestically and globally**

   • Health impacts and co-benefits to be fully evaluated, costed and reflected in all domestic, regional and global climate decisions on both mitigation and adaptation;
   • Health and environmental costs to be reflected in corporate and national accounts;

(continued)

- Assessment of loss and damage from climate change to include impacts on human health, wellbeing and community resilience, as well as impacts to health care infrastructure and systems;

2. **Investment in climate mitigation and adaptation to be significantly increased on a rapid timescale**

- Priority given to decarbonisation of national and global energy supplies;
- Cessation of fossil fuel subsidies globally and greater funding for renewable and clean technologies;
- Funding for programs to support and protect health in vulnerable countries to be significantly increased;
- Investment in adaptation and mitigation programs that can demonstrate health benefits to be substantially increased;

3. **The health sector and the community to be engaged and informed on climate action**

- The health sector to be engaged and included in the processes of designing and leading climate mitigation and adaptation worldwide;
- National and global education programs to increase public awareness of the health effects of climate change and promote the health co-benefits of low-carbon pathways; and
- More inclusive consultation processes in global climate negotiations to reflect the views of young people, women and indigenous people.

More information: http://dohadeclaration.weebly.com/

14.2 Positive Health Benefits of Climate Change

There is no doubt that tackling climate change will also have overall positive benefits in individual and community health and wellbeing. Here are some simple examples.

14.2.1 Cooking

Over 50 % of the world's population use inefficient indoor stoves with associated high GHG and black carbon emissions. They also cause significant indoor pollution leading to respiratory disease especially pneumonia. It has been estimated that more than a 1,000,000 deaths each year could be associated with indoor pollution from such stoves. Many aid agencies have focussed on providing modified and new low-emission stoves that not only save GHG emissions but also reduce health impacts. There is a 10-year programme to introduce 150 million low-emission stoves in India which is hoped will prevent about two million premature deaths.

14.2.2   Electricity Generation

Airborne particulates and gases from coal fired power stations cause respiratory and heart disease, as well as lung cancer. This is a major problem in China and India. This is solved by low emission power generation and clean coal fired stations using best available pollution control technology and carbon capture.

14.2.3   Transport

Better public transport infrastructure reduces car use in cities which in turn reduces emissions and particulates associated with respiratory and heart disease and possible cancers. Encouraging cycling and walking has huge heath benefits such as less heart disease, loss in weight, reduction in diabetes, reduction in depression, reduced risk from dementia and breast cancer; as well as creating cleaner and safer streets with less car related accidents especially around schools.

14.2.4   Eating Less Meat and Dairy

A better diet results in less obesity, reduces heart disease, and reduces cancer risk as well as really reducing your food associated carbon emissions.

Dealing with climate change will provide huge heath benefits at the personal and community level, reduce pressure on health services, and simply make us fitter, healthier, have longer active lives and most importantly happier. All these positive outputs really offset the cost of mitigation strategies and help us deal with these very serious health issues.

What is good for planet health is good for human health as well

14.3   Wellbeing and Sustainability

Human wellbeing is such an important issue when dealing with environmental problems but is something that is very often overlooked. Yet with depression at an all-time high, it is so important that actions to deal with global warming take our wellbeing very much into account. The reality is the concerns and doomsday predictions about climate change have not helped people to face the future with a hopeful and positive attitude which is why we need to take control individually, and also as small groups and communities to tackle this issue and thereby support each other. Wellbeing is dependent on lots of factors such as living in a secure and safe environment, being with others, having enough to eat and much more. In some ways this book will have touched on many areas that you may have felt will change your current lifestyle and that such changes would in fact damage your wellbeing. It is important that we see that change is not always negative but can also be positive and life enhancing.

Human wellbeing is defined by academics as being dependent on the preservation and enhancement of three conditions and processes:

- *Economic conditions and processes*, such as production, employment, income, wealth, markets, trade, and the technologies that facilitate all of these.
- *Socio-political conditions and processes*, such as national and personal security, liberty, justice, the rule of law, education, health care, the pursuit of science and the arts, and other aspects of civil society and culture.
- *Environmental conditions and processes*, including our planet's air, water, soils, mineral resources, biota, and climate, and all of the natural and anthropogenic processes that affect them.

So is personal wellbeing and dealing successfully with global warming compatible?

John Holdren, the former president of the American Association for the Advancement of Science defined sustainable wellbeing as: '**pursuing sustainable development to achieve wellbeing where it is now most conspicuously absent, as well as converting to a sustainable basis the maintenance and expansion of wellbeing where it already exists but is being provided by unsustainable means**.'

Source: http://www.sciencemag.org/content/319/5862/424.full

I want to state clearly that I genuinely believe that **it is absolutely possible to maintain personal, family and community wellbeing within a one planet Earth economy**. This has to be our ultimate goal. It is no good tackling climate change if we are unable to live full and happy lives afterwards. Yet why is wellbeing so difficult to achieve?

Let start by listing the main shortfalls to wellbeing. These are:

- Poverty
- Preventable disease
- Pervasiveness of organized violence
- Oppression of human rights
- Wastage of human potential

The driving forces and aggravating factors that are driving these shortfalls are:

- Non-use, ineffective use, and misuse of science and technology
- Maldistribution of consumption and investment
- Incompetence, mismanagement, and corruption

- Continuing population growth
- Ignorance, apathy, and denial

The response to global warming will in fact require us to tackle and overcome many of these shortfalls to wellbeing, handing back personal responsibility in many cases, and giving us all a shared and unifying purpose. What is interesting is that there is no mention of consumerism here. Tackling climate change will require tackling those socio-political issues that currently effect human wellbeing as well as tackling unsustainable consumerism. All these issues are primarily local requiring individuals to act both locally and regionally, **so change is in the grasp of all of us if we really want it**. For me consumerism is a breakdown of the idea of individually, which is moving us towards a collective almost monocultural society. This is so well illustrated by the massive investment being made in developing data mining, which at one level is simply exploiting personal data to sell you what you need according to the retailer, rather than you deciding what you want?

The key underlying principal of personal action is summarized by Jonathan Porritt, in *Redefining Prosperity*, published by the Sustainable Development Commission.

> *…people who have reached a certain level of material comfort and security can (and should) be persuaded that their future quality of life resides in freeing themselves of the trappings of consumerism and in opting instead for low-maintenance, low-throughput, low-stress patterns of work, recreation and home life…*

More information: http://www.sd-commission.org.uk/presslist.php/21/redefining-prosperity

Consumerism is often self-perpetuating, constantly requiring its participants to upgrade and mimic. Consumerism disengages us from the life outdoors, in preserving and enjoying our locality, interacting with neighbours, creating a better, safer and more sustainable place in which to live and work. For example, we can be so busy planning and investing in travel to take us to unspoilt and hopefully safe places that we forget the desire and possibility to have that where we live. Old people afraid to leave their homes, young women intimidated on their way to work on public transport, soiled and littered open spaces, all this happens because we let it happen. Our intention to create a sustainable planet must also include outside our very own door and to do that we need to act collectively to create the environment in which we want to live. It is all part of tackling global warming and it isn't going to be easy, but it's going to be very rewarding and a lot of fun trying.

> The *hedonic treadmill* is our desire for more constantly outstripping what we already have.
>
> Richard Easterlin

Most research indicates that peoples' quality of life is determined far more by the quality of their working life, their family life and their overall social relationships, then anything else, all these factors seem to be relatively more important than the amount of consumption they are able to enjoy (Barry 2009). If that consumption is increasingly eroding the quality of those other aspects of overall wellbeing, then it is clearly far less beneficial than it might at first sight appear. De-coupling of consumption and the improvement of wellbeing is possible and offers a way of moving towards a low-energy economy at the same time. **Wealthy nations must not only consume in more socially and environmentally responsible ways, they must also be persuaded to consume less**. For us to achieve a sustainable society we will have to find ways of making the economy independent of growth (Jackson 2009).

So are happiness, wellbeing, life satisfaction largely independent of wealth and consumerism? I am not naive and clearly an improvement in our income clearly does make a difference, especially at the lower end of the salary scale, but the immediate reward in terms of wellbeing are quickly lost as disposable income rises above a certain threshold. Indeed the recent study published in 2013 shows that while life satisfaction increases with GDP in poor countries, in richer countries life satisfaction remains more or less constant and even declines after a salary threshold of $30,000 per annum (Proto and Rustichini 2013).

More information: Proto, E. and Rustichini, A. (2013) A reassessment of the relationship between GDP and life satisfaction. PLOS One. http://www.plosone.org/article/info%3Adoi%2F10.1371%2Fjournal.pone.0079358

"We can no longer depend on our growth-obsessed model of progress to generate the improvements in quality of life and personal wellbeing that people are now so hungry for. The evidence shows that even as they get richer people aren't getting any happier. Yet our entire macro-economic strategy is still dedicated to a set of policies that demonstrably are not delivering the goods".

Jonathon Porritt,
Chairman of the Sustainable Development Commission UK.

Most researchers agree that external factors such as consumerism accounts for only a small percentage of what generates happiness and that the effects of consumerism tend to be temporary anyway. The primary driver of happiness is in fact genetic, but the major determinant is engaging in activities both mental and physical that lead to long term improvements to our wellbeing or how we experience the quality of our lives. So the difficulty is deciding what lifestyle you actually want, and how you can achieve this with the constraints that climate change will place on us. There is no right or wrong lifestyle, only the one that is best for you. There are no right or wrong lifestyles, that is for you to decide. But whatever route you choose it has to fit into your share of the planet's bioresources, that is a one Earth lifestyle. Coping with global warming is changing your approach to life and making those adjustments to your lifestyle and your community right now.

The last but one step is understanding that if we do not tackle global warming it will have an increasingly negative effect on human health through water and food scarcity, spread of diseases and increased severity of weather patterns.

Wellbeing is largely independent of wealth and so each of us should strive to tackle those socio-political issues that affect wellbeing as well as addressing unsustainable consumerism which often becomes manipulative and addictive.

Ask yourself what lifestyle you really want for yourself and your family and plan how to achieve this sustainably.

14.4   Conclusions

- Climate change affects human health, and, if no action is taken, problems such as malnutrition, deaths and injury due to extreme weather conditions, and change in geographical distribution of disease vectors will worsen.
- Also overall benefits in individual and community health and wellbeing help offset cost of mitigating global warming.
- Without effective responses, climate change will compromise:

  - **Water quality and quantity**: Contributing to a doubling of people living in water-stressed areas by 2050.
  - **Food security**: In some African countries, yields from rain-fed agriculture are predicted to halve by 2020.
  - **Control of infectious disease**: Increasing population at risk of malaria in Africa by 170 million by 2030, and at risk of dengue by two billion by 2080s.
  - **Protection from disasters**: Increasing exposure to coastal flooding by a factor of 10, and land area in extreme drought by a factor of 10–30.

- Tackling climate change will require tackling those socio-political issues that currently effect human wellbeing as well as tackling unsustainable consumerism.

Homework !

This is the most difficult task of all. Decide on what kind of career and lifestyle that you want and how to achieve this within the constraints of living sustainably. This is explored in detail in the final chapter.

How can you get involved with increasing your own biocapacity? Some ideas were given in Chap. 5, but look out for community action projects. For example, Kensington Borough Council in the heart of London helped to convert a derelict tennis courts into 42 kitchen gardens involving 1000 volunteers. People are cultivating and planting even the smallest areas in the built up areas.

For more inspiration look at these links; http://www.incredible-edible-todmorden. co.uk/; http://growsheffield.com/; http://www.incredible-edible-wilmslow.co.uk/ home; http://www.grownyc.org/openspace; http://brooklynfoodcoalition.org/.

References and Further Reading

Barry, J. (2009). Choose life' not economic growth: Critical social theory for people, planet and flourishing in the age of nature. In H. F. Dahms (Ed.), *Nature, knowledge and negation. Current perspectives in social theory* (Vol. 26, pp. 93–113). Bingley, England: Emerald Group.

Checkley, W., Epstein, L. D., Gilman, R., Figueroa, D., Cama, R. I., Patz, J. A., & Blck, R. E. (2000). Effects of El Nino and ambient temperature on hospital admissions for diarrheal diseases in Peruvian children. *Lancet, 355*(9202), 442–450.

IPCC. (2007). *Climate change 2007: Working group II: Impacts, adaptation and vulnerability, Chapter 8: Human health.* IPCC fourth assessment report: Climate change 2007. Geneva, Switzerland: Intergovernmental Panel on Climate Change. Retrieved from http://www.ipcc.ch/ publications_and_data/ar4/wg2/en/ch8.html

Jackson, T. (2009). *Prosperity without growth? The transition to a sustainable economy.* London, England: Sustainable Development Commission. Retrieved from http://www.sd-commission. org.uk/data/files/publications/prosperity_without_growth_report.pdf

Patz, J. A., McGeehin, M. A., Bernard, S. M., Ebi, K. L., Epstein, P. R., Grambsch, A., … , Trtanj, J. (2000). The potential health impacts of climate variability and change for the United States: Executive summary of the report of the health sector of the U.S. National Assessment. *Environmental Health Perspectives, 108*(4), 367–376.

Proto, E., & Rustichini, A. (2013). A reassessment of the relationship between GDP and life satisfaction. *PLOS One.* Retrieved from http://www.plosone.org/article/info%3Adoi%2F10.1371% 2Fjournal.pone.0079358

Rogers, D. J., & Randolph, S. E. (2000). The global spread of malaria in a future, warmer world. *Science, 289*(5485), 1763–1766.

Health

McMichael, A. J., Campbell-Lendrum, D. H., Corvalán, C. F., Ebi, K. L., Githeko, A., Scheraga, J. D., & Woodward, A. (2003). *Climate change and human health—Risks and responses.* Geneva, Switzerland: World Health Organization. Retrieved from http://www.who.int/global-change/publications/cchhbook/en/

Checkley, W., Epstein, L. D., Gilman, R., Figueroa, D., Cama, R. I., Patz, J. A., & Blck, R. E. (2000). Effects of El Nino and ambient temperature on hospital admissions for diarrheal diseases in Peruvian children. *Lancet, 355*(9202), 442–450.

Climate and Health Council

http://www.climateandhealth.org/
http://epa.gov/climatechange/effects/health.html
http://www.imn.ie/features/3424-climate-change-and-health-in-ireland

http://www.ria.ie/getmedia/f3890d6e-a756-4d97-96b8-54a41878d6aa/climate-change-7.pdf.aspx
EEA (European Environment Agency). (2004). *Impacts of Europe's changing climate*. Brussels,
    Belgium: OPOCE (Office for Official Publications of the European Communities). Retrieved
    from http://www.eea.europa.eu/publications/climate_report_2_2004

Wellbeing

Brey, P., Briggle, A., & Spence, E. (Eds). (2012). *The good life in a technological age*. London,
    England: Routledge.
Scott, K. (2012). *Measuring wellbeing: Towards sustainability?* London, England: Routledge.

Zero Economic Growth

Jackson, T. (2009). *Prosperity without growth? The transition to a sustainable economy*. London,
    England: Sustainable Development Commission. Retrieved from http://www.sd-commission.
    org.uk/data/files/publications/prosperity_without_growth_report.pdf

Consumerism

http://www.consume.bbk.ac.uk/

Chapter 15
In Your Hands!

So is global warming induced climate change real? In this chapter we review the evidence and show how you really can make a difference through personal action. Climate Summit by Joel Petts. *Source*: *Image* The Lexington Herald, http://www.kentucky.com/joel-pett-cartoon/. Reproduced with permission of Joel Petts and the Lexington Herald

15.1 Introduction

The other day a friend of mine made a general observation which made me sit up and rethink a few things. She said '**people live as if there is no tomorrow**', and I suddenly realized that she was absolutely right. There is a kind of manic intensity in our lives that is driven by a need to succeed and be seen as successful. I suppose it all started with the idea of empowerment which is a great thing, you know the

© Springer International Publishing Switzerland 2015
N.F. Gray, *Facing Up to Global Warming*, DOI 10.1007/978-3-319-20146-7_15

phrase '*if you believe in yourself enough you can achieve anything.*' Then hijacked by the advertisers…remember the slogan '*because you are worth it*', a concept subsequently driven by social media until we have ended up at the end of 2013 with the most widely used new word to be included in the 2014 edition of the Oxford English Dictionary, *Selfism.*

We have become obsessed with ourselves and now perceive the world, not as being a part of a complex system but the centre of our own universe, unable possibly to see ourselves in any other context. Everything we do is documented, in fact on average people under the age of 30 take more photos of themselves than of anything else. It has made us very insular and less able to see ourselves as an important part of something much larger and complex. With this has developed a need for conformity, to be seen wearing the right branded clothes, to travel to the right places, and to document all this for the approval of others.

We have lost basic skills of making and mending clothes, growing our food, the knowledge of how to store and manage food stuffs, and to know when something is safe or not to eat. We are obsessed with cooking programmes and buy more cookery books than any other genre of book, and indeed will often prepare elaborate and rich meals on the weekends; yet often do not have the skill or knowledge to cook all our meals on a daily basis and provide ourselves and our families with a healthy diet. We have allowed a whole generation to finally lose all the important life skills that my parents, and especially my mother, and also my grandparents, and in this case especially my grandfather, had to use in order to survive. This transition has been going on for a long time and started in earnest in the 1960s. But the good news is things are changing. TV Cooks such as Delia Smith, Nigella Lawson and Mary Berry have all caught the Nation's imagination and through their series have encouraged new generations to cook and bake; with similar revivals happening with growing vegetables, knitting, crotchet, dress making skills, DIY and many more. These are real life skills, and in developing these skills something important happens, we start to reconnect with the physical world, by sourcing and using basic ingredients, by growing our own food, and through this we begin to see the world in a different way, we are finding connections again with our planet. Reconnecting in this way and developing these important life skills are key elements in wellbeing as well tackling the larger issues associated with global warming.

15.1.1 Global Warming

Over 95 % of scientists and engineers agree that greenhouse gas mediated global warming is a major catastrophe of global proportions. Not only is it creating unique problems of its own as we have seen, it also intensifies the damage being caused by many of the other major environmental problems (e.g. ozone layer). So if we simply leave this problem to sort itself out?….well we can't, because our lives and those of our children and in fact the survival and wellbeing of all species on planet Earth depends on us to tackling this problem now…that's not just you and me that is everyone.

Now is the time to accept that global warming is real and is negatively impacting on us so that we can finally move on and engage in dealing with this major global problem in a meaningful way. Scepticism is stalling the need to take action. Of course we don't want to alter our lifestyles or the way in which do business, but we are living in a fool's paradise if we believe that we can continue as we are (i.e. business as usual). Numerous surveys has shown that the vast majority of us know that we can't sustain this level of consumption and that global warming requires us to take action, but we just don't want to admit it and have to start taking action ourselves.

Our lifestyle has been derived from centuries of access to cheap resources which we have continued to this very day. That was the past, we now have to make the large mental leap and accept that global warming and the need to reduce our use of energy either directly or indirectly must take precedence in all aspects of our future economic, political and social futures. This is something we have to do, if we don't do it ourselves then Governments will have to impose it on us anyway. So we need to put aside the doom and gloom and start to see the positive side of all this and get on with the job of tackling climate change and creating a sustainable global society.

Doom and Gloom. Image by Rupert Besley. Reproduced under licence

So we need to put aside the doom and gloom and start to see the positive side of all this and get on with the job of tackling global warming and creating a sustainable global society.

A recent headline in the Times Newspaper (19th September, 2013), '***Number of climate change sceptics soars as support for alternative energy wanes***,' highlights the growing problem of negativity towards taking positive action to mitigate global warming. A UK Government survey published by the UK Energy Research Centre showed that those who do not believe in climate change had quadrupled from just 4 % in 2005 to 19 % in 2013. A recent IPSOS MORI poll of 1000 people has shown a similar trend with 91 % accepting that the climate was changing due to global warming in 2005 compared to just 72 % in 2013. The level of concern has also fallen from 74 % being very or fairly concerned in 2012 to 60 % in 2013. So we are seeing the general public choosing not to believe the overwhelming consensus of scientists, which has been strengthened by the recent IPCC report published in September 2013. But rather choosing to believe the sceptics who often have vested interests in maintaining a high-energy society.

The first task we have to tackle is asking ourselves honestly where we stand in all this. Each and every one of us needs to identify what our stance on global warming and planet health really is? Below are four options. Which one do you honestly belong to and which one would you like to be in?

- Total denial

  - Just natural perturbations in climate
  - Do nothing—business as usual

- Acceptance, but will not affect me

  - Idea that this is nothing to do with you personally
  - Let's see how this one pans out
  - Scientists will find a way
  - Governments will find a way
  - Meanwhile do nothing—business as usual
  - Good time to make some money of out of people's paranoia
  - There will always be somewhere else I can move to and be safe and happy

- Acceptance, but nothing can be done

  - What will happen will happen
  - Deal with problems as they arise
  - Be philosophical and prepare yourselves mentally

- Acceptance, but prepared to act personally

  - Act now to slow down global warming
  - Try and prepare yourself for a low-energy future

The answer I am hoping you will select is the last one… *Acceptance but prepared to act personally*. It is important that we all accept what is happening, but that global warming can be tackled, and that we can as individuals do something to mitigate the problem. If we don't then we leave no future security or hope for those who come after us and most likely for some of you actually reading this book in 2015.

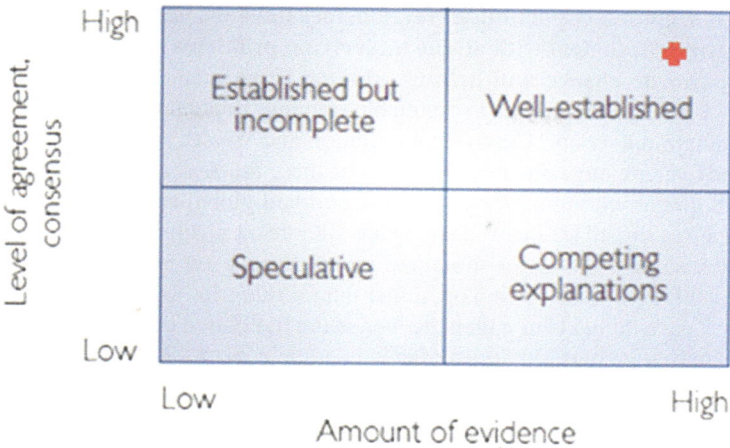

Fig. 15.1 Uncertainty analysis is used to test the validity of evidence based theories

Is global warming and all those effects we have discussed earlier actually real? Uncertainty analysis is used to establish the certainty of something being valid and is based on the level of agreement or the consensus between experts and the volume of evidence to support the theory. More scientists, engineers and other disciplines have been involved in the study of global warming both through the work of the IPCC and independently than any other issue in history. Using this approach then global warming, with a high level of consensus and a huge amount of evidence, places global warming in the top right hand corner of Fig. 15.1. So uncertainty assessment reinforces that global warming is occurring and that the effects predicted are occurring at a rate closely related to atmospheric GHG concentration. So we have to accept that this is happening to us right now.

> *Uncertainty Assessment reinforces that global warming is occurring and that the effects predicted are occurring at a rate closely related to atmospheric GHG concentration.*

Even if we were to be wrong about climate change caused by global warming, then we are still in the same position, meaning that we cannot continue as we are with a high population growth rate, dwindling renewable and non-renewable resources, and increasing pollution. Planet Earth simply can't support a human population much in excess of nine billion without ecosystem processes and services collapsing, which means that our ability to sustain all the people will not be possible. Our current path is leading to inevitable consequences that can only end in massive migration, increasing conflict, the potential death from starvation for tens of millions, and the possible disintegration of the global society that we enjoy today and the relative peaceful world in which we live.

This is a golden opportunity. Never before have we been in such a position of such strength to deal with all the underlying problems of our global society. Tackling climate change will help us address: Hunger and poverty; Equality of opportunity; Water scarcity and sustainable farming; Population control and health; Maintaining biodiversity; Controlling pollution and Wealth distribution. In fact we are already seeing huge advances in many of these areas as a response to action to deal with global warming. We live in a networked global society, so it has never been easier to organize, share ideas, make significant changes, and to act as individuals but as part of a larger structured society. So we are now able to act against climate change from the bottom up, rather than waiting for top down solutions. The longer we leave taking action then the increasing likelihood of those top down decisions becoming increasingly draconian is inevitable. So we have gone past the tipping point where change can be prevented and are entering a new era in terms of climate. We must take action ourselves to deal and prepare ourselves for the future which will be an exciting and challenging period for everyone.

> *Tackling climate change is a global revolution not only in how we live our lives on a daily basis, but how we interact with others, and how we nurture our planet, ourselves and each other.*

15.2   Revisiting the Previous Chapters

15.2.1   Defining the Problem

The global population is expanding very rapidly and for every death there are 2.4 births. We have already exceeded the carrying capacity of the planet in terms of resources and waste assimilation. Global warming will put more pressure on dwindling resources and our very survival as the climate changes. The dilemma of dwindling planet health and our survival on planet earth is due to three factors that must be tackled: The underlying problem of **population size; sustainability of lifestyles** that are adopted or desired which inevitably leads to inequality; and finally the way in which we use **natural resources**. The magazine New Scientist listed smaller families at the top of their ten steps to reducing greenhouse gas emissions. **Each of us must see ourselves in context with the 7,282,385,400** (as of the 20th December, 2014) **other people all with an equal share in the biocapacity of the planet**.

Update global population: http://www.worldometers.info/world-population/

> *The first step was to accept that our climate is changing and that there* **is a finite global population that is sustainable.**

15.2.2 What Is Sustainability?

It is important to see ourselves as part of the natural system. We cannot exclude humanity in our vision of planet Earth nor must we see humanity in isolation. Any resolution of the environmental crisis must ensure continued economic stability, otherwise society will break down and we will enter a global dark age caused by famine and conflict. Sustainability is a complex interaction between social, economic and environmental goals. But **no matter how complex and ill define it is, the concept of sustainability is the best mechanism that we have to ensure global stability and fairness**.

By creating your own definition of sustainability you made it clear that you understood the basic underlying principle that we have to live with our own share of the planet's biocapacity. Many of you will have accepted this to be a moral duty.

The second step was accepting Solow's definition of sustainability as 'an obligation to conduct ourselves so that we leave to the future the options and the capacity to be as well off as we are, not to satisfy ourselves by impoverishing our successors' and personally agreeing to act to help achieve this.

15.2.3 The Concept of Resources

Production is linked to demand and with a continuously growing economy, demand is not falling. Without constraint then exploitation of increasingly inaccessible resources from increasingly fragile environments will continue with the potential for severe environmental consequences as we have seen with oil extraction in the Gulf of Mexico and Alaska. New resources such as oil sands and shales require more complex and costly extraction, processing and transportation, resulting in higher greenhouse gas emissions per unit of energy supplied.

The list of non-renewable resources shouldn't be simply seen as materials that are limiting our development. It is more important to consider that these are non-renewable resources, which means that **when they are gone there will be no more**. When the CO_2 we emit today is finally removed from the atmosphere by the oceans and weathering processes, nearly all non-renewable resources really will be exhausted. So avoid products that exploit and/or waste non-renewable resources, be aware of what products contain, and view these resources as your children's inheritance.

> *The third step was to accept that all resources are potentially finite and that
> we must use them sensibly and sparingly, preserving and recycling them
> whenever possible.*
> *We can do this by developing simple strategies to maximize our use of non-
> renewables by careful initial product selection, maximizing product use
> and making an extra effort at the end of life of products to prevent the
> loss of non-renewables into the waste stream.*

15.2.4 Global Warming and CO_2

The greenhouse effect and global warming are real physico-chemical processes that
are occurring right now. The energy balance which controls warming is complex but
has been conclusively shown to be affected by man's activities in terms of burning
fossil fuel, and also industrial and agricultural practices. The rate of greenhouse gas
emissions is continuing to rise each year and **we need a significant reduction in
emissions to stabilize greenhouse gas levels in the atmosphere**. What you emit
today in terms of CO_2e will affect people for at least 100 years to come. **Action to
reduce emissions is urgent and the time to start is right now**.

> *The fourth step was to accept that greenhouse gases actually control the
> surface temperature of the planet and that man's activities are releasing
> new GHGs, above those emitted by natural processes, at an ever increasing
> rate. Action is urgent because of the longevity of global warming potential
> (GWP) of the GHGs emitted, and that the targets set by the IPCC are real-
> istic and necessary.*

15.2.5 Measuring and Offsetting CO_2 Emissions

**Every action uses energy and has a measurable carbon dioxide equivalent
(CO_2e) footprint**. Footprint models allow us to effectively measure and manage our
own emissions. If you followed the tasks in this chapter then you have already measured your own footprint, but remember that these calculators do not take into
account embedded CO_2e (i.e. they measure your direct primary footprint and not the
secondary footprint which is energy used on your behalf). Such models are only of
any value if you **accept that an individual's behaviour or lifestyle is a source of
global CO_2e emissions**.

To reduce your emissions you first need to know what is being emitted by which
actions. You have explored which models or calculators are best, so now you need
to commit yourself to one, analyze your emissions regularly and explore ways of

reducing unnecessary wastage. In Ireland personal (primary) emissions are 5.7 t of CO_2e per person each year with the total emissions (both primary and secondary) 13.6 t of CO_2e per person each year.

Ideally we should all strive to reduce total our emissions to the current global average of 3.8 t per person per year, although this will require a huge personal and national effort. We have until 2050 to reach the IPCC target of an 80 % reduction in our emissions which would require our primary footprint to be just 1.06 t of CO_2e per person each year. **Your choices control the energy used and CO_2e emitted by you…no-one else makes or should make that decision for you**.

> *The fifth step was to accept that an individual's behaviour or lifestyle is a source of global CO_2e emissions and that carbon footprinting enables individuals, households and companies to measure and manage their emissions.*
> *This also involves you agreeing to carrying out regular analysis of your own CO_2e emissions and agreeing to set yourself personal emissions targets, goals, or limits.*

15.2.6 The Real Cost of Carbon and Offsetting

The impact of GHG emissions can only be reduced by using either less fossil fuels or actually removing CO_2 from the atmosphere. Both alternative energy and mitigation strategies need a long term fixed price for carbon to encourage investment, to make alternatives viable and to reduce energy use. Carbon is currently trading at €7.11 per tonne of CO_2 (November, 2014). We have to accept that the real cost of CO_2 and other greenhouse gases have to be high enough to stimulate and support emissions reduction. However, to achieve this it needs to be fixed at around €100 per tonne or a minimum trading price fixed. We also need to ensure that carbon trading is equitable and that carbon credits are allocated in such a way as to genuinely reduce carbon emissions overall. **The challenge is to allocate a realistic charge to our own personal carbon emissions and use money or time equivalent to offset carbon activities by investing in the reduction of your own or community GHG emissions**.

The burning of fossil fuel releases carbon that has not been in the carbon cycle for millions of years, and therefore creates a net increase of carbon mainly as carbon dioxide in the biosphere. It's not possible to truly remove (offset) carbon emissions or become 'carbon neutral' even by planting trees. So promoting the concepts of 'carbon offsetting' and of being 'carbon neutral' runs the risk of providing an apparent justification for continuing with a fossil-fuel intensive lifestyle and culture, whereas **it is a drastic reduction in fossil fuel usage that is required now**. Offsetting can and does invest in future emission reduction technology and helps those in developing countries to improve their lifestyle through the more efficient use of energy, but can also drive extra growth, prosperity and so inadvertently result in increased emissions.

> *The sixth step was to accept that the price of carbon should be set at a cred-ible value in order to be incentive for both carbon reduction and the innovation of low-carbon technologies.*
>
> *Step five was also accepting that offsetting can be a positive action against climate change but not in reducing existing GHG emissions. As individ-uals we can't buy ourselves out of our personal moral responsibility to act for all those people and other species we share our planet by offset-ting. So the only certain way to reduce CO_2 emissions is to use less fixed carbon.*
>
> *Use money or time equivalent to offset carbon activities by investing in the reduction of your own or community GHG emissions.*

15.2.7 Ecological Footprint

Whatever the reason, the longer we delay in tackling greenhouse gas emissions then the more severe climate change will be. Although not perfect, ecological footprint-ing (EF) gives us a far more holistic approach to tackling global warming than just using carbon or water footprint models. Governments, especially within Europe, now have the technical and management tools to start to tackle global warming seri-ously but they need a strong mandate! So we all need to put pressure on European, National and local politians and send a clear message that we need action.

Ecological footprinting gives us the realistic universal target of one planet which is currently in 2014 equivalent to 1.8 global hectares per capita, although this is get-ting less each year. This approach gives us clear goals requiring us to stabilize and increase bio-capacity, reduce the ecological footprint through reduced consumerism and greater efficiency, and to stabilize and eventually reduce population.

> *The seventh step was accepting the concept of ecological footprinting as a holistic approach to achieving sustainability and dealing with global warming. This involves us all looking at the three key mechanisms and how we can make positive contributions.*
>
> *Your aim should be*:
>
> - **To do twice as much with half of the resources (Factor Four Reduction) and thereby reduce your ecological footprint;**
> - **To stabilize and increase bio-capacity by using your space, garden and land better and by becoming involved with conserving and transforming your immediate environment;**
> - **Thinking about population and sustainable family size.**

15.2.8   Energy: Green or Otherwise

It is remarkable that since the industrial revolution more than half of the world's energy has been consumed in the last two decades, despite advances in efficiency and sustainability. So there is an urgent need to bridge the gap between current usage and the required reductions in order to meet International greenhouse gas emission targets. This can only be done by using less energy, using energy more effectively, and by replacing traditional fuels with cleaner and renewable energy alternatives. For the individual the major uses of energy are for transportation, household space heating and water heating, so it is in these areas that investment can yield significant savings and emission reductions. As energy becomes increasingly expensive and pressure to reduce emissions to meet international targets becomes acute then every one of us will have to address the issues that will arise from living within a slowly reducing energy budget. Two major challenges face us, the adoption of a low-energy (carbon) lifestyle and the move towards a low-energy (carbon) economy. This requires replacing supply-side management to demand-side management by setting ourselves strict energy budgets or limits. This can be only be achieved through a combination of structural (e.g. smart lighting/heating, CFC lights, etc.) and behavioural actions (e.g. switching off appliances rather than leaving on standby, turning off lights when not in use, etc.).

> *The eighth step was reducing the use of energy by adopting a proactive low-energy lifestyle.*
> *Start by setting personal reduction targets for electricity used in the home, taking time to select and investment in the best energy efficient appliances, turning lights and appliances off when not required, and always selecting the most efficient energy sources. Use this ethos to influence others in your family and work.*

15.2.9   Travelling Here, There, Everywhere

Transportation is the largest component of the Irish personal carbon footprint (56 %) and is growing. This is the same in most developed countries. Cheap flights have exacerbated aviation use with the subsequent proliferation of weekend breaks, multiple holidays, second homes, etc. Ideally aviation should be considered a luxury and subject to high carbon taxation; however, should be minimized and offset against other emissions within your own budget (i.e. personal trading). So if travel is really important to you, and travelling is something that is very special, then you should try to find those carbon credits by economizing elsewhere. Alternatively spread travel emissions over a longer period of several years to incorporate that once in a lifetime trip or studying abroad, or offset by investing your money or time in personal or community projects that really do reduce GHG emissions.

> *The ninth step was to accept that travel is a major factor in personal and business GHG emissions and that you should try to minimize emissions from transportation and build them into your own personal carbon budgets.*
> *Always ask the questions … Is this trip necessary? Is there a more efficient mode of transport I could take? Can I make long trips more effective (**e.g.** by staying longer)? Can I combine trips? Can I find alternatives to the need to travel (**i.e.** why are you going there)?*

15.2.10   Having Enough to Eat

It is more than likely that if you are reading this book that you probably do have enough to eat. In fact in the Western world we do tend to eat far too much protein, carbohydrates and fat which has lead to rising concerns about obesity and the rise in diabetes in the under 40s. However, 35 % of the global population is living in food poverty and global warming will cause this figure to rise year on year.

The type of food you eat and how it's produced is the single largest component in your meal's carbon footprint. Although it often seems that we are obsessed with how far our food has travelled, in general food miles associated with air travel are a very small component and are typically associated with expensive perishable items such as out of season strawberries. That is not to say that much of our staple foods do not travel long distances, but the most significant carbon footprint in terms of transport is your trip to the shop or supermarket to fetch it. We waste up to a third of our food by cooking too much or never cooking purchased food at all; so less waste means more food to go around and less demand means lower prices globally. So if you eat what you buy; reduce meat and dairy or not eating meat at least one day per week; be more seasonal and avoid out of season hothouse grown or air freighted food, you will *half* your carbon footprint emissions for the food that you or your family eat.

> *Step ten was accepting that food waste is a critical and unnecessary factor in carbon emissions and that by buying only what we need, careful selection of what we eat, and eating everything that we buy, that we can significantly reduce emissions as well as increase global food reserves.*

15.2.11   Where Does Water Fit in?

How climate change will effect society will largely focus around water availability which not only controls agricultural production but is the key cause for conflict and migration. Increased migration increases urbanization which subsequently leads to an increased risk of water shortages leading to disease and poverty.

Increasing scarcity of water is exacerbated by global warming with renewable resources becoming threatened by over exploitation. All products contain embedded water with our direct water use (i.e. the portion that is piped to our home) representing on average only 3 % of our total water footprint. As water becomes scarcer we need to adopt the principles of water demand management, where consumers work within an adequate but fixed budget of water, with emphasis on conservation of water, use of water saving appliances, reduction in leakage and metered based charging.

> *The eleventh step was to accept that water is a limited resource and so should be used thoughtfully.*
> *Reduced consumption and its careful use in the home helps to reduce carbon emissions from heating water, and also where water has come from expensive sources such as desalinization, and also reduces the impact of water abstraction on ecosystems.*

15.2.12   Waste Not Want Not

Recycling … does it matter? It has been shown that recycling does significantly reduce personal and business carbon emissions. Recycling also preserves and maintains non-renewable resources and reduces environmental degradation. So recycling is good for the environment, for preserving resources and reducing emissions. But we must learn not to create waste in the first place. So we each need to **develop a waste minimization strategy**. Our new mantra must be to prevent and reduce waste in all areas of our life. However, we all need to buy things from food to computers, so remember the three Rs: **Reduce**—buy less and use less with the basic concept here to use all you buy, wear it out and reduce the amount of material discarded; **Reuse**—discarded items or components of the item should be used again before finally recycling and upcycle waste into more valuable items whenever possible; and finally **Recycle**—ensure that discarded items are dismantled and sorted so that all the key resources can be recovered and reused.

> *Step twelve was acknowledging that recycling can make a significant difference both to GHG emissions and preserving non-renewable resources. This involves you changing your lifestyle by seriously adopting the three Rs and to*:
>
> - **Buy less and buy better quality**
> - **Consume all your food**
> - **Use consumables carefully**
> - **Wear things out**
> - **Keep electronic devices longer before replacing them**
> - **Become an active and predatory recycler**

15.2.13 The Planet's Health

The health of the planet is totally subjective as it is based on our own view of what it should be, and to a great extent it is a romantic ideal based on a manmade eighteenth century English landscape. The planet is continuing to evolve, and while we currently enjoy a period of global stability there is in reality no correct bioplan for planet Earth. Planet health is really about sustaining us within the biocapacity available. For that **we need other species of plants and animals…we can't survive on our own**. As global temperature rises we are approaching some irreversible tipping points that will alter planet Earth forever.

To achieve a sustainable balance between us and planet Earth we have to act now, not only on global warming, but on a large array of environmental, political and social issues. It is a very big and complex problem. However, we need to avoid whole planet engineering and management solutions as quick fix. Our planet Earth is so fragile that we should not interfere with its natural evolution on a global scale as we have no idea of the long-term consequences. Such approaches should only be used as a very last result.

We all need to understand our planet, to live as closely as we can with it, be a part of it, and understand its seasons and rhythms. We must preserve and care for it, because quite simply it's our home. So step lightly, use only what you need, and waste nothing.

> *Step thirteen was realizing that the planet Earth is at a tipping point inorganically and that our existence as a species is dependent on all of us realizing that we have exploited the planet too far and that we must act now to prevent these fundamental changes occurring.*
> *It was also realizing that we can still prevent the worse scenarios occurring if each of us chooses to act right now.*

15.2.14 Your Health and Wellbeing

Climate change can directly as well as indirectly affect human health, and if no action is taken, then problems such as malnutrition, deaths and injury due to extreme weather conditions, and change in geographical distribution of disease vectors, will worsen. Set against this are significant potential benefits in both individual and community health and wellbeing from implementing carbon mitigation strategies which help offset their cost. So, for example, the reduction in city traffic density lowers air pollution levels leading to improved health.

Peoples' quality of life is determined far more by the quality of their working life, their family life and their overall social relationships than the amount of consumption or wealth they are able to enjoy. So it is important to ask yourself honestly what lifestyle you really want for yourself and your family, and plan how to achieve this sustainably. Numerous studies have shown that the de-coupling of consumption and the improvement of wellbeing is possible. Tackling climate change will require tackling those socio-political issues that currently effect human wellbeing as well as tackling unsustainable consumerism. It has become the norm to go away to find the environment and lifestyle you want, and this is reflected by the very high ownership of second homes. We are often amazed how safe it is and how relaxed it is to walk the streets at night in Italy or Greece, something we would never do in the UK or in some parts of Ireland or the USA. Yet an important factor in sustainable living must be creating a place where you want to live, where you feel safe and unstressed. For example, many people who would like to use public transport don't because they feel threatened and unsafe. So tackling social problems is at the core in creating a sustainable society.

> *The penultimate step was to understand that if we do not tackle global warming it will have an increasingly negative effect on human health through water and food scarcity, spread of diseases and increased severity of weather patterns.*
> *Wellbeing is largely independent of wealth and so each of us should strive to tackle those socio-political issues that affect wellbeing as well as addressing unsustainable consumerism which often becomes manipulative and addictive.*
> *Ask yourself what life style you really want for yourself and your family and plan how to achieve this sustainably.*

15.3   The Next Step?

Global warming is real and is happening to us all right now. This is causing our climate to change far more quickly than we have experienced before, and certainly since records began, resulting for some drought and desertification and for other increasingly violent and unpredictable storm events leading to flooding. **So global warming doesn't always mean hotter, it means unpredictability**. So without doubt our climate is changing which means that we need to take action to meet the new challenges that will face us all in the coming decades. The cartoon at the start of this final chapter is by the Pulitzer Prize winner editorial cartoonist Joel Petts of the Lexington Herald. He has produced numerous climate related cartoons but this one sums up exactly our current dilemma. Even if we are wrong about climate change, and that is extremely unlikely, then taking actions now will solve many other pressing and indeed equally damaging problems.

It is often stated that any person on the planet can be connected to any other person through a nexus of just six relationships. Therefore the individual, through modern media can connect and influence potentially every other human on the planet. So we as individual's have the power to influence, lead by example and cause change. The cause and solution to climate change lies with the individual, that's you and me.

I said at the outset that global warming is a major global catastrophe inflicted by Humankind on itself. A **catastrophic failure** is a sudden and total failure of some system from which **recovery is impossible**. Catastrophic failures nearly always lead to cascading systems failure. However, this is not going to happen in relation to climate change because politians will respond with increasingly draconian measures in order to deal with the consequences of global warming such as disastrous storm events, crop failures, major migration etc. Some of these events could severely test our very social and economic stability. Therefore we must pre-empt such measures by taking the lead now and not leaving it until uncontrollable events take over. Our climate is changing and we need to be able to stop the changes becoming too severe and we can only do that by all of us significantly reducing our emissions. **Governments can't tackle global warming alone and they need us to help and we can do that by raising the issue whenever possible and ensuring that it is always on the agenda no matter what other economic emergency there may be. This is done by giving the issue relevance and urgency and by giving actions validation and support; most importantly by altering our own behaviour and finally by encouraging those around us to do the same.**

15.4 Implementing Personal Action

We saw in Chap. 5 that the average Irish carbon footprint could be broken down into the primary personal emissions of 5.7 t CO_2e per year and the secondary or embedded personal emissions equivalent to 7.9 t CO_2e per year which gives us our total carbon footprint 13.6 t CO_2e per person each year. The first stage is to identify what your primary footprint is, although some calculators give you a total footprint as well. Once we have identified what our emissions are and how we are currently using them (i.e. household heating, travel, food, consumables etc.) we can begin to take control. To do this most effectively we need to develop a personal action plan.

15.4.1   The Personal Plan

The idea of a personal plan is not a new one and has been used extensively for career and personal development. The same approach can be used to create a personal plan to achieve a sustainable lifestyle. Let me say at the outset that many people manage pretty well without a plan and simply use common sense. Others are going to live their lives exactly as they have always done but will take on board the problems and try to reduce and mitigate their actions as best as they can. So to a great extent the need for a personal plan really depends on the type of person you are, the level of commitment you are making, and finally how much help you need to achieve those goals. Others simply get a thrill at setting goals and seeing if they achieve them. To be honest with you, **the level of reductions that our Governments have agreed to by the years 2020 and 2050, will really require all of us to reduce our emissions to quite a low level**; **so a personal plan will help to ensure that we sustain the lifestyle that we want within those limits**.

The mechanism of creating a personal plan starts with **recognizing the need to change**. This requires you to ask some very fundamental questions about your stance on climate change and how you are going to deal with the problems that may arise in the future as a result of global warming.

- Do I want to be proactive in dealing with climate change?
- Do I want to be prepared for the challenges that climate change will bring?
- Do I have the skills that I need for a more dynamic future?
- Do I want to develop the skills that will make me more prepared for the future?

You can see straight away that these fundamental questions are also about personal lifestyle and career. The next step is to **accept change**. This is always going to be the most difficult step, and that is why when you come to set goals it is important that they are achievable within the limits of your commitment. You can only change things that are under your control so it is important not waste time trying to achieve unobtainable or unrealistic goals. Acceptance of change is being willing to alter behaviour and at the same time being willing to grow both in terms of being a self determining individual and also in personal independence. Once you have recognized and accepted the need to change you can then start to **create your personal plan** which will list your skills, measure your carbon, water and ecological footprints, and explore your personal and career goals. The plan will also contain details of your short-medium and long term targets for CO_2e emissions, water use and ecological footprint. The difficult part of course is going to be how you accomplish these targets, which require you to identify and detail each step to achieve your objectives. Finally we put words into action and slowly **implement** our personal plan.

This is not a race, and what we hope to achieve is a slow but sustainable reduction in energy use and ecological footprint coinciding with Government targets. Then comes the rest of your life in which to slowly change to a sustainable, safer and hopefully happier lifestyle. Of course you will need to **measure your progress** by rechecking your personal footprints at regular intervals to ensure that you on are target, but also as circumstances change you may need to **review goals** and if necessary readjust timescales or interim targets (Table 15.1). But by making this change you will be helping to slow global warming, to prepare yourself for the changes in our climate which are already in progress and improve your lifestyle and wellbeing.

You have already begun your plan by working through the simple exercises in the book. By creating your own personal definition of sustainability you have recognized the need to change. You have looked at different ways of measuring your impact and have established your current primary and secondary footprint. Next you need to set a target. Whether this is based on one Earth or a specific emission goal (remember we all have been set an 80 % reduction of our greenhouse gas emissions by 2050), or perhaps both. However, it is important that it is whatever you feel happiest with. Your immediate short term goal should be to get to the national average carbon footprint, although many of you will already be well below this. Short term (12 months) targets should be easily achievable very quickly by carrying out simple energy saving changes, by minimizing waste and introducing more effective recycling. Medium term goals to reach lower emission targets will require more effort and changes in your behaviour, with some things only achievable with significant investment over time in structural changes such as replacing worn out appliances with more energy efficient ones, or replacing household heating, or investment in more advanced household insulation. Long term targets may require significant changes in lifestyle, so it is important that these changes are made over a significant period of 10 years or more. All these changes will save you money in the long term which you can reinvest to create your sustainable dream, whether it's a city flat with travel at the heart of your life, a small house and allotment, a larger house with a garden, or a small holding in the country and a commute. All five options are possible within one Earth lifestyle; it just requires careful management, imagination, innovation and commitment.

What will the plan physically look like? Well that depends on you. It can be an Excel spread sheet on your computer, a detailed plan bound in a folder, or a couple of sheets stuck to the fridge…whatever works for you. But it is important to keep a record and to ensure that you do have a long term plan which you stick to. Remember to reward yourselves when you reach targets.

Whatever your goals, whether they are simply to make small behavioural changes or more structured emission reductions, **every little helps and more importantly will change attitudes both within your family, among your friends and work colleagues**. One of the great successes in Ireland has been the green flag initiative to make schools more sustainable. At the core of this initiative has been recycling and this has been so successful that our new generation of young people going to university and entering employment recycle as a matter of course. It was drummed into them at primary and secondary level that recycling was important…and also

Table 15.1 Overview of main components of the personal plan

Recognition of the need to change

- Performance review by carrying out footprint analyses
- Identify career objectives
- Identify lifestyle objectives
- List skills
- Identify critical skills to acquire

Accept change

- While the process is not personal in that we all need to reach common targets, we do all start from different initial points in terms of our current emissions. There are no bad or good guys, only those who have decided to do something about making planet Earth sustainable
- Everyone is scared of change and of the unknown. What is scarier is being aware that things are going to alter for the worse and doing nothing about it. So the challenge will seem daunting, even impossible at first, and this will make acceptance difficult. But this is an exciting opportunity to embrace the future with renewed hope, so it is really important that you accept the challenge willingly and positively
- Try and image what change will be like, what the new lifestyle will give you and the advantages and renewed hope for the future
- Instead of watching global warming slowly create havoc, you will be doing something positive to mitigate its effects and prepare yourself and your children to understand, accept and react positively to the challenges ahead

Create personal plan

- Set targets/goals (i.e. carbon emissions, water usage, Earth equivalents)
- Break targets down into key footprint categories such as transportation, aviation, home energy, consumables, clothes, food, waste, etc.
- Create a realistic time scale to achieve reduction targets (e.g. 10 % per annum)
- Plans should have short (less the 1 year), medium (1–5 years) and long term targets/goals (greater than 5 years) and actions that may coincide with life stages (career, house purchase, having a family, retirement, etc.). Remember there will be periods in your life when you will have to use more energy than you would like, so this must be built into your plan
- Identify what actions are needed to achieve each interim target/goal
- Set about gaining the skills you have identified that will be useful in the future
- Invest in both your sustainability and lifestyle which should be one of the same
- Use personal trading as a mechanism to learn how to manage your emissions
- Explore how your plan can be integrated with family, friends and the wider community (e.g. combined actions such as car sharing or car pooling)
- Set interim dates to assess progress
- Create a reward system

Implementation

- Once you have recognized the need to change and accepted the challenge, with your plan conceived you need to begin to implement the activities listed
- Start small at first so that you can achieve your goals and targets. This will give you the confidence to continue and achieve greater things
- Use skill development to meet other like minded people who will support your efforts
- Talk about what you are doing and try to show by example what can be done

Measure progress and review

- A critical part of the plan is to review progress regularly and where necessary reassess targets and goals, and also the actions needed to achieve them

carried a detention if not done. They have also been introduced to a wide range of life skills, which often have been lost during their parent's generation, such as a greater understanding and appreciation of food through gardening and cooking, sewing, knitting, first aid, lifesaving and much more. By adopting a sustainable lifestyle we will give ourselves and our children the skills and understanding to live sustainably on planet Earth. At the heart of acting individually to address climate change is becoming a more independent and self-determining person.

15.4.2 More on Setting Targets

What should be our individual target? We know from experience that as the household size increases the average individual footprint falls because you are sharing heating, food preparation and often transport. So managing, and hence reducing, emissions often work best at family or community level. It also allows you to trade emission allowances between family members, so that emissions from holidays by one member can be absorbed by the others. The first interim target should be getting your primary footprint down to the national average which in Ireland is 5.7 t per person per year. This is a very generous allowance (equivalent to 3.5 planets) and should be achieved as quickly as possible and will be quite easy by making simple behavioural changes such as switching off lights, recycling, not wasting food, etc. For many of you your average footprint will be well below this already, so what is a reasonable target? Well we have quite a few to choose from. You can use ecological foot printing and aim for one Earth which is the simplest approach and in many ways the most logical. It also allows you to build biocapacity, such as growing your own food, planting and managing woodlands, etc. Alternatively you can adopt a more direct approach that gives you far more control in how you manage your greenhouse gas emissions. Perhaps the most sensible is to follow the IPCC reduction targets of 20 and 80 % reduction by the years 2020 and 2050 respectively. Another approach is to aim for the global average which would require us all to reduce our annual personal primary emissions to just 1.55 t. This allows everyone in the planet to have the same emissions, permitting developing countries to grow and improve their lifestyles while putting a ceiling on their emissions, while at the same time allowing developed countries to use structural and behavioural strategies to significantly reduce their emissions. This is a huge reduction and cannot be realistically achieved quickly by most of us, although I know many people who have even lower emission footprints than the global average and have, what appears to be, completely normal and certainly very happy and fulfilled lives...so this is not impossible by any means. Everyone will have a different strategy for achieving targets. Some will want to achieve their final goal quickly and will perhaps make large financial investments to help them achieve them. Others, like me, I suspect will have to do this over longer timescales. The emissions reduction system I like best is the adoption of a 10 % reduction per year (Table 15.2). This appears a very relaxed approach, but it does allow people to make changes in a more sustained way

Table 15.2 Step reduction plan where emissions are reduced by 10 % each year

Year	Percentage of national average footprint (%)	Reduction in national average footprint (%)	Primary footprint allowance tCO₂e/year	Total footprint allowance tCO₂e/year	Milestone
1	100	0	5.70	13.6	
2	90	10	5.13	12.2	
3	81	19	4.62	11.0	2020 IPCC 20 % target
4	73	27	4.16	9.9	
5	66	34	3.74	8.9	
6	59	41	3.37	8.0	
7	53	47	3.03	7.2	
8	48	52	2.73	6.5	Halved emissions!!!
9	43	57	2.46	5.9	
10	39	61	2.21	5.3	
11	35	65	1.99	4.8	
12	32	68	1.79	4.3	
13	28	72	1.61	3.9	
14	25	75	1.45	3.5	Global average (1.55/3.8)
15	23	77	1.31	3.2	
16	21	79	1.18	2.9	
17	19	81	1.06	2.6	2050 IPCC 80 % target

Each year the actual reduction is less but becomes more difficult to achieve as the target threshold falls. Allowance for primary footprint and total (primary and secondary footprint) are given as tonnes CO_2e per person per year

that is probably more likely to succeed. It gets tougher as time goes on to make those reductions even though actual annual reductions do get less each year. So it would take 14 years (the long term goal) to get to our global average primary footprint average and would depend on a similar reduction of the secondary footprint from the current 7.9–2.25 t per person per year. To achieve the 2050 IPCC 80 % reduction target which is below the global average, would take 17 years if you are starting at the National average footprint. It is important that you start on this process where your current emissions are. So if you current footprint is 3.7 t per annum, then you would come in at that point in Table 15.2 giving you just 9 years to reach the ultimate target, even though the yearly reduction in actual tonnes of CO_2e are relatively small (e.g. about a third of a tonne in the first year and slowly falling each year after that). We need to be able to prioritize and use greenhouse gas emissions as we want to use them and personal emission reduction targets allows us to do just that. You can't offset your energy use, you have to live within your own carbon budget, and what is brilliant, is that you make up your mind how you are going to use that energy.

There are going to be times when you will exceed your targets due to unforeseen circumstances, the need to travel to an important event, having a baby (although he or she will also have their own emissions allowance), moving house, etc. How do we deal with that. Ideally we would absorb the extra over a couple of years within our own plan by making savings elsewhere, or alternatively we may have to use a carbon credit system to offset. However, make sure that you offset at a realistic cost (Sect. 6.4) and use that money to invest in your own personal emissions reduction plan or that of your local area.

15.4.3 Checklist

I have given you quite a bit of advice during the book but I would like to finish by giving a quick check list of what I think is important:

- Reconnect with the living planet and be a part of it, learn to respect and love it.
- Be educated as global warming may throw up challenges that will require you to act more independently.
- Be flexible in all things.
- Be as fit and healthy as you can. Invest in both your physical and mental health.
- Plan ahead
- Invest in longevity and high quality consumer goods
- Always incorporate climate change mitigation and impact into your long term investments (car, housing, family)
- Always minimize your carbon footprint in every action you take
- Don't offset via a third party but do it through direct or personal action
- Create a personal plan…stick to it…review it frequently…persuade others
- Learn to live in a low-carbon way

15.5 In Conclusion

Every one of us is now aware of the problems facing the planet, especially the problem of global warming and population growth. Also we all know that we will have to act anyway whether prompted by Government or by taking the initiative. We have the means and the technology to do it, so all we have to do is get on with it and stop waiting for someone else to tell us what to do and when. There are no easy solutions to tackling climate change and meeting the challenges that it will eventually present to us all. We need to find sustainable outcomes and we, as individuals, are the solution. We need you to be part of this exciting new period…to act as leaders, advocates, role models, or simply to be proactive. This is your time, your planet, and you can decide what kind of world you want to create.

Probably without realizing it you are already making a difference. The majority of us recycle, pay carbon tax on fuel, use our cars less and public transport more, replacing incandescent light bulbs with long life low energy bulbs and much more. The fact that you are reading this shows your willingness to engage and take positive action. I have found that people in general are overwhelmingly positive about tackling global warming but are genuinely at a loss to know what to do. As a species we have both the intellectual understanding and empathy that gives us all the ability to do wonderful things and to make the seemingly impossible happen.

Whatever the climate has in hold for us it is possible to survive and to live full, complete and happy lives. But we must each and every one of us play our part and we need to start altering our lifestyle now by making small and considered changes. Now is the time to back off from high-energy consumerism and to live within the limits placed on the planet that needs to support 7.3 billion people just like you and me (and still growing) and countless other species. If you do it, and your friends do it, and their friends do it then suddenly things will have changed. **It really does only take one person to make a difference and that person is you**.

This is your one planet…take care of it…pass it on at the end of your lifetime in better shape than I've left it to you.

References and Further Reading

- Your own portfolio
- Your own common sense
- Your belief that you can create a sustainable planet

Index

A

Acceptance of change, 391
Access to water, 45, 281
Agriculture, 14, 55, 70–73, 83, 139, 146,
    149–153, 192, 243–251, 253, 256,
    264, 266, 269, 276, 279, 281, 288,
    292, 297, 335, 372
Air pollution-related health effects,
    358, 364–365
Albedo, 85, **89–90**, 339
Angle of tilt of the Earth, 21
Antarctic, 3, 18, 344, 346
Arctic, 3, 18, 22, 103, 342–344, 347,
    350, 353
Atmosphere, 19–21, 24, 25, 36, 46, 81–83,
    85–91, 94–97, 107, 115, 141, 143,
    147, 149, 151–153, 229, 336,
    338–340, 348, 350, 351, 355,
    381–383
Atmospheric CO2 concentration, 95–101
Aviation, 112, 129, 135, **217–224**, 229, 230,
    238, 385, 393

B

*Beyond scarcity: Power, Poverty, the Global
    Water Crisis*, 280
Bike hire, 237
Biodiversity, 5, 13, 15, 35, 38, 56, 63, 70, 71,
    153, 170, 195, 247, 253, 255, 335,
    336, 340, 349, 354, 379
Biomass, 54, 69–71, 73, 83, 149, 152, 162,
    184, 188, 193, 195, 196, 248,
    250, 336

C

Cap and trade mechanism, 136, 137, 140
Capture and storage, 143, 144, 154, 289
Carbon cycle, 70, 85, 94, 149, 151, 336–339,
    356, 383
Carbon dioxide (CO_2), 17, 36, 62, 70, **81–101**,
    104, 107, 113, 115, 116, 118, 119,
    124, 131, 138, 141, 143, 144, 146–155,
    157, 179, 180, 184, 187, 211, 222,
    225, 226, 232, 244, 252, 256, 259,
    263, 286, 293, 318, 335, 336,
    338–342, 348–352, 355, 381–383
    emissions (*see* Emissions)
Carbon dioxide equivalent (CO_2e), 46, 61,
    83, 104, 136, 182, 216, 251, 286,
    304, 339, 382
Carbon footprint
    aviation, 112, 217, 223, 224, 230, 238, 385
    calculators, 109–116, 261–262
    carrier bags, 127–129
    commuting, 113, 223, 224, 230
    driving, 119
    food, 255–262
    household, 121–123
    internet, 110, 115, 126
    mobile communication, 124–126
    total carbon footprint, 109
    transport fuel, 118
    travel, 216–217
Carbon offsetting, 147–154
Carbon pricing, 47, 107, 142–144, 154–155
Carbon sink, 149, 336–338
Carbon taxation, 47, 110, 111, 144–147, 154,
    156, 238, 256, 385

Cars
    efficiency, 109, **119–121**, 230, 231, 235, 285
    emissions, 109, 111, 112, 114, 118–121,
            127, 129, 130, 155, 219, 225–238,
            285, 316, 325, 368
        from driving, 119–121, 124, 127, 219,
            221, 231, 234
        by engine size, 111, 112, 114, 118,
            120, 235
    ownership, 226–229
    pooling, 230, 236–238, 393
    recyclable components, 326
    sharing schemes, 237
    travel, 120, 148, 169
CCX. *See* Chicago climate exchange (CCX)
CDM. *See* Clean development mechanisms
            (CDM)
CER. *See* Certified emissions reduction (CER)
            unit
Certified emissions reduction unit (CER),
            139, 142
Changing attitudes, 173, 392
Charcoal, 70, 144, 154, 156
Chicago climate exchange (CCX), 142
Clean development mechanisms (CDM),
            137–140, 142, 148, 149, 152, 155
Climate, 3, 17–35, 55, 81, 106, 135, 160, 187,
            225, 243, 274, 307, 334, 357, 375
    another ice age?, 21
    change sceptics, 378
    El Niño, 19–21
    health, 358, 366–367
    La Niña, 19, 20
    precipitation, 24–27
    records, 17–21, 23, 139
    sensitive diseases, 362
    temperature, **18–23**, 29, 46, 365
    wind, 26–29
Climate smart agriculture (CSA), 72, 248
Coal, 19, 20, 36, 57, **62–66**, 75, 82, 85, 97,
            107, 109, 112, 145–147, 152, 182,
            184, 185, 187, 188, 190, 198, 199,
            201, 211, 339, 368
CO₂e. *See* Carbon dioxide equivalent (CO₂e)
Commuting, 113, 223, 224, **229–231**
Consumerism, 33, 37, 38, 54, 99, 163, 167,
            174, 182, 217, 294, 315, 319, 327,
            370–372, 384, 389, 397
Contrails, 86, 87, 215, **222**
Conversion factors
    electricity, 114, 123, 132, 155, 184, 186,
            187, 202, 203, 207, 211
    food, 115, 262
    gas, 203, 211

liquid fuels, 211
    solid fuels, 211
    transport, 235
Cooling effects, 339
Cost of carbon, 135–157, 383–384
CSA. *See* Climate smart agriculture (CSA)

D
Decarbonization of energy, 198, 201, 367
Deforestation, 70–73, 83
Demand-side management, 283–287
Desalination, 252, 293–294
Direct taxation of emissions, 136
Drought, 72, 289–293

E
The Earth Charter, 48
Ecocentrism, 43, 44
Ecological debtors, 164–167
Ecological footprint, 12, 14, 35, **159–174**,
            256, 384, 391, 392
    calculation, 163–170
    China, 162, 167
    creditors, 164, 165, 168
    debtors, 164–168
    global hectares, 12, 162, 163, 165–170, 384
    India, 167
    Ireland, 35, 165, 166, 173, 174, 256
    London, 35, 169, 170, 173
    per capita, 162, 163, 165–170, 384
    USA, 35, 162–168, 170
Ecological services, 13, 14, 69, 161, 290
The Economics of Climate Change, 44, 250
Economic sustainability, 38–41, 43–49, 94, 152,
            171, 193, 290, 369, 371, 377, 381
Economy-ecology-social nexus, 40–42
ECX. *See* European climate exchange (ECX)
Electricity
    all fuels, 187–192
    biomass, 152, 184, 188, 193, 195, 196
    coal, 62–65, 107, 109, 112, 145–147,
            182, 184–185, 187, 188, 198,
            199, 211, 368
    conversion factors, 114, 123, 155, 184,
            186, 187, 202, 203, 206, 211
    energy flows in generation, 188
    Europe, 25, 107, 152, 155, 182, 184, 186,
            187, 195, 197, 198, 200
    fuel mix, 109, 114, 160, 184–188, 210
    global consumption, 277
    hydroelectric, 109, 193
    imported energy, 184, 191

Ireland, 123, 184, 185, 188, 196, 206
  nuclear, 160, 182, 186–188, 198–201
  renewable energy, 145, 147, 155, 160, 182,
       184, 185, 187, 192–198
  solar energy, 21, 82, 83, 89, 222, 274
  standby, 204–206, 286
  USA, 187, 196
  wind power, 28, 179, 196, 197
Electronic waste, 321
El Niño, 19–21
Embedded energy in carbon footprints, 117–118
Embedded footprint of car manufacture, 236
Emission reduction Units (ERU), 138–140, 142
Emissions
  allowance, 112, 138, 140–142, 202, 220,
       394, 396
  carbon footprints, 65, 104, 106, **109–117**,
       122, 126, 127, 129, 131, 148,
       216–217, 222–224, 230, 236, 251,
       253, 255, 258–261, 265, 286, 314,
       383, 385, 386, 390, 392
  car manufacture, 225, 236, 238
  carrier bags, 129, 130, 320
  cars, 109, 111, 112, 114, 118–121, 127,
       129, 130, 155, 219, 225–238, 285,
       316, 325, 368
  commuting, 113, 223, 224, 229–231
  by country, 105–106
  developed countries, 98, 99, 107, 138,
       140, 155, 182, 204, 209, 225,
       265, 385, 394
  developing countries, 47, 71, 99, 104, 130,
       138–140, 152, 155, 156, 253, 265,
       384, 394
  diet, 251, 253, 256–258, 262, 265, 270, 368
  driving, 116, 119–121, 124, 127, 130, 136,
       219, 221, 232, 234
  efficiency, 232–236
  embedded, 113, 114, 117–118, 128, 129,
       131, 217, 236, 322, 325, 382, 390
  flying, 116, 219–222
  food, 257, 259, 263
  global average, 104, 130, 265, 345, 383,
       394, 395
  internet, 110, 115, 124, 141
  Ireland, 65, 106, 107, 111, 113, 114, 121,
       123, 125, 128, 130, 132, 138, 141,
       142, 145, 146, 183–188, 196, 202,
       204, 206, 208, 210, 217, 230, 256,
       267, 268, 286, 293, 313, 316, 322,
       383, 392, 394
  lights, 65, 109, 110, 115, 119, 121–124,
       129, 194, 201, 202, 210, 230, 232,
       234, 325, 327, 385

limits, 4, 87, 98, 104, 131, 136–138, 140,
       141, 152, 153, 198, 209, 225, 264,
       353, 383, 391
  measuring, **103–132**, 219, 382–383
  methane hydrate, 84, 355
  mobile communication, 124–126
  offsetting, 103–132, 144, 147–149,
       152–156, 217, 223, 319, 382–384
  permafrost, 194, 349, 353, 355
  personal, 83, 97, 113, 114, 117, 124, 131,
       223, 238, 258, 383, 390, 395, 396
  personal targets, 174, 209
  predicting, 93–95
  primary, 46, 109–111, 113, 117, 157, 316,
       382, 383, 390, 392, 394, 395
  rebound effect, 129–130
  recycling, 111, 112, 114, 125, 126,
       128–130, 236, 263, 304, 313, 314,
       316–320, 323–325, 327, 328, 387,
       392, 394
  secondary, 109–114, 117–118, 340,
       382–383, 390
  sector, 107, 109, 114, 141, 145, 151, 153,
       186, 201, 216, 220, 224, 251, 256,
       314, 316
  by sector, 108, 243
  stakeholder share, 71
  sulphur dioxide, 198, 339, 340
  trading, 111, 130, **136–143**, 146, 149, 155,
       220, 223
  travel, 112–114, 116, 118–121, 129, 132,
       148, 153, 216–239, 251, 252, 385,
       386, 391, 392, 396
  USA, 23, 83, 98, 100, 125, 142, 196, 222,
       251, 257, 264, 265, 342
  volcanoes, 338–339
  water, 14, 63, 123, 126, 194, 201, 202, 222,
       252, 263, 265, 284–287, 293, 294,
       299, 313, 316, 351, 384, 385, 387, 391
  wildfires, 339–342
Emissions Trading Registry, 139
Emissions trading scheme (ETS), 136,
       138–142, 149, 220
Energy
  consumption by fuel type, 189
  conversion factors, 123, 155, 184, 186,
       203, 211
  decarbonization, 198, 201, 367
  desktop PCs, 206–207
  efficient house, 204
  electricity, 184–188
  flows in electricity generation, 188
  global consumption, 277
  household use, 114, 201–211, 214, 224

Energy (*cont.*)
    measuring usage, 103–132, 203–211, 294,
        382–383
    non-thermal uses, 201, 202, 204
    nuclear, 66–67, 187, 198, 200, 201, 210
    personal targets, 174, 209
    renewable, 47, 66, 67, 124, 145, 147, 149,
        155, 160, 179, 182, 184, 185, 187,
        192–198, 218, 294, 385
    sources, 188–192
    targets, 183, 192, 195, 200, 207–209
    thermal uses, 201
    trilemma index, 192
    usage, 180–183
Environmental sustainability, 14, 15, 34, 35,
        38, **43–49**, 55, 69, 72, 73, 94, 136,
        144, 153, 171, 173, 174, 192, 232,
        251, 253, 314–315, 368–371, 381,
        388–389
ERU. *See* Emission reduction units (ERUs)
Ethanol, 152, 195
ETS. *See* Emissions trading scheme (ETS)
European Climate Exchange (ECX), 142–144
EUV. *See* Extreme ultraviolet radiation (EUV)
Extreme ultraviolet radiation (EUV), 20
Extreme weather-related health effects,
        357, 363–364, 372, 388

F
Finite resources
    land, 69–74
Flooding, 17, 19, 22, 25, 26, 28, 29, 47, 55,
        138, 243–245, 249, 282, 287, 346,
        363, 365, 372, 390
Food, 5, 35, 54, 112, 152, 162, 195, 217,
        241–274, 304, 335, 357, 376
    calculators, 261–262
    conversion factors, 262
    drought, 245–249
    emissions, 257, 259, 263
    footprint, 115, 252, **255–262**
    local, 73, 251
    miles, 251–252, 256, 386
    organic, 253–255
    reducing emissions, 253, 264, 394
    scarcity, **243–251**, 255, 372, 389
    supply chain, 259, 264–266
    waste, 169, 255, **264–269**, 318, 319, 328,
        386, 393
The Food miles report, 256
Forestry, 70–73, 139, **149–153**
Fracking, 39, 64–66
Frugality, 269, 303–307
Fuel mixes, 109, 114, 160, 184–188, 210, 232

G
Gaia, **36, 37**, 353
Gaia: a New Look at Life on Earth, 36
Geothermal, 54, 188, 194
GHG. *See* Greenhouse gases (GHG)
GHGs. *See* Greenhouse gases (GHGs)
Global CO_2
    emissions, 84, 99, 126, 130
    Kyoto Protocol, 87, 98, 136, 137, 148
    predictions, 8, 16, 19, 68, 93–95, 161,
        183, 193, 244, 245, 340, 353,
        355, 357, 360
    proposed limits, 95–100
Global Footprint Network, 12, 161–168, 170
Global hectares (gha), definition, 162
Global justice, 35, 130
Global living planet index (LPI), 170, 171
Global temperature, 17, 18, 25, 29, 46, 83, 90,
        94, 96, 97, 353, 354, 359, 366, 388
Global warming potential (GWP), **87–90**, 100,
        104, 382
Government action, 4, 111, 160, 174, 207,
        209, 318, 378, 384, 390, 391
Green bin, 320, 321
Greenhouse effect, **81–85**, 100, 104, 274, 382
Greenhouse gases (GHGs), 46–48, 54, 58,
        60–65, 67, 69–72, 75, 81–97, 99,
        100, 104, 106–109, 112, 115, 117,
        118, 120, 121, 124–127, 136–138,
        140, 141, 143, 146–148, 150, 153,
        154, 156, 157, 160, 163, 174,
        184–188, 195, 196, 201–203,
        205–207, 210, 216, 217, 221–223,
        225, 232, 236, 238, 243, 248, 249,
        251–259, 261, 264, 265, 270, 284,
        286, 287, 293, 313, 314, 316, 318,
        319, 322–325, 327, 328, 336,
        338–342, 353–355, 367, 376,
        379–387, 392, 394, 395
    carbon dioxide (CO_2), 62, 70, **81–100**, 104,
        106–109, 112, 115, 117, 118, 120,
        121, 124–127, 136, 143, 146–148,
        153, 184, 187, 201, 225, 256, 259,
        286, 293, 336, 340, 342, 355, 382–384
    emissions (*see* Emissions)
    emissions by sector, 108, 243
    Kyoto Protocol, 87, 136, 137, 148
    methane (CH_4), 64, 83–85, 88, 104, 107,
        113, 115, 243, 251, 256, 313, 314,
        318, 336, 340
    models, 90–95
    nitrous oxide (N_2O), 83–85, 88, 104, 115,
        243–244, 340
    predictions, 340, 353, 355
    proposed limits, 95–100

Greenland, 343–346, 353
The Growth of World Population, 14
GWP. *See* Global warming potential (GWP)

H
HDI. *See* Human Development Index (HDI)
Health, 7, 11, 15, 35, 37, 45, 69, 70, 84,
　　　127, 160, 225, 239, 242, 248, 251,
　　　254, 255, 258, 261, 262, 264, 270,
　　　276, 278, 280, 284, 290, 333–355,
　　　357–373, 376, 378–380, 388–389,
　　　396
　air pollution, 357, 364–365, 388
　effect of temperature, 95
　extreme weather, 17, 70, 357, 358,
　　　363–365, 372, 388
　malaria, 358, 360–363, 365, 366, 372
　positive health benefits, 367–368
　rodent-borne diseases, 358–363
　vector-borne diseases, 358–363
　waterborne diseases, 363
Helium, 56, 57, 66
Home energy measurements, 203–204
Household electricity consumption, 319
Household lighting, 122
How the Other Half Dies, 35
Hubbert peak theory, 57
Human Development Index (HDI), 170
Human water cycle, 276
Hurricane Christian, 26
Hurricane Katrina, 27, 28, 364
Hurricane Sandy, 26, 28
Hurricane Xavier, 26
Hydroelectric, 109, 193
Hydrogen, 194, 195, 351
Hydrological cycle, 274

I
Ice cover, 17, 274, **342–344**, 355
Imported energy, 184, 191
Intergovernmental Panel on Climate Change
　　　(IPCC), 7, 8, 18, 20, 22, 40, 83, 86,
　　　91–98, 150, 173, 207, 225, 243,
　　　264, 337, 344, 345, 350, 352, 365,
　　　378, 379, 382, 383, 394, 395
Internet–carbon footprint, 110, 126
IPCC. *See* Intergonmental Panel on Climate
　　　Change (IPCC)
IPCC-Fifth assessment Report (AR5), 22, 91–93
IPCC-Fourth assessment Report (AR4), 7, 8,
　　　18, 86, 91, 93, 243
Irish climate: the road ahead, 22, 23
Is Global Collapse Imminent, 16, 17

J
JI. *See* Joint implementation scheme
Joint implementation scheme (JI), 137–140

K
Kyoto Protocol, 87, 98, 136, 137, 139, 141,
　　　148, 151

L
La Niña, 19, 20
Lignite, 53, **62–64**, 147, 338
Limits for emissions, 4, 87, 98–100, 104, 131,
　　　136–138, 140, 141, 152, 153, 198,
　　　209, 225, 264, 353, 383, 391
Limits to growth, 15–17, 41
Low energy light bulbs, 121, 129
LPI. *See* Global Living Planet Index
LULUCF, 139

M
Making food poverty history, 243
Malaria, 358, 360–363, 365, 366, 372
Malthusian catastrophe, 13–17
Mauna Loa CO_2 concentrations, 96
Maximum sustainable yield (MSY), 55, 56
Meatless Mondays, 264, 307
Metals, 14, 15, 56, 57, 61, **66–69**, 74–76, 263,
　　　304, 315, 321, 323, 325, 326
Methane (CH_4), 21, 64, 83–88, 104, 107, 115,
　　　149, 243, 251, 256, 313, 314, 318,
　　　328, 340, 348, 349
Methane hydrate, 84, 336, 350, 355
Million tonnes of oil equivalent (Mtoe),
　　　181–183, 189
Mobile communication–carbon footprint,
　　　124–126
MSY. *See* Maximum sustainable yield (MSY)
Mtoe. *See* Million tonnes of oil equivalent (Mtoe)

N
Natural gas, 36, 57, **63–66**, 82, 83, 85, 97,
　　　109, 112–114, 146, 147, 169, 182,
　　　184, 187–190, 202, 203, 211, 286
Natural gas conversion factors, 114, 184, 187,
　　　202, 203, 211
Neo-Malthusianism, 14, 15, 43
New opportunities, 44–47, 126, 131, 379
Newspapers, 3, 4, 116, 239, 304, 378
Nitrous oxide (N_2O), 83–85, 88, 104, 113,
　　　115, 149, 225, 243–244, 339, 340
N_2O. *See* Nitrous oxide (N_2O)

Non-renewable resources, 35, 40, 43, 53–58,
          62, 63, 69, 74–76, 155, 321, 328,
          379, 381, 387
Nuclear energy, 66–67, 187, 198–201, 210

O
Offsetting
     cost, 144, 145, 147–154, 383–384
     emissions, 103–132, 382–383
     mechanisms, 110, 111
Oil peak, 60
Oil sands, **60–62**, 75, 189, 381
Oil shales, 39, **60–62**, 66, 75, 381
One Planet Economy Network, 171–173
Organic food, 253–255

P
Panic, 159
Paper bags, 127–129
Peak ecological water, 289
Peak non-renewable water, 289
Peak oil, 58
Peak renewable water, 289
Peak theory, 57
Peak water, 287–294, 299
Perihelion, 21
Permafrost, 84, 194, 336, 342, **347–350**,
          353, 355
Personal action, 4, 111, 168, 192, 355, 370,
          390–396
     food, 391, 393, 394
     food waste, 393
     recycling, 111, 319–327, 392, 394, 397
     travel, 391, 393, 394
     waste minimization, 392
Personal emissions, 83, 97, 113, 114, 117,
          124, 131, 223, 238, 258, 383, 390,
          395, 396
Personal plan, 5, 50, 131, 355, 391–396
Personal targets for energy use, 209, 382, 383,
          385, 392–395
Personal targets for transport, 209, 381, 385,
          386, 389, 393, 394, 397
Petroleum, 58–62, 188, 211
Planet health, 84, 160, 290, 333–355, 368,
          378, 380, 388
     biodiversity, 335, 336, 340, 349, 354
     carbon cycle, 336–339
     ice cover, 342–344, 355
     methane hydrate, 84, 336, 350, 355
     organic balance, 335–339
     permafrost, 84, 336, 342, 347–350, 353, 355

     sea acidification, 351–353
     sea level, 342–347, 355
     sulphur dioxide, 339, 340
     tipping points, 353–355, 388
     volcanoes, 338–339
     wildfires, 84, 339–342, 355
Plastic bags, 128, 129, 263
Population, **5–17**, 29, 30, 38, 44, 46, 49, 50,
          56, 69, 72, 75, 93, 94, 98–100, 125,
          136, 161–163, 165, 167–170, 173,
          174, 182, 184, 189, 226, 228, 242,
          245, 247, 249–251, 255, 269, 270,
          273–277, 288, 289, 311, 312, 317,
          325, 358, 363, 367, 369, 372, 379,
          380, 384, 386, 396
     consequences, 11–13, 15, 379
     growth rate, 6–11, 14, 29, 379
     hockey stick, 7
     predictions, 8
     total fertility rate, 10, 11, 29
The Population bomb, 14, 15
Positive action, 156, 335, 378, 384, 397
Positive health benefits of global warming,
          367–368
Precipitation, 17, 22, **24–27**, 29, 90, 91, 244,
          245, 274, 287, 359, 363
     flooding, 17, 22, 25, 26, 29, 244, 245, 287,
          363
     Sahel, 24, 25
Primary carbon footprint, 109–115,
          224, 238

R
Radiative forcing (RF), **85–87**, 104, 144,
          148, 222
Rainfall. *See* Precipitation
Rare earth metals, 57, 67, 68, 74, 76, 325
Rebound effect, 129–130
Recognition of the need to change, 391–393
Recycling
     car, 114, 129, 174, 236–237, 316, 320,
          325, 326
     contents of green bin, 320, 321
     contents of waste bin, 304, 305, 321,
          323–324
     electronic waste, 321
     paper, 125, 126, 128–130, 304, 308, 309,
          315, 316, 321, 327
     personal action, 74, 392
     signs, 304–311
     USA, 264, 317
     waste hierarchy, 314, 315, 319, 327
Removal unit (RMU), 139

Renewable energy, 47, 66, 67, 124, 145, 147, 149, 155, 160, 179, 182, 184, 185, 187, **192–198**, 218, 294, 385
Renewable resources, 35, 43, 54–57, 76, 193, 288, 299, 387
Resources
    coal, 57, **62–66**, 75, 85, 182
    concept of peak, 57
    crude oil, 58–62
    forestry, 70–73, 139, **149–153**
    fracking, 39, 64–66
    helium, 56, 57, 66
    land, **69–74**, 162, 248, 249, 289
    lignite, 53, **62–64**
    maximum sustainable yield, 55
    metals, 14, 15, 56, 57, 61, **66–69**, 74–76
    natural gas, 57, 59, 63–66, 85, 182, 189
    non-renewable, 35, 40, 43, **53–58**, 62, 63, 69, 74–76, 321, 328, 379, 381, 387
    oil sands, **60–62**, 75, 189, 381
    oil shales, 39, **60–62**, 66, 75, 381
    petroleum, 58–62, 65, 67
    rare earth metals, 57, 67, 68, 74, 76
    renewable, 17, 35, 40, 43, 54–57, 66–67, 76, 167, 182, 185, 189, 192, 193, 288, 299, 379, 387
    shale gas, 39, 64–66, 189, 381
    sustainable yield, 55, 56
    uranium, 57, 66–67
RF. See Radiative forcing (RF)
River basin management, 281
RMU. See Removal unit (RMU)
Rodent-borne diseases, 358–363

S
Sahel precipitation, 24, 25
Sceptics, 21, 45, 160, 377, 378
Sea acidification, 351–353
Sea level, 17, 63, 90, 91, 245, 249, 344–347, 355, 363
Secondary carbon footprint, 109, 117–118
Sequestration, 69, 70, 72, 139, 143–145, 149–154, 166, 336, 351, 353
    agriculture, 70, 72, 139, 149–153
    capture and storage, 143–145, 154
    charcoal, 70, 144, 154
    cost, 139, 143–145, 149–154
    forests, 70, 73, 139, 144, 149–153, 336
    oceans, 166, 350, 351
    soils, 69–70, 72, 73, 149–150, 350
Setting emission targets, 100, 104, 141, 209, 385, 392
Shale gas, 64–66, 189

Shape of Earth's orbit, 21
Small is Beautiful, 44
SO_2. See Sulphur dioxide (SO_2)
Social justice, 42, 369
Social sustainability, 40
Solar energy, 21, 82, 83, 89, 222, 274
Solar radiation, 82, 85, 89, 339
Space heating, 147, 196, 202, 210, 385
Standby electricity use, 204–206
Step reduction plans, 395
Stern Committee, 44–48
Storms, 26–29
Sulphur dioxide (SO_2), 19, 20, 198, 339, 340
Surface temperatures (USA), 19–21
Sustainability, 15, **33–50**, 54, 55, 71, 75, 93, 94, 129, 145, 160, 161, 168, 172–174, 192, 207, 210, 254, 292, 304, 306, 307, 314–315, 332, 366, 368–372, 380, 381, 384, 385, 392, 393
    ecocentrism, 43, 44
    economic, 38–41, 43–49, 94, 152, 171, 193, 290, 369, 371, 377, 381
    environmental, 14, 15, 34, 35, **38–49**, 55, 69, 72, 73, 94, 136, 144, 153, 171, 173, 174, 192, 232, 251, 253, 314–315, 368–371, 381, 388–389
    definition, **37–40**, 49
    the future, 14, 37–40, 48, 49, 55, 56, 153, 207, 251, 304, 306, 335, 368, 377, 381
    nexus, 40–42
    social, 14, 39–42, 47, 72, 173, 370, 371, 377, 381, 388, 389
    Stern, 44–48, 172
    sustaincentrism, 43
    targets, 42, 46, 94, 138, 154, 173–174, 207, 316, 385, 392, 394–395
    technocentrism, 43
    wellbeing, 37–38, 43, 366, 368–372, 389, 392
Sustainable development, 38, 39, 48, 49, 67, 69, 139, 171, 369–371
Sustainable yield, 55, 56
Sustaincentrism, 43

T
Technocentrism, 43
Temperature. See Climate
Temperature related illness, 357, 359
TFR. See Total fertility rate (TFR)
Tipping points, 349, **353–355**, 379, 388
Total carbon footprint, 109–116, 118, 260, 390
Total fertility rate (TFR), 10, 11, 29

Transportation, 14, 59, 63–64, 73, 75, 85, 107,
        109, 111–115, 117, 118, 125, 151,
        182–183, 189, 192, 195, 196, 202,
        207, 209, 210, 215–220, 224–226,
        229–234, 236–239, 251, 252, 255,
        256, 259, 267, 293, 314–316, 328,
        368, 370, 381, 385, 386, 389, 393,
        394, 397
Transport fuel footprint, 118
Travel, 5–6, 111–114, 116, 118–121, 129, 130,
        132, 146–148, 153, 169, **215–239**,
        251, 252, 348, 370, 376, 385–386,
        391, 392, 396
    carbon footprint, 118, 132
Typhoon Haiyan, 27, 28

U
Uncertainty analysis, 379, 380
Upcycling, 387
Uranium, 57, 66–67

V
Vector-borne diseases, 358–363
Vehicle emissions, 119, 120, 232, 236
Virtual water, 294–297
Volcanoes, 19, 20, 84, 91, 336, **338–339**

W
Waste, 11, 36, 61, 85, 107, 144, 161, 186, 217,
        251, 274, 304, 363, 380
    bin, 304, 305, 320, 321, 323–324
    disposal hierarchy, 314

generation, 311, 312, 316, 317
    minimization, 267, 268, 314–315, 319,
        327, 387, 392
Water, 5, 35, 53, 82, 123, 147, 162, 194, 222,
        248, 273–300, 318, 335, 357, 379
    agriculture, 276, 279, 281, 288, 289, 292,
        296–299
    conflict, 281–282, 299
    conservation, 42, 283–284, 286, 290, 298,
        299, 387
    demand management, **283–287**, 291,
        299, 387
    diary, 294, 298–300
    footprint, 172, 174, **294–300**, 384, 387
    metering, 26, 283, 284, 286, 294, 387
    peak. *See* Peak water
    poverty threshold, 277
    scarcity, 29, 249, 273, **277–282**, 288, 290,
        293, 357, 372, 379, 386–387, 389
    setting limits, 283, 289, 299
    stress threshold, 278, 279
    use, 277, 281, 283, 284, 286–287, 290,
        294–296, 298–300, 387, 391
    virtual, 294–297
Waterborne diseases, 363
Water efficiency labelling (WELS), 284–286
Wealth, 15, 37, 44, 93, 97, 104, 155, 167, 215,
        250, 277, 285, 315, 347, 365, 369,
        371, 372, 379, 388, 389
Wellbeing, 5, 11, 15, 37–38, 43, 69, 127, 131,
        357, **368–373**, 376, 388–389
WELS. *See* Water efficiency labelling (WELS)
Wildfires, 17, 29, 55, 84, 150, 165, 336,
        339–342, 355
Wind power, 28, 179, 194, 196